IET ENERGY ENGINEERING 549

Non-Redundant Near-Field to Far-Field Transformation Techniques

Other volumes in this series:

Volume 1	**Geometrical theory of diffraction for electromagnetic waves, 3rd edition** G.L. James
Volume 10	**Aperture antennas and diffraction theory** E.V. Jull
Volume 11	**Adaptive array principles** J.E. Hudson
Volume 12	**Microstrip antenna theory and design** J.R. James, P.S. Hall and C. Wood
Volume 15	**The handbook of antenna design, volume 1** A.W. Rudge, K. Milne, A.D. Oliver and P. Knight (Editors)
Volume 16	**The handbook of antenna design, volume 2** A.W. Rudge, K. Milne, A.D. Oliver and P. Knight (Editors)
Volume 18	**Corrugated horns for microwave antennas** P.J.B. Clarricoats and A.D. Oliver
Volume 19	**Microwave antenna theory and design** S. Silver (Editor)
Volume 21	**Waveguide handbook** N. Marcuvitz
Volume 23	**Ferrites at microwave frequencies** A.J. Baden Fuller
Volume 24	**Propagation of short radio waves** D.E. Kerr (Editor)
Volume 25	**Principles of microwave circuits** C.G. Montgomery, R.H. Dicke and E.M. Purcell (Editors)
Volume 26	**Spherical near-field antenna measurements** J.E. Hansen (Editor)
Volume 28	**Handbook of microstrip antennas, 2 volumes** J.R. James and P.S. Hall (Editors)
Volume 31	**Ionospheric radio** K. Davies
Volume 32	**Electromagnetic waveguides: theory and applications** S.F. Mahmoud
Volume 33	**Radio direction finding and superresolution, 2nd edition** P.J.D. Gething
Volume 34	**Electrodynamic theory of superconductors** S.A. Zhou
Volume 35	**VHF and UHF antennas** R.A. Burberry
Volume 36	**Propagation, scattering and diffraction of electromagnetic waves** A.S. Ilyinski, G. Ya.Slepyan and A. Ya.Slepyan
Volume 37	**Geometrical theory of diffraction** V.A. Borovikov and B.Ye. Kinber
Volume 38	**Analysis of metallic antenna and scatterers** B.D. Popovic and B.M. Kolundzija
Volume 39	**Microwave horns and feeds** A.D. Olver, P.J.B. Clarricoats, A.A. Kishk and L. Shafai
Volume 41	**Approximate boundary conditions in electromagnetics** T.B.A. Senior and J.L. Volakis
Volume 42	**Spectral theory and excitation of open structures** V.P. Shestopalov and Y. Shestopalov
Volume 43	**Open electromagnetic waveguides** T. Rozzi and M. Mongiardo
Volume 44	**Theory of nonuniform waveguides: the cross-section method** B.Z. Katsenelenbaum, L. Mercader Del Rio, M. Pereyaslavets, M. Sorella Ayza and M.K.A. Thumm
Volume 45	**Parabolic equation methods for electromagnetic wave propagation** M. Levy
Volume 46	**Advanced electromagnetic analysis of passive and active planar structures** T. Rozzi and M. Farinai
Volume 47	**Electromagnetic mixing formulas and applications** A. Sihvola
Volume 48	**Theory and design of microwave filters** I.C. Hunter
Volume 49	**Handbook of ridge waveguides and passive components** J. Helszajn
Volume 50	**Channels, propagation and antennas for mobile communications** R. Vaughan and J. Bach-Anderson
Volume 51	**Asymptotic and hybrid methods in electromagnetics** F. Molinet, I. Andronov and D. Bouche
Volume 52	**Thermal microwave radiation: applications for remote sensing** C. Matzler (Editor)
Volume 53	**Principles of planar near-field antenna measurements** S. Gregson, J. McCormick and C. Parini
Volume 54	**Satellite-to-ground radiowave propagation, 2nd edition** J.E. Allnutt
Volume 502	**Propagation of radiowaves, 2nd edition** L.W. Barclay (Editor)
Volume 534	**Emerging Evolutionary Algorithms for Antennas and Wireless Communications** S. Goudos
Volume 555	**Bioelectromagnetics in Healthcare: Advanced sensing and communication applications** W Whittow

Non-Redundant Near-Field to Far-Field Transformation Techniques

Claudio Gennarelli, Flaminio Ferrara, Rocco Guerriero, and Francesco D'Agostino

The Institution of Engineering and Technology

Published by SciTech Publishing, an imprint of The Institution of Engineering and Technology, London, United Kingdom

The Institution of Engineering and Technology is registered as a Charity in England & Wales (no. 211014) and Scotland (no. SC038698).

First published 2022

The Institution of Engineering and Technology
Futures Place
Kings Way, Stevenage
Herts, SG1 2UA, United Kingdom

www.theiet.org

British Library Cataloguing in Publication Data
A catalogue record for this product is available from the British Library

ISBN 978-1-83953-141-5 (hardback)
ISBN 978-1-83953-142-2 (PDF)

Typeset in India by Exeter Premedia Services Private Limited
Printed in the UK by CPI Group (UK) Ltd, Croydon
Cover Image: Alexandr Gnezdilov Light Painting via Getty Images

To Catello
Scientist, mentor and friend
and
to our families

The perfect paper will never be written. One step toward perfection is better than no step at all.

Leopold B. Felsen

Contents

About the Authors

Claudio Gennarelli was born in Avellino, Italy, in 1953. He received the Laurea degree (*summa cum laude*) in Electronic Engineering from the University of Naples, Italy, in 1978. From 1978 to 1983, he worked with the Research Group in Electromagnetics at the Electronic Engineering Department of the University "Federico II" of Naples. In 1983, he became Assistant Professor at the Istituto Universitario Navale (IUN), Naples. In 1987, he was appointed Associate Professor of Antennas, formerly at the Engineering Faculty of Ancona University and subsequently at the Engineering Faculty of Salerno University. In 1999, he has been appointed Full Professor at the same University. The main topics of his scientific activity are reflector antennas analysis, antenna measurements, diffraction problems, radar cross section evaluations, scattering from surface impedances, application of sampling techniques to electromagnetics and to near-field to far-field (NF–FF) transformation techniques, NF–FF transformations with spiral scannings, and compensation of the probe positioning errors in the NF–FF transformations. Dr. Gennarelli is co-author of over 450 scientific papers, mainly in international journals and conference proceedings. In particular, he is co-author of five books on NF–FF transformation techniques and co-author of the chapter "Near-field antenna measurement techniques" of the *Handbook of Antenna Technologies*. He is a Senior Member of the IEEE since 2002 and member of the Editorial board *of the International Journal of Antennas and Propagation*. He has also been member of the Editorial board of the *Open Electrical and Electronic Engineering Journal* from 2007 to 2019.

Flaminio Ferrara was born near Salerno, Italy, in 1972. He received his degree in Electronic Engineering from the University of Salerno in 1999. Since the same year, he has been with the Research Group in Applied Electromagnetics at the University of Salerno. He received the Ph.D. degree in Information Engineering at the same University, where he is presently an Associate Professor of Electromagnetic Fields. His scientific interests include application of sampling techniques to the efficient reconstruction of electromagnetic fields and to NF–FF transformation techniques, monostatic radar cross section evaluations of corner reflectors, electromagnetic wave diffraction from canonical and complex structures. Dr. Ferrara is co-author of more than 280 scientific papers, mainly in international journals and conference proceedings. In particular, he is co-author of four books on NF–FF transformation techniques and co-author of the chapter "Near-field antenna measurement techniques" of the *Handbook of Antenna Technologies*. He is reviewer for several international journals and member of the Editorial board of the *International Journal of Antennas and Propagation* and of *Electronics*. He is member of the IEEE society.

Rocco Guerriero received his degree in Electronic Engineering and the Ph.D. degree in Information Engineering from the University of Salerno in 2003 and 2007, respectively. Since 2003, he has been with the Research Group in Applied Electromagnetics of University of Salerno, where he is currently an assistant professor of Electromagnetic Fields. His scientific interests include application of sampling techniques to the efficient reconstruction of electromagnetic fields and to NF–FF transformation techniques; antenna measurements; inversion of ill-posed electromagnetic problems; analysis of microstrip reflectarrays; diffraction problems. Dr. Guerriero is co-author of more than 240 scientific papers, mainly in international journals and conference proceedings. In particular, he has co-authored three books on NF–FF transformation techniques and is co-author of the chapter "Near-field antenna measurement techniques" of the *Handbook of Antenna Technologies*. He is reviewer for several international journals and conferences and is member of the Editorial board of the *International Journal of Antennas and Propagation* and of *Electronics*. Since 2015, he is member of IEEE.

Francesco D'Agostino received the Laurea degree in Electronic Engineering from the University of Salerno in 1994. Since the same year, he has been with the Research Group in Applied Electromagnetics at the same University, where in 2001 he received the Ph.D. degree in Information Engineering. From 2002 to 2005, he was Assistant Professor at the Engineering Faculty of the University of Salerno where, in October 2005, he was appointed Associate Professor of Electromagnetics and joined the Department of Industrial Engineering, where he is currently working. His research activity includes application of sampling techniques to electromagnetics and to innovative NF–FF transformations, diffraction problems, radar cross section evaluations, Electromagnetic Compatibility. In this area, Dr. D'Agostino has co-authored 5 books and over 270 scientific papers, published in peer-reviewed international journals and conference proceedings. He is a member of the Editorial Boards of the *International Journal of Antenna and Propagation* and of *Microwave and Wireless Communications.* He serves as a regular reviewer for several journals and conferences and has chaired many international events and conferences. Dr. D'Agostino is a fellow of Antenna Measurement Techniques Association (AMTA), senior member of the Institute of Electrical and Electronics Engineers (IEEE), member of the European Association on Antennas and Propagation (EurAAP), and of the Italian Electromagnetic Society (SIEM).

Preface

The near-field (NF) measurement techniques and the related NF to far-field (FF) transformations are gaining an ever growing interest for their own peculiarity to allow a very accurate evaluation of the radiation characteristics of those antennas, whose electric sizes do not make it possible to perform direct FF measurements in a controlled and reflection-free environment, such as the anechoic chamber.

This book aims to provide a comprehensive treatment of the classical NF–FF transformation techniques and to describe the significant improvements achieved in their performance by correctly applying the non-redundant sampling representations of antenna radiated electromagnetic (EM) fields. These progresses arise from the strict collaboration of Professor Claudio Gennarelli and his dear friend Professor Catello Savarese under the precious scientific direction of Professor Ovidio Mario Bucci. This collaboration began in the mid-1980s and continued involving first Professor Giovanni Riccio and then the other co-authors. After the death of Professor Savarese, this scientific activity moved definitively to the University of Salerno under the scientific direction of Professor Gennarelli.

The book is designed to meet the needs of students approaching the antenna NF measurements, as well as of engineers and physicists working in such an area. It has been written keeping in mind the fulfilment of two main objectives. The former is to provide all analytical details on the derivation of the classical NF–FF transformation techniques without and with probe compensation, which are not easily available in literature. The latter is to give a comprehensive description of effective representations of the EM fields radiated over arbitrary rotational surfaces, which make use of a non-redundant (i.e, minimum) number of samples collected on these surfaces or along proper spirals wrapping them. To this end, the theoretical bases of the non-redundant sampling representations of EM fields and of spiral scans are described in Chapters 2 and 3, respectively, and optimal sampling interpolation (OSI) formulas, which allow the fast and accurate reconstruction of the EM field or of the response of the measuring probe over a rotational surface from its samples acquired on such a surface, or on the spiral wrapping it, are presented. These sampling representations have been properly exploited to develop probe-compensated non-redundant NF–FF transformations with plane-polar, bi-polar, cylindrical, and spherical scannings and those employing the innovative planar spiral, helicoidal, and spherical spiral scannings. Note that there is no comprehensive source covering all topics relevant to these NF–FF transformations. The description of these transformation techniques and some numerical and experimental results assessing them are collected in the chapters from 4 to 6, depending on the kind of the scanning geometry, together with all the analytical details on the derivation of the corresponding classical NF–FF transformations without and with probe compensation.

To improve the book readability, some long and non-trivial analytical details supporting the derivation of formulas are collected in the appendices to the chapters or to the book.

An overview of the state of art about the different techniques available in the open literature for the antenna characterization from NF measurements completes the book.

The authors are grateful to Professor Ovidio Mario Bucci, whose scientific ideas have made possible the writing of this book, and are indebted to Professor Giovanni Riccio, who profitably cooperates with them on these topics. Moreover, they would like to thank Dr. Massimo Migliozzi for his precious contribution in performing the experimental tests, whose results are reported throughout the book. At last, they dedicate the book to Professor Catello Savarese, who greatly contributed until his early death to obtain most of the here presented results.

Claudio Gennarelli, Flaminio Ferrara, Rocco Guerriero,
and Francesco D'Agostino
April 2022

Chapter 1
Introduction

1.1 Motivation and background

The large spreading of high-performing antennas, adopted in satellite communication systems and radar equipment, has given rise to an increasingly demand for accurate measurements of the antenna radiation characteristics, which, as well known, are defined in the antenna far-field (FF) region. To ensure that the reflections from the surroundings and incoming electromagnetic (EM) interferences do not affect the accuracy, as in the case of the outdoor FF ranges, the measurements are performed in controlled indoor environments able to emulate the free-space propagation conditions, i.e., shielded anechoic chambers. However, due to the limited sizes of an anechoic chamber, the direct FF measurements are possible only for antennas under test (AUTs) whose electric dimensions allow the fulfilment of the FF distance requirements. When these requirements are not satisfied, only near-field (NF) measurements are made possible there and, accordingly, the evaluation the antenna radiation characteristic can be obtained by resorting to a NF–FF transformation technique [1–13], which allows the accurate computation of the amplitude, phase, and polarization of the antenna far field from the knowledge of the acquired NF data. Besides the evaluation of the complete antenna FF pattern, NF measurements can be employed to obtain, through a back transformation, the field at the antenna surface (microwave holography [14, 15]) and this information can be exploited for diagnostic purposes to detect, e.g., faulty elements in an array or surface deformations in a reflector antenna.

A brief discussion about the field behaviour in the space surrounding a radiating antenna on increasing the distance (see Figure 1.1) is provided in the following in order to make more clear the framework wherein the NF–FF transformation techniques are applied. The region closest to the antenna, extending up to a distance of $\lambda/2\pi$ (with λ being the wavelength), is the *reactive NF region*, wherein the reactive field components are predominant.* These reactive field components

*This boundary can be easily determined by considering the field radiated by an elementary electric dipole [16], However, λ is a more reasonable boundary [2].

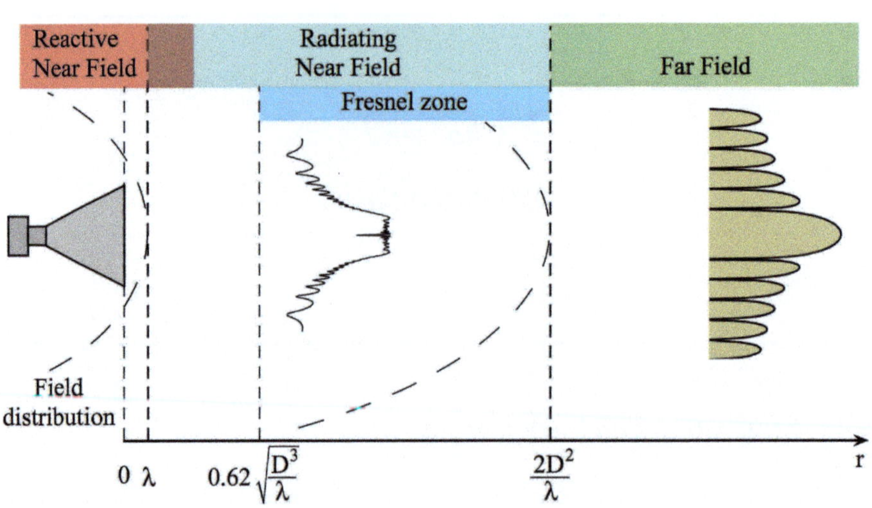

Figure 1.1 *Field regions surrounding electrically large antennas*

decay rapidly on increasing the distance, becoming negligible at a distance of few wavelengths, which is the inner boundary of the *radiating NF region*. In this region, the radiative field components predominate, even though the angular distribution of the field depends on the distance and the field does not exhibit the typical FF dependence $e^{-j\beta r}/r$, where β denotes the wavenumber and the time dependence $e^{j\omega t}$ is assumed. This time dependence is assumed and suppressed throughout the book. The region outer to the antenna is the *FF region,* which is characterized by an angular field distribution independent of the distance and by the dependence $e^{-j\beta r}/r$ of the field components. For antennas having large dimensions with respect to the wavelength, the inner boundary of such a region is usually fixed to $2D^2/\lambda$, D being the antenna maximum dimension and is determined by considering tolerable a maximum phase error of $\pi/8$, when using a linear approximation in the integral expression of the vector potential [17, 18]. For antennas focusing at infinity, the FF region can be also denoted by the optical term “Fraunhofer region” and the optical term “Fresnel region” can be used to indicate a sub-region of the radiating NF region. For electrically large antennas, the inner boundary of the Fresnel region is set at $0.62\sqrt{D^3/\lambda}$ and is determined [2, 18] by considering also the quadratic phase term in the approximation of the vector potential expression.

It must be stressed that the distance constraints are determined by the level of first sidelobe and the required accuracy [19, 20]. As a matter of fact, negligible pattern errors at the Rayleigh $2D^2/\lambda$ distance can be obtained only for antennas with a moderate first sidelobe level (–25 dB). Much larger measurement distances are needed in the case of antennas with low (–30 ÷ –40 dB) and ultralow (below –40 dB) sidelobe levels [21–23]. As shown in Figure 1.2, when measuring an antenna with low sidelobes at the Rayleigh distance, the first sidelobe level increases

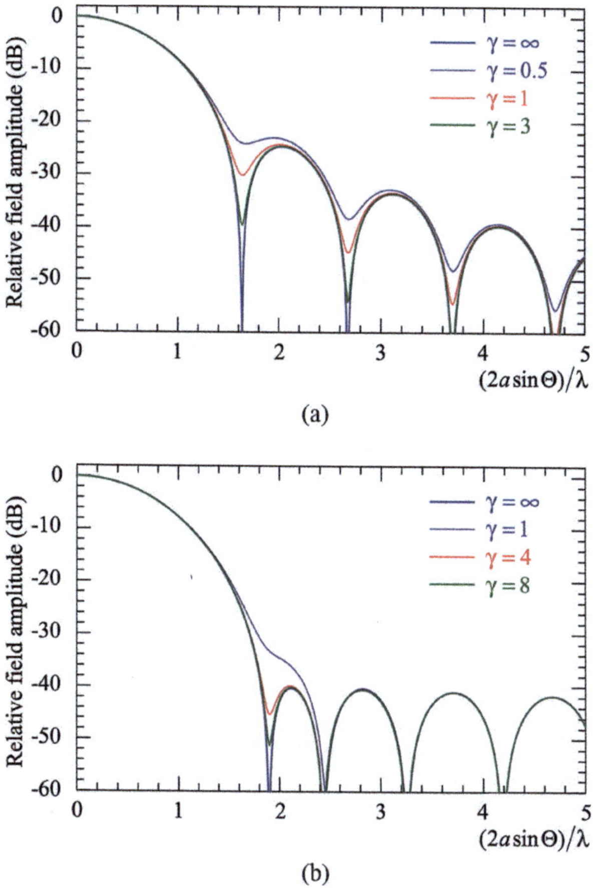

Figure 1.2 *Measured pattern at different normalized distances* $\gamma = r/(2D^2/\lambda)$.
(a) Circular aperture with radial dependence $f(\rho) = (1 - \rho^2)$.
(b) Circular aperture with Taylor distribution.

and the null between it and the main lobe considerably raises. For instance, the first sidelobe of a circular aperture antenna having a Taylor aperture distribution with a 40 dB sidelobe ratio can be measured with an accuracy of 1 dB only at a distance greater than $6D^2/\lambda$ [23].

The NF–FF transformation techniques can make use of complex or only amplitude NF data. Most of these techniques employ complex, i.e., amplitude and phase data, which are usually collected by moving the measuring probe over a planar, cylindrical, or spherical scanning surface and stored with their positions (Figure 1.3). Then, the antenna far field is evaluated by properly expanding the field in terms of modes, which are elementary solutions of the vector wave equation in a source-free region. Spherical, cylindrical, or plane waves can be employed depending on whether a planar, a cylindrical, or a spherical scanning is adopted. The choice of the

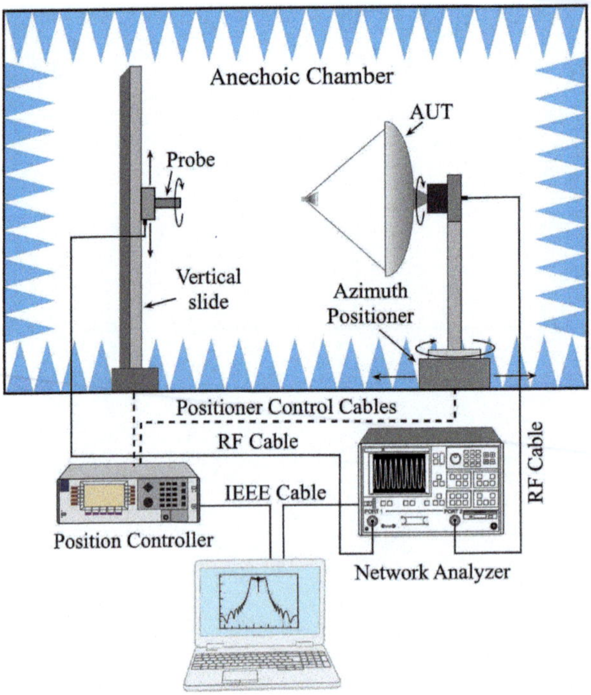

Figure 1.3 NF measurement facility

scanning type depends on the kind of the antennas, the measurement requirements, and the analytical and mechanical complexity. From two sets of measurements of the voltage detected by the probe in two different orientations (the latter is obtained by rotating the probe by 90° around its axis) on a proper grid of the scanning surface, it is possible to get the unknown modal expansion coefficients, by exploiting the orthogonality properties of the modes on these surfaces and taking into account the probe effects. These coefficients, once substituted in the expansion valid in the FF region, allow the accurate evaluation of both the copolar and cross polar components of the AUT far field [2, 3, 11]. In any case, no preliminary information about the antenna polarisation is required, but this information can be properly obtained from the recovered far field. It must be stressed that the probe must be electrically small to introduce minimum disturbance into the measured NF data and its characteristics need to be stable over time and not affected by changes of its orientations and of the environmental conditions [9].

Among the NF–FF transformation techniques, those adopting the planar scannings are the simplest ones from the analytical and computational viewpoints. They represent the better choice when characterizing highly directive antennas radiating pencil beam patterns, since they allow the accurate AUT pattern reconstruction only within the solid angle specified by the edges of the AUT and those of the scanning area. The conventional planar scannings can be accomplished by means of a

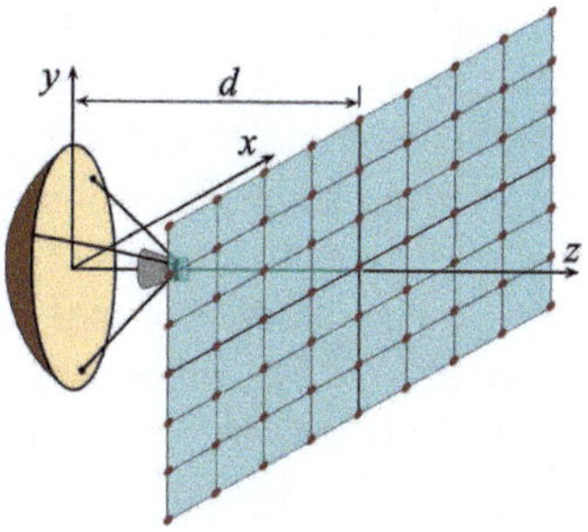

Figure 1.4 *NF–FF transformation with plane-rectangular scanning*

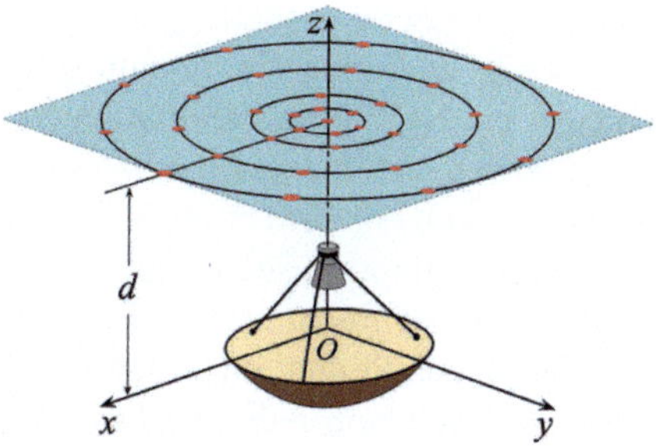

Figure 1.5 *NF–FF transformation with plane-polar scanning*

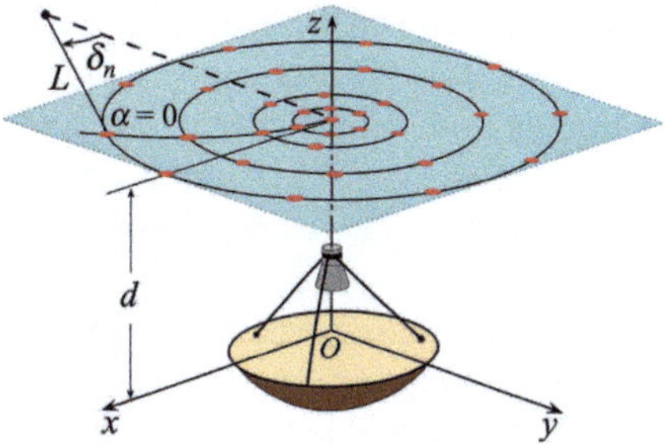

Figure 1.6 *NF–FF transformation with bi-polar scanning*

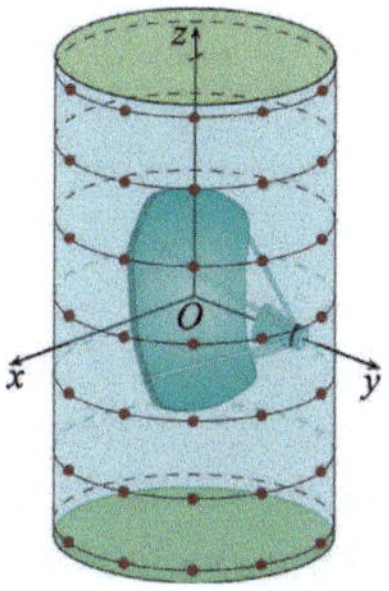

Figure 1.7 *NF–FF transformation with cylindrical scanning*

plane-rectangular [24–31] (see Figure 1.4), a plane-polar [32–45] (Figure 1.5), and a bi-polar [15, 46–51] (Figure 1.6) NF measurement facility.

The analytical and computational complexity increases in the case of the NF–FF transformation with cylindrical scanning [52–65] (see Figure 1.7), which makes possible the whole reconstruction of the antenna far field except for the regions nearby the spherical poles. Accordingly, it results to be particularly suitable for AUTs radiating mainly in the horizontal plane, as e.g., the antennas employed in the broadcasting communications.

At the cost of a significant increase of the analytical complexity and computational effort, the NF–FF transformation with spherical scanning [64–101] (see Figure 1.8) can allow the complete reconstruction of the far field radiated by the AUT avoiding the error arising from the truncation of the scanning area characterizing the NF–FF transformations with planar and cylindrical scannings. Accordingly, it is usually adopted for the characterization of non-directive or multi-beam antennas.

The issue of the reduction of the time required to acquire the NF data has been, and still is, particularly felt by the antenna measurement community, since this time is much longer than that needed to execute the NF–FF transformation, which, anyway, is usually performed offline. Such a goal can be certainly achieved by lowering the amount of the NF measurements, which, when using the classical NF–FF transformation techniques, can become larger and larger on increasing the AUT electric dimensions. It is clear that the acquisition time saving has a beneficial effect on the measurements accuracy, that otherwise could be severely compromised by long acquisition times, owing to the thermal drifts of the involved instrumentation. To reduce the number of NF measurements, the spatial bandlimitation properties [102, 103] of radiated or scattered EM fields and their non-redundant sampling representations [104, 105] have been suitably exploited to develop optimal sampling interpolation (OSI) expansions of central type [39, 106, 107], which allow one to accurately reconstruct the massive data needed by a classical probe-compensated NF–FF transformation from a minimum number of NF samples collected through the corresponding non-redundant scan (Figure 1.9). It has been so possible to develop probe-compensated non-redundant NF–FF transformations with plane-polar [40–45], bi-polar [49–51], cylindrical [59–63], and spherical [93–101] scannings, which remarkably reduce, even by several orders of magnitude, the number of the required samples and the related acquisition time. These results have been achieved by taking into account [104] that the EM fields radiated by antennas or the voltage acquired by a non-directive probe [41] on a rotational surface $\mathcal{M}$ can be very well approximated by spatially bandlimited functions, once these antennas are considered as contained in a convex domain bounded by a surface Σ with the same rotational symmetry of $\mathcal{M}$, a proper phase factor is extracted from the field (or voltage) expression, and proper parameterizations are adopted to represent $\mathcal{M}$. The non-redundant sampling representations have been also applied to develop a new effective NF–FF transformation with a non-conventional plane-rectangular scan [108, 109], which has been named planar wide-mesh scanning (PWMS), since its sample grid has meshes becoming larger and larger when moving away from the scanning plane centre (Figure 1.10).

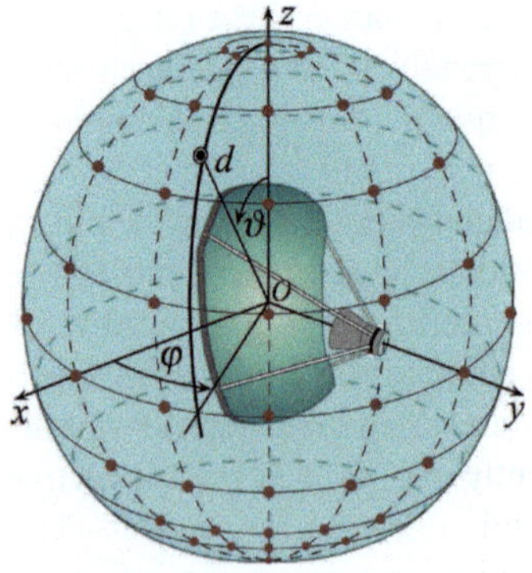

Figure 1.8 NF–FF transformation with spherical scanning

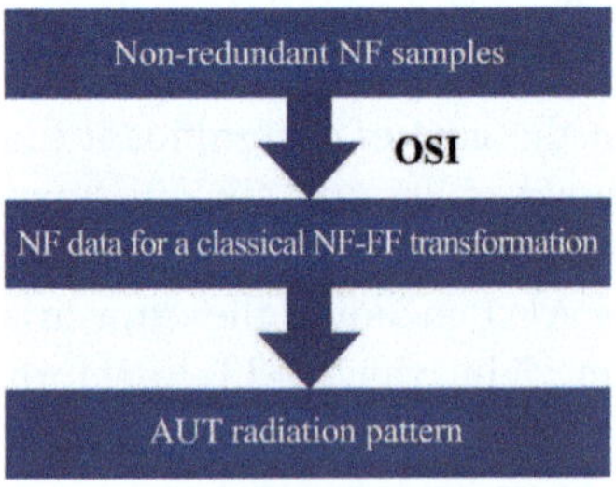

Figure 1.9 Flowchart of the non-redundant NF–FF transformations

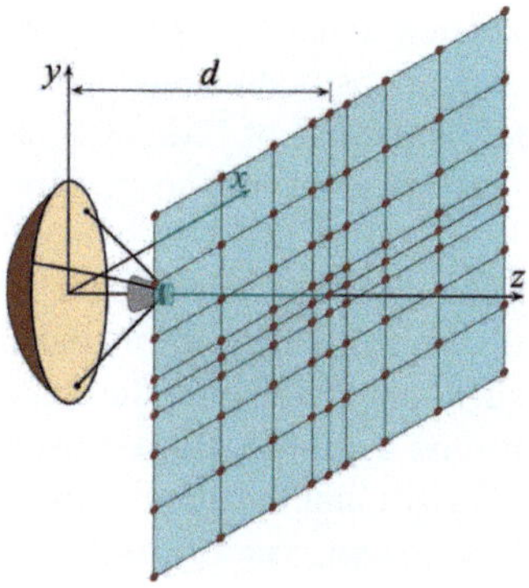

Figure 1.10 *NF–FF transformation with PWMS*

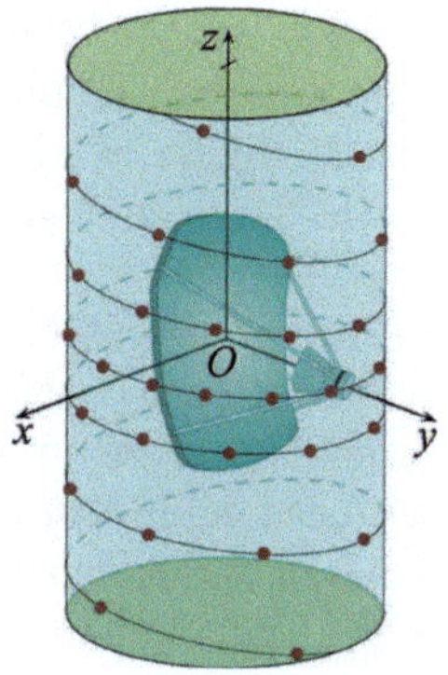

Figure 1.11 *NF–FF transformation with helicoidal scanning*

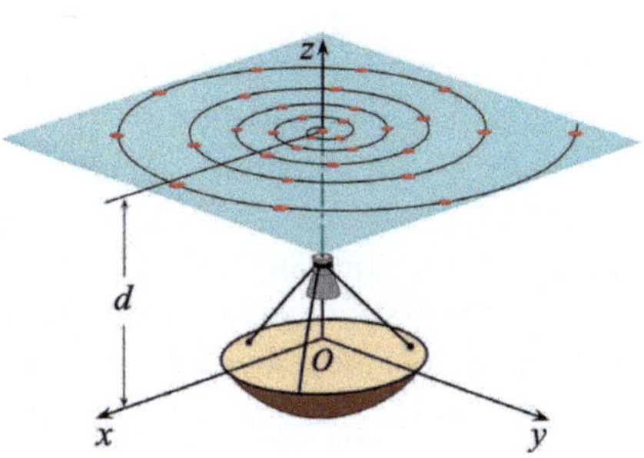

Figure 1.12 *NF–FF transformation with planar spiral scanning*

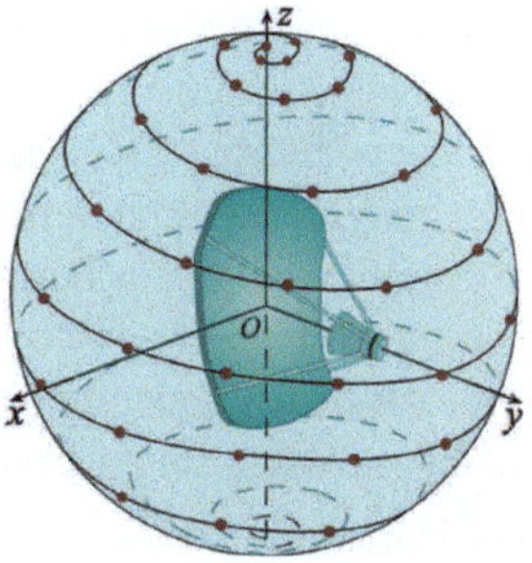

Figure 1.13 *NF–FF transformation with spherical spiral scanning*

It is easy to recognize that the measurement time saving can be achieved not only by reducing the number of needed NF samples, but also by making faster their acquisition. To this end, the modulated scattering technique, which uses arrays of scattering probes in order to realize a very fast electronic scanning, has been proposed in [110, 111]. In any case, apart from the issues related to the measurement accuracy, NF facilities adopting such a technique are not very flexible. The acquisition of the NF samples can be more conveniently sped up by gathering them on fly along spirals drawn through continuous and synchronized motions of the positioners of the probe and AUT, as suggested by Rahmat-Samii *et al.* in [112]. Accordingly, effective non-redundant NF–FF transformations with helicoidal scanning [113–126] (see Figure 1.11), planar [122–130] (see Figure 1.12) and spherical [122–126, 131–141] (see Figure 1.13) spiral scans have been recently proposed. They are based on the non-redundant sampling representations and recover the NF data required by the standard NF–FF transformation employing the corresponding classical scan (cylindrical, plane-rectangular, and spherical), by interpolating, via suitable

OSI expansions, the non-redundant samples gathered along the spiral. For completeness of information, other NF–FF transformations with spiral scannings, which, not making use of the non-redundant representations of EM fields, require an unnecessary large amount of NF measurements, have been also developed in [112, 142, 143].

NF–FF transformation techniques based on the source reconstruction method have recently developed [144–158], as an alternative to those using the modal expansion approach. They exploit Love's equivalence theorem and require the evaluation of a proper set of equivalent electric and/or magnetic currents on a reconstruction surface enclosing the antenna to obtain the field at any point outside the equivalent source domain and, in particular, at those in the FF region. The equivalent current distribution can be determined by properly solving a set of integral equations relating the unknown equivalent currents to the NF data acquired on the scanning surface. The main feature of these techniques is that, unlike those based on the modal expansions, they can be applied to arbitrary scanning surfaces. Their main disadvantage is the considerably large computational effort required to solve the related inverse problem, which involves the solution of an equations system with a very great number of unknowns. Since such a system is usually ill-posed, proper regularization techniques must be applied. In order to reduce the computational time, the use of a graphics processor has been proposed in [155], whereas the multilevel fast multipole method has been exploited in [156, 157]. It is noteworthy that, in these NF–FF transformation techniques, the probe effects can be compensated by considering the FF pattern of the probe as a weighting function in the integral equations to be solved [157, 158].

As already underlined, the NF–FF transformation techniques usually require the acquisition of both amplitude and phase NF data. However, when characterizing antennas working at the millimetre and submillimetre frequency range, accurate phase measurements become increasingly difficult and expensive. Therefore, proper phaseless approaches [159–187] are required to recover the missing phase information. Several phase recovery techniques, which can be arranged in three main groups, are available in the open literature. The first group [159–164] makes use of interferometric or holographic techniques, which require the knowledge of a measured or synthesized interfering signal to retrieve the lacking phase information. A hybrid interferometric/phase-difference approach, which uses two identical probes and a simple microstrip circuit, has been also developed in [142, 165–168]. The second group [169–180] refers to the techniques exploiting the functional relationship which exists between two sets of phaseless NF data gathered by a single probe on two measurement surfaces [169–179] or by two different probes on the same measurement surface [180]. The third group [181–187] refers to the techniques which exploit the source reconstruction method and, accordingly, can be applied also to an arbitrary scanning surface.

Probe-compensated NF–FF transformation techniques, which are suitable for dealing with electrically large antennas and quite arbitrary measurement surface, have been developed in [188–190], by using the plane wave expansion and diagonal translation operators [188]. In particular, the plane waves are related to the

NF values using diagonalized translation operators obtained by efficiently solving the integral equations through fast multipole methods [191, 192] from the NF data acquired at arbitrary sampling points.

An efficient approach to characterize aperture antennas from NF measurements has been developed in [177, 178, 193–199]. In particular, such an approach formulates the NF–FF transformation as a linear inverse problem, wherein the number and the locations of the NF samples are determined, by means of a singular-value optimization procedure, as those leading to the best-conditioned inversion. To this end, it exploits a proper representation of the radiating source in terms of prolate spheroidal wave functions to provide the minimum sized subspace embedding the aperture field.

An algorithm, which properly hybridizes measurements and simulations, has been proposed in [200, 201] to remarkably reduce the number of required NF samples with respect to the conventional Nyquist criteria. The underlying idea is that the simulations, performed through advanced numerical modelling tools and exploiting a-priori information on the geometry and materials of the AUT, can provide the missing quantity of information in the NF sampling. Since the approach is based on the numerical simulations of the AUT, the accuracy of the reconstructions obtained from the reduced set of NF samples may obviously depend on the available information on the AUT in the form of computer aided design and on the quality of the performed simulations.

In the last few years, the antennas and propagation community has shown a growing interest in compressive sensing [202, 203]. Such a technique originates in the area of information theory and concerns the reconstruction of a signal from incomplete frequency information. It can be viewed as a new sampling theorem, since it allows one to recover a signal from a significantly fewer number of data than that required by the traditional sampling techniques. Its application involves the solution of an undetermined system of linear equations, properly built by incorporating a-priori information on the structure of the vector of unknowns, properly represented in a selected basis, and on the system matrix, which has to be sparse and satisfy proper sufficient and/or necessary conditions in order to allow a perfect recovery of the signal. In the antenna measurement framework, the compressive sensing has been applied to inverse problems [204], sparse array synthesis from NF and FF measurements [205–210], and antenna characterization from NF and FF measurements [210–214]. In any case, the application of the compressive sensing to the solution of electromagnetic problems is still at the beginning and the related theoretical issues, mathematical implementations, and numerical features still need to be carefully examined and tackled in progress and future research activities.

In order to reconstruct accurately the FF pattern of the AUT, it is mandatory that the employed positioning systems be able to acquire the NF data exactly at the positions prescribed by the adopted NF–FF transformation technique. Unfortunately, their cost can become prohibitively expensive on increasing the operating frequency. To overcome such a drawback, effective approaches, allowing the accurate retrieval of the data required by the considered NF–FF transformation from the positioning error affected ones, whose locations are determined by means of inexpensive laser interferometric techniques, have been proposed over the years. The early simple and approximated approach to correct three-dimensional positioning errors in the classical NF–FF transformations is the so-called "k-correction technique" [215, 216], which practically consists in adjusting the phase term of the measured NF data. Other more accurate approaches employ iterative techniques [217–221], which result to be rapidly convergent even in presence of rather large positioning errors. More recently, an approach, relying on the conjugate gradient iteration technique and using the fast Fourier transform for nonequispaced data [222], has been applied in [223] and [224] to correct known three-dimensional position errors altering the NF data required by the classical NF–FF transformations with plane-rectangular and spherical scannings, respectively. An interpolation algorithm [225], based on the Yen's technique [226], has been exploited to correct small two-dimensional positioning errors affecting the NF measurements in the case of the classical NF–FF transformation with plane-rectangular scanning. It must be stressed that all the above positioning errors correction techniques are not suitable for the non-redundant NF–FF transformations, wherein the NF data required by the corresponding classical NF–FF transformation are reconstructed through a two-dimensional OSI algorithm from the collected non-redundant samples. As pointed out in [227, 228], a direct reconstruction of these NF data from the position error affected samples is not convenient. In such a case, it is more convenient [227, 228] to first retrieve the non-redundant samples from the acquired ones corrupted by position errors and then determine the needed NF data through an effective OSI algorithm. To this end, two approaches have been developed. The former relies on the singular value decomposition (SVD) method [229] and has been applied when the two-dimensional problem can be split in the solution of two independent one-dimensional problem, otherwise, a very large computational effort results owing to the growth of the sizes of the involved matrices. The latter relies on an iterative technique which has been found convergent only if it is possible to build a one-to-one correspondence, linking each non-redundant sampling point to the nearest position error affected one. Accordingly, non-redundant NF–FF transformations from position errors affected samples in planar [230–239], cylindrical [240–245], and spherical [246–251] scanning geometries have been developed.

In many cases, spurious reflections within antenna NF measurement facilities represent the largest source of error in the evaluation of the error budget [252]. Recently, in order to identify and then suppress the effects of these spurious reflections, a frequency domain (FD) measurement and post-processing technique consisting in a combination of modal filtering with an out-of-centre antenna measurement, named mathematical absorber reflection suppression (MARS), (see [13] for the detailed treatment of the technique and its application) has been successfully applied in planar [253, 254], cylindrical [255, 256], and spherical [256–258] scanning NF facilities, as well as in direct FF measurements [259, 260]. Other techniques to mitigate the effects of spurious reflections on the reconstructed FF pattern are based on the spatial filtering of the measured NF data and exploit the a-priori information on the antenna dimensions. The first approach [12, 261] takes advantage of the low-pass spatial characteristics of the cylindrical and spherical-wave expansions of the field radiated by the AUT, which are related to the minimum cylinder and sphere, respectively, able to contain it. The second approach [12, 261, 262] relies on the reconstruction of equivalent currents and, in such a case, the spatial filtering depends on the size and shape of the chosen reconstruction surface. The third approach exploits the bandlimitation properties of antenna radiated fields and their related non-redundant sampling representations. As a matter of fact, once a suitable phase factor is extracted from the expression of the voltage acquired by the probe and proper parameterizations are adopted for describing the observation surface, the so obtained "reduced voltage" is a spatially quasi-bandlimited function, whose bandwidth depends only on the convex surface adopted to model the antenna [104, 105]. In this framework, the filtering strategy shown in [263] can be applied. In any case, as widely testified from the experimental results reported throughout this book, an effective filtering of the spatial harmonics of the noise exceeding the AUT spatial bandwidth is directly obtained when the NF data required by a classical NF–FF transformation technique are reconstructed from the noisy non-redundant NF samples.

Another serious source of error, which can affect the FF pattern reconstruction obtained from NF measurements, is the truncation of the scanning surface. This error is unavoidable in the NF facilities with planar and cylindrical scannings, but may occur also in that adopting the spherical scanning, since the presence of the AUT positioner can prevent the possibility to perform the measurements on the whole sphere. It can be easily realized that the NF–FF transformation techniques based on the source reconstruction method, wherein the AUT far field is obtained by considering the radiation of reconstructed equivalent electric and/or magnetic currents, do not suffer in principle from this error. As a matter of fact, if the scanning surface is enough large, the equivalent currents are correctly determined and, then, the AUT far field can be accurately evaluated in the whole angular range. As already stressed, the main limitation of such an approach is the considerably large computational effort for the solution of the resulting equations system, usually ill-conditioned and involving a very great number of unknowns. The estimation (extrapolation) of some NF samples falling outside the scanning area can surely allow the reduction of the truncation error. However, since only very few external samples can be estimated with a reasonable accuracy, this strategy results to be useless when applying the classical $\lambda/2$ interpolation techniques. On the contrary, when applying the non-redundant sampling representations [104, 105], the samples spacing increases remarkably when moving far away from the centre of the scanning region. Accordingly, even the estimation of very few samples external to such a region allows a significant extension of the zone wherein the near field is known, thus remarkably enlarging the angular region where an accurate FF reconstruction is achieved. In this framework, a procedure to reduce the truncation error in the NF–FF transformations with plane-polar and cylindrical scannings has been developed in [264] and [265], respectively, by using the cardinal series (CS) expansions and the SVD method [229] for the recovery of the outside data. A more convenient extrapolation procedure employing, on the contrary, the OSI expansions has been proposed in [266, 267] and applied to the plane-polar [268], bi-polar [269], planar wide-mesh [270], cylindrical [271], and spherical [272] scanning cases. A different approach to reduce the truncation error affecting the FF pattern reconstruction obtained from planar NF measurements has been proposed in [273] in the case of aperture antennas. It takes into account that the reliable portion of visible region of the plane wave spectrum is related to the extension of the scanning surface and makes use of the iterative technique proposed by Gerchberg [274] and further investigated by Papoulis [275] to extrapolate the remaining portion of the visible region, subject to the condition that the inverse Fourier transform of the plane wave spectrum is spatially limited to the aperture, i.e., that the fields are different from zero only on the AUT aperture. Such an extrapolation approach has been extended in [276] also to the cylindrical and truncated spherical scanning cases. Another approach, which allows a significant reduction of the truncation error arising when a small antenna is characterized in a NF spherical facility from the NF data collected only on the forward hemisphere, has been proposed in [277]. It exploits a least squares minimization procedure to evaluate the spherical wave expansion coefficients from this set of incomplete NF data.

The ultra wide band (UWB) antennas, widely used in radars and modern telecommunication systems to transmit and/or receive short pulsed signals, must be characterized over a wide frequency band and, accordingly, the use of the NF–FF transformations in the frequency domain are no longer convenient, since these techniques allow to evaluate the AUT far field at a single frequency at time. As a matter of fact, in such a case, a large number of FD NF measurements must be carried out to cover the whole frequency band of interest, so that the acquisition time becomes prohibitively large. Thus, a suitable extension of these transformation techniques to the time domain (TD) is becoming more and more appealing. In this framework, a probe-uncompensated TD NF–FF transformation technique with plane-rectangular scanning has been proposed in [278, 279]. It allows to evaluate the far field of an UWB antenna from the knowledge of the time derivative of its near field at the points of a rectangular grid, whose spacings are smaller than $\lambda_{min}/2$, with λ_{min} being the wavelength corresponding to the maximum frequency of interest. At each of these points, the data must be acquired at a time interval lower than $\lambda_{min}/2c$, with c being the light velocity in the free space. Accordingly, by using a short pulsed signal, it is possible to evaluate the AUT far field over the entire frequency band of interest from a single set of NF measurements. The probe-corrected version of such a TD NF–FF transformation has been also developed in [280] (see also [281]). A TD probe-compensated NF–FF transformation with spherical scanning has been also proposed in [282]. The study on the influence of the NF spatial sampling spacing, the measurement distance, and the scanning area truncation has been presented in [283] in the case of the TD NF–FF transformation with plane-rectangular scanning, whose experimental assessment has been also provided in [284]. A NF–FF transformation with cylindrical scanning from TD NF measurements has been proposed in [285]. It exploits the Fourier transform to obtain the NF data at each frequency of interest and then evaluates the FF pattern through the classical FD NF–FF transformation with cylindrical scanning. TD gating techniques have been profitably exploited in [286, 287] to remove multiple reflections from the measured NF data in the case of the TD NF–FF transformation with spherical scanning, thus enabling the use of a non-perfectly anechoic chamber. Effective OSI algorithms allowing a fast and accurate reconstruction of the TD NF data required by the NF–FF transformations [278, 279] from a minimum number of TD NF samples collected by a plane-polar, bi-polar, or PWMS NF facility have been developed in [288, 289], [290], and [291], respectively.

1.2 The versatile NF facility at the UNISA antenna characterization lab

The versatile NF facility (see Figure 1.14) available in the anechoic chamber of the Antenna Characterization Lab of the UNIversity of SAlerno (UNISA) is described in this section, since it has been employed to carry out all the experimental results shown throughout the book. The chamber is 8 m × 5 m × 4 m sized, with a 1m × 1m × 1m (–50 dB) quiet zone, working up to 38 GHz. Its walls, ceiling, and floor are

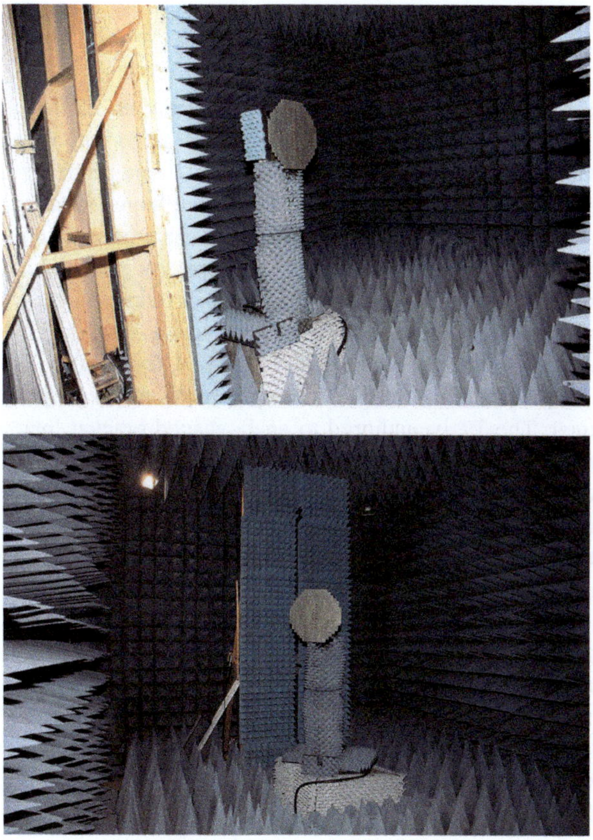

Figure 1.14 Photos of the versatile NF facility available at the UNISA Antenna Characterization Lab

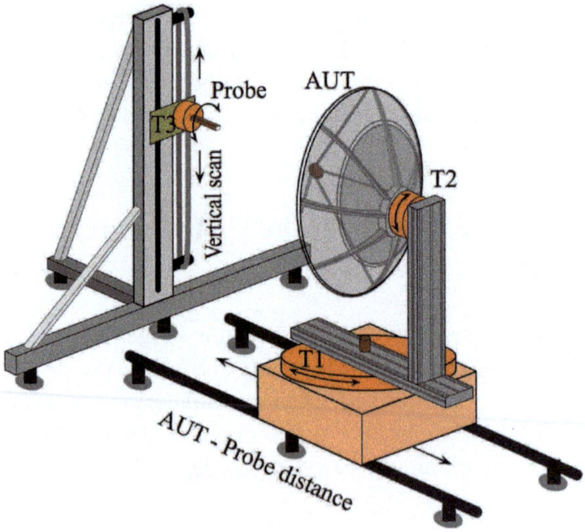

Figure 1.15 Sketch of the versatile NF facility

covered by pyramidal absorbers, which assure a reflection level lower than −40 dB. The NF measurement system is equipped (see Figure 1.15) with several turntables (Ts) and a vertical linear positioner, having height 270 cm and characterized by a linear precision of ±0.005 cm. They can be properly arranged in order to collect the NF data as they would be acquired in a cylindrical, spherical, plane-polar, bi-polar, and plane-rectangular scanning NF facility. Moreover, by means of continuous and synchronized movements of the probe and AUT positioning systems, it is also possible to perform the helicoidal, as well as the planar and spherical spiral scans. In this NF acquisition system, the probe is fixed to the T3, mounted on the linear vertical positioner, and the AUT on the T2, having its rotation axis perpendicular to the vertical positioner. The T2 is anchored to an L–shaped bracket, placed on the T1 (see Figure 1.16), which is in turn mounted on a horizontal slide, so that the distance between the AUT and the probe can be suitably modified. The T1 has an angular precision of ±0.09°, whereas T2 and T3 are characterized by a precision of ±0.05°. The cylindrical and the helicoidal scans are performed by using the vertical linear positioner and the T1. The plane-polar and the planar spiral scannings are realized by means of the vertical linear positioner and the T2. The spherical and the spherical spiral scannings are achieved by exploiting the Ts 1 and 2. In particular, the T2 allows the scanning along a parallel, whereas the T1 allows the change of the scanning parallel, thus realizing a roll-over-azimuth spherical NF system. The T3 (see Figure 1.17) is exploited in all the scannings to allow the gathering of the voltages measured by the probe and the rotated probe. Moreover, it allows the acquisition of the classical plane-rectangular NF data and of the PWMS ones, as well as of the bi-polar ones, by using the same arrangement employed for the plane-polar scanning. In fact, the positions, wherein the related NF data must be acquired, are reached by means of the plane-polar scanning arrangement and the corresponding correct

Figure 1.16 *Photo with the detail of the turntable T1 mounted on the horizontal slide*

Figure 1.17 *Photo with the detail of the turntable T3 mounted on the vertical slide by using a proper counterweight system*

orientations of the probe axes are achieved by means of the T3. Apart from the positioners, the high-performance RF cables, the controller, and the encoders supplied by MI Technologies, the measurement setup includes the vector network analyzer Anritsu 37 247C having high sensitivity, wide dynamic range, and linearity over the range from 40 MHz to 20 GHz, which is employed to perform the measurements of both the amplitude and the phase of the voltage revealed by the probe, a standard open-ended rectangular waveguide. Such a versatile NF facility has been realized by modifying over the years the cylindrical setup previously equipping the UNISA Antenna Characterization Lab. In particular, it has been added a horizontal slide, over which the turntable T1 has been mounted, to allow the change of the distance between the AUT and the probe. Then, the earliest T1 has been substituted with a greater one, which has made possible to mount, on the L–shaped bracket, the table T2 needed to provide the roll axis. The addition of the turntable T3, not present in the supplied NF cylindrical facility, has required to ensure the mechanical balance and stability of the vertical positioner by means of a proper counterweight system (see Figure 1.17). At last, a further turntable, placed in the quiet zone of the anechoic chamber, can be employed to perform direct far-field measurements in the case of electrically small antennas.

1.3 Structure of the book

After this introductive section, wherein an overview of the NF–FF transformation techniques has been presented and a brief description of the versatile NF facility available at the UNISA Antenna Characterization Lab has been provided, the book content is summarized.

Chapter 2 gives a comprehensive description of effective sampling representations of the EM fields, radiated by sources, contained in an arbitrary convex domain bounded by a surface with rotational symmetry, on a surface having the same rotational symmetry. These representations are very appealing, since they require a non-redundant, i.e., minimum, number of field samples, which is always finite also for an unbounded observation surface and depends only on the geometry of the radiating source. The problem of the choice of the optimal phase factor to be extracted from the expression of the field (or of the voltage measured by a probe) and of the proper parameterizations to be adopted to describe the observation surface is tackled and their explicit formulas for quite arbitrary rotational observation and modelling surfaces are determined. These formulas are then particularized to the cases of spherical, spheroidal, and flexible modelling surfaces. The use of CS and OSI expansions for the efficient interpolation of the field (voltage) on the observation surface is discussed, by highlighting the advantages of the OSI ones. At last, the application of the related formulas to the FF interpolation, in the case of spheroidal modellings of the source, is fully described.

In Chapter 3, a unified theory of the NF–FF transformation techniques with spiral scannings for volumetric antennas modelled by a sphere is rigorously derived. This has been accomplished by introducing a sampling representation of the voltage acquired by the measuring probe on a rotational surface from the knowledge of a non-redundant and always finite number of its samples on a spiral wrapping the surface. The obtained results are general, since they are valid for spirals wrapping on quite arbitrary rotational surfaces, and can be directly applied, as shown in the next chapters, to the pattern reconstruction via NF–FF transformation techniques. The unified theory of the NF spiral scannings is then heuristically extended in the same chapter to the case of non-spherical antennas, considered as enclosed by spheroidal and flexible modelling surfaces, which allow a better fitting of the actual geometry of the AUT. This gives a formidable tool to overcome the drawbacks related to the spherical AUT modelling, thus making the spiral scannings more and more appealing from the practical viewpoint. The application of these results to the case of elongated antennas and a spiral lying on a prolate spheroid is explicitly considered in the numerical simulations. It is worthy to note that, although such a case has been tackled for its generality, the related OSI algorithm allows one to recover the NF data required by the NF–FF transformation process [292].

Chapter 4 deals with the NF–FF transformations with planar scannings. The classical plane-rectangular NF–FF transformation without probe compensation is first detailed and the effect of the sizes of the scanning region on the accuracy of the recovered far field is outlined. A two-dimensional OSI scheme for evaluating the FF components from the knowledge of the data obtained via the NF–FF transformation is also presented. The key steps of the probe-compensated NF–FF transformation with plane-rectangular scanning, as proposed by Paris *et al.*, are then described, by highlighting the need of the probe compensation. A NF–FF transformation technique using the innovative planar PWMS is also presented. Its main advantage is to make use of a reduced number of NF data with respect to that employing the classical plane-rectangular scanning. Non-redundant NF–FF transformations with plane-polar and bi-polar scannings are then described. In both the cases, the massive NF data required by the classical probe-compensated plane-rectangular NF–FF transformation are obtained by interpolating by means of proper OSI formulas the acquired NF samples. The described techniques reduce significantly the number of needed NF data without losing the efficiency of previous approaches. Moreover, it is shown that the use of a probe, whose radiated far field exhibits a first-order azimuthal dependence, allows one to avoid the probe co-rotation with the AUT. At last, the unified theory of spiral scannings for non-volumetric AUTs is applied to develop effective non-redundant NF–FF transformations with planar spiral scanning. Numerical and experimental results validating the effectiveness of the described NF–FF transformations are shown. Analytical details on the plane wave spectrum representation of the field radiated by an antenna and on its relation with the corresponding far field are also reported for reader's convenience.

Chapter 5 is concerned with the cylindrical scanning. Analytical details on the derivation of the classical cylindrical NF–FF transformation without probe compensation are first reported and an OSI algorithm to evaluate the FF components from the FF data reconstructed by means of the NF–FF transformation is also described. The main steps of the basic theory of the probe-compensated NF–FF transformation with cylindrical scanning as developed by Paris *et al.* are then reported for reader's convenience. Also for such a NF–FF transformation, the need of the probe compensation is highlighted and the effect of the scanning cylinder height on the accuracy of the recovered far field is discussed. Non-redundant NF–FF transformations with cylindrical scanning are then described. They exploit a two-dimensional OSI expansion to accurately reconstruct the voltage values at the points needed to perform the classical probe-compensated NF–FF transformation from a non-redundant amount of NF samples. At last, effective non-redundant NF–FF transformations with helicoidal scanning, developed by properly applying the unified theory of spiral scannings for non-volumetric AUTs, are presented. Representative numerical and experimental reconstruction examples assessing the effectiveness of the described NF–FF transformations are reported. Analytical details on the cylindrical wave expansion of the antenna field and on the steps to get its expression valid in the FF region conclude the chapter.

The last chapter addresses the problem of the FF reconstruction from spherical NF measurements. The classical spherical NF–FF transformation without probe compensation and its modified version, which takes into account the spatial band-limitation properties of the EM fields, are first detailed and an effective OSI formula to determine the FF components from the FF data recovered through the NF–FF transformation is also presented. The key steps of the classical and modified probe-compensated NF–FF transformations with spherical scanning are then described, by highlighting the need of the probe compensation, which, in such a case, is less significant than in planar and cylindrical NF–FF transformations, since the probe is physically pointed toward the AUT centre during the scanning. Non-redundant sampling representations, using suitable modellings for elongated and quasi-planar AUTs, and the related OSI algorithms, allowing to accurately reconstruct from the probe voltage samples the massive NF data required by the classical spherical NF–FF transformation, are presented. At last, the use of the unified theory of spiral scannings for non-volumetric AUTs to develop efficient non-redundant NF–FF transformations with spherical spiral scanning for elongated and quasi-planar antennas is described. Numerical and experimental results are also presented in order to show the effectiveness of these NF–FF transformation techniques. Analytical details on the spherical wave expansion of the field radiated by an antenna are also provided and its expression valid in the antenna FF region is derived.

To improve the book readability, some fundamental background results are collected in the Appendices to the book, whereas long and non-trivial analytical details supporting the derivation of the formulas relevant to a particular chapter are reported in an appendix to the chapter.

Chapter 2

Non-redundant sampling representations of electromagnetic fields

Aim of this chapter is to give a comprehensive description of effective sampling representations of the electromagnetic (EM) fields, radiated by sources contained in an arbitrary convex domain bounded by a surface with rotational symmetry on a surface having the same rotational symmetry. These representations are very appealing since they require a non-redundant, i.e., minimum, number of field samples. It will be shown that such a number is always finite also for an unbounded observation surface, depends only on the geometry of the source, and essentially coincides with the degrees of freedom of the field.

It is useful to note that the sampling representations of the EM fields are typically more convenient than those using asymptotic or modal expansions, because the related coefficients are the field samples, namely, the directly available computed or measured values at the sampling points, and the basis functions are universal and simple. Therefore, they can be exploited to obtain very efficient representations of the radiated EM fields on quite arbitrary surfaces. To achieve these results, a suitable phase factor is extracted from the field expressions and proper parameterizations are adopted for describing the observation surface. Then, optimal sampling expansions of central type are used to accurately reconstruct the radiated EM field on such a surface from the knowledge of its samples.

2.1 The choice of the optimal parameterization and phase factor

Let $\boldsymbol{E}$ be the field radiated by finite size sources S, enclosed in a convex domain $\mathcal{D}$ bounded by a surface Σ with rotational symmetry and observed on a regular surface $\mathcal{M}$ external to $\mathcal{D}$ and with the same rotational symmetry (Figure 2.1). In the following, it is dealt with the representation of the field over a regular curve C described by a parameterization $\boldsymbol{r} = \boldsymbol{r}(\xi)$, because $\mathcal{M}$ can be described by two families of coordinate curves, that is to say, meridian curves and azimuthal circumferences. As shown in [102, 104], the "bandlimitation" error, arising when the "reduced electric field"

$$\boldsymbol{F}(\xi) = \boldsymbol{E}(\xi)\, \mathrm{e}^{\mathrm{j}\gamma(\xi)} \tag{2.1}$$

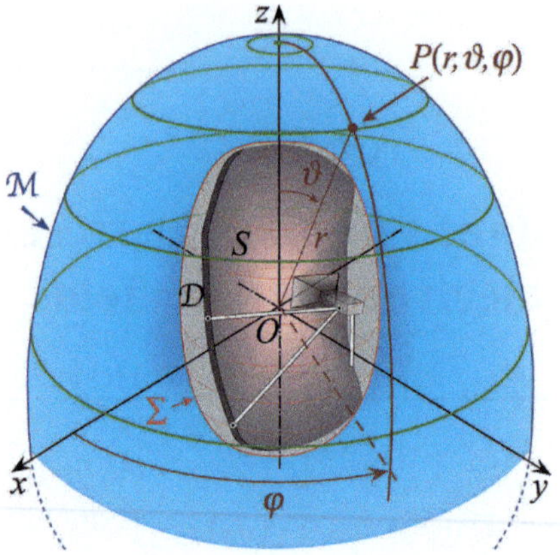

Figure 2.1 Geometry of the problem

is approximated by a bandlimited function, exhibits a step-like behaviour becoming negligible as the spatial bandwidth is greater than the threshold value

$$W_\xi = \max_\xi \left[w(\xi) \right] = \max_\xi \left[\max_{r'} \left| \frac{\mathrm{d}\gamma(\xi)}{\mathrm{d}\xi} - \beta \frac{\partial R(\xi, \boldsymbol{r}')}{\partial \xi} \right| \right] \tag{2.2}$$

where $w(\xi)$ is the so called local bandwidth, $\gamma(\xi)$ is an appropriate phase function to be determined, β is the free-space wavenumber, $\boldsymbol{r}'$ denotes the source point, and $R = |\boldsymbol{r}(\xi) - \boldsymbol{r}'|$. Thus, such an error can be controlled in an effective way by approximating the reduced field with a function having bandwidth equal to $\chi' W_\xi$, where χ' is an enlargement bandwidth factor (slightly greater than unity for antennas whose sizes are large with respect to the wavelength λ).

In order to get a non-redundant (NR) representation, the local bandwidth $w(\xi)$ has to be minimized for each ξ. This is achieved [104, 105] by choosing γ such that its derivative is equal to the average between the minimum and maximum value of $\beta\, \partial R/\partial \xi$, when $\boldsymbol{r}'$ varies in the source domain $\mathcal{D}$, namely,

$$\frac{\mathrm{d}\gamma(\xi)}{\mathrm{d}\xi} = \frac{\beta}{2} \left[\max_{r'} \frac{\partial R}{\partial s} + \min_{r'} \frac{\partial R}{\partial s} \right] \frac{\mathrm{d}s}{\mathrm{d}\xi} \tag{2.3}$$

where s is the curvilinear abscissa along the curve $\mathcal{C}$. As a consequence, the appropriate phase function is:

$$\gamma(\xi) = \frac{\beta}{2} \int_0^{s(\xi)} \left[\max_{r'} \frac{\partial R}{\partial s} + \min_{r'} \frac{\partial R}{\partial s} \right] \mathrm{d}s = \frac{\beta}{2} \int_0^{s} \left[\max_{r'} \hat{R} \cdot \hat{t} + \min_{r'} \hat{R} \cdot \hat{t} \right] \mathrm{d}s^{*} \tag{2.4}$$

*By differentiating the relation $R^2 = \boldsymbol{R} \cdot \boldsymbol{R}$ with respect to s, it results $\partial R / \partial s = \hat{R} \cdot \partial \boldsymbol{R} / \partial s$. Since $\boldsymbol{r} = \boldsymbol{r}' + \boldsymbol{R}$, it follows that $\partial \boldsymbol{R} / \partial s = \mathrm{d}\boldsymbol{r}/\mathrm{d}s = \hat{t}$ and, accordingly, $\partial R / \partial s = \hat{R} \cdot \hat{t}$.

where $\hat{t}$ is the unit vector tangent to $\mathcal{C}$ at the observation point P, $\hat{R}$ is the unit vector from the source point to P, the symbol $(\cdot)$ indicates the scalar product, and it has been assumed that $\gamma(0) = 0$. It can be easily verified that, according to the choice done for $\mathrm{d}\gamma/\mathrm{d}\xi$, it results:

$$w(\xi) = \frac{\beta}{2}\left[\max_{r'}\frac{\partial R}{\partial s} - \min_{r'}\frac{\partial R}{\partial s}\right]\frac{\mathrm{d}s}{\mathrm{d}\xi} = \frac{\beta}{2}\left[\max_{r'}\hat{R}\cdot\hat{t} - \min_{r'}\hat{R}\cdot\hat{t}\right]\frac{\mathrm{d}s}{\mathrm{d}\xi} \tag{2.5}$$

To avoid redundancy, the parameterization $\boldsymbol{r}(\xi)$ must be determined [104] by imposing that the local bandwidth is constant and, as consequence, equal to W_ξ. In fact, if $w(\xi)$ varies with ξ, the sample spacing, which is fixed by the bandwidth W_ξ, becomes uselessly small in the zones where the local bandwidth is smaller than its maximum value W_ξ, thus causing a redundancy in the sampling representation. Accordingly, by enforcing the constancy of $w(\xi)$ and assuming the origin for ξ coincident with that for the curvilinear abscissa, it follows that:

$$\xi = \frac{\beta}{2W_\xi}\int_0^{s(\xi)}\left[\max_{r'}\frac{\partial R}{\partial s} - \min_{r'}\frac{\partial R}{\partial s}\right]\mathrm{d}s = \frac{\beta}{2W_\xi}\int_0^{s}\left[\max_{r'}\hat{R}\cdot\hat{t} - \min_{r'}\hat{R}\cdot\hat{t}\right]\mathrm{d}s \tag{2.6}$$

It is worthy to observe that a change of the bandwidth W_ξ implies only a simple change of scale in the parameterization $\boldsymbol{r}(\xi)$.

The case of a meridian observation curve $\mathcal{C}$, i.e., a curve lying on a plane containing the rotation axis z, is now considered. It can be recognized through very simple geometrical considerations that, if this curve is such that the unit vector $\hat{t}$ is always external to the cone of vertex P tangent to the surface Σ, the minimum and maximum values of the scalar product $\hat{R}\cdot\hat{t}$ take place at the two tangency points $P_{1,2}$ on the intersection curve C' between the meridian plane Π and Σ (see Figure 2.2). By taking into account that in such a case [104, 105]

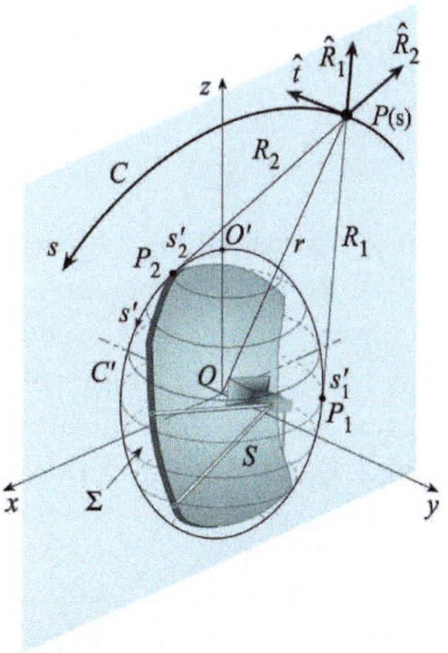

Figure 2.2 Relevant to the representation along a meridian curve

$$\left.\frac{\partial R}{\partial s}\right|_{s'_{1,2}} = \frac{\mathrm{d}R_{1,2}}{\mathrm{d}s} \pm \frac{\mathrm{d}s'_{1,2}}{\mathrm{d}s}^{\dagger} \tag{2.7}$$

where $s'_{1,2}$ are the curvilinear abscissas of $P_{1,2}$ and $R_{1,2}$ are the distances from P to $P_{1,2}$ (Figure 2.2), relations (2.4) and (2.6) become

$$\gamma = \frac{\beta}{2}\left[R_1 + R_2 + s'_1 - s'_2\right] \tag{2.8}$$

$$\xi = \frac{\beta}{2W_\xi}\left[R_1 - R_2 + s'_1 + s'_2\right] \tag{2.9}$$

Since the optimal parameter ξ is an angle like coordinate, whose variation is equal to $\beta\ell'/W_\xi$ when P, moving on a closed meridian curve, makes a complete turn around the source, it is convenient to choose W_ξ in such a way that this variation equals 2π. Accordingly,

$$W_\xi = \beta\ell'/2\pi \tag{2.10}$$

where ℓ' is the length of the curve C'. Then the expression of ξ becomes

$$\xi = \frac{\pi}{\ell'}\left[R_1 - R_2 + s'_1 + s'_2\right] \tag{2.11}$$

According to relations (2.8) and (2.11), the phase function γ and the parameter ξ depend only on the point P and are independent of the particular meridian curve passing through it, moreover, their evaluation is reduced to that of quantities with very simple geometric meanings. It must be stressed that the number of samples at Nyquist spacing ($\Delta\xi = \pi/W_\xi$) on any closed meridian curve C (also unbounded) is always finite and equal to $N_\xi = 2\pi/\Delta\xi = \ell'/(\lambda/2)$.

By differentiating relations (2.8) and (2.11) with respect to s and taking into account (2.7), it follows:

$$\frac{\mathrm{d}\gamma}{\mathrm{d}s} = \frac{\beta}{2}\frac{\mathrm{d}}{\mathrm{d}s}\left[R_1 + R_2 + s'_1 - s'_2\right] = \frac{\beta}{2}\frac{\partial}{\partial s}\left(R_1 + R_2\right) = \frac{\beta}{2}\left(\hat{R}_1 + \hat{R}_2\right)\cdot\hat{t} \tag{2.12}$$

$$\frac{\mathrm{d}\xi}{\mathrm{d}s} = \frac{\pi}{\ell'}\frac{\mathrm{d}}{\mathrm{d}s}\left[R_1 - R_2 + s'_1 + s'_2\right] = \frac{\pi}{\ell'}\frac{\partial}{\partial s}\left(R_1 - R_2\right) = \frac{\pi}{\ell'}\left(\hat{R}_1 - \hat{R}_2\right)\cdot\hat{t} \tag{2.13}$$

Accordingly, the curves at γ = constant and ξ= constant passing through P are perpendicular to $\hat{R}_1 + \hat{R}_2$ and $\hat{R}_1 - \hat{R}_2$, respectively, and, therefore, are orthogonal.

The case of an azimuthal observation circumference C at z of radius ρ is now considered (Figure 2.3). In this case, because of the rotational symmetry of Σ, the minimum and maximum values of $\hat{R}\cdot\hat{t}$ are opposite and constant along the circumference. Hence, from relations (2.4) and (2.6), it follows that the phase function is

†By differentiating the relation $R_{1,2} = |\mathbf{r}(s) - \mathbf{r}'(s'_{1,2})| = R(s, s'_{1,2}) = R[s, s'_{1,2}(s)]$ with respect to s, it follows that: $\frac{\mathrm{d}R_{1,2}}{\mathrm{d}s} = \left.\frac{\partial R}{\partial s}\right|_{s'_{1,2}} + \frac{\partial R}{\partial s'_{1,2}}\frac{\mathrm{d}s'_{1,2}}{\mathrm{d}s} = \left.\frac{\partial R}{\partial s}\right|_{s'_{1,2}} \mp \frac{\mathrm{d}s'_{1,2}}{\mathrm{d}s}$ since, being $P_{1,2}$ tangency points, $\partial R/\partial s'_{1,2} = \mp 1$. Accordingly, (2.7) follows.

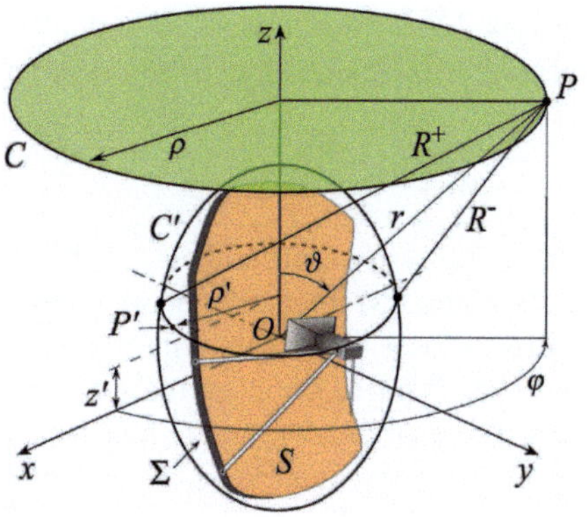

Figure 2.3 Relevant to the representation along an azimuthal circumference

constant and any parameter proportional to the curvilinear abscissa s is optimal. Therefore, the azimuthal angle φ can be conveniently adopted as parameter and the phase function relevant to any meridian curve passing through the observation point on C can be chosen as function γ. As shown in Appendix 2A.1, the corresponding bandwidth is [104]:

$$W_\varphi = \frac{\beta}{2}\max_{z'}\left(R^+ - R^-\right)$$

$$= \frac{\beta}{2}\max_{z'}\left(\sqrt{(z-z')^2 + (\rho+\rho'(z'))^2} - \sqrt{(z-z')^2 + (\rho-\rho'(z'))^2}\right) \quad (2.14)$$

It is worthy to note that the maximum in (2.14) is attained [104] on the part of the surface Σ which lies on the same side of C with respect to its maximum transverse circumference (see Appendix 2A.1).

In the limiting case of an observation circumference C whose radius approaches infinity, $R^+ = R^- + 2\rho'\sin\vartheta$ and, as a consequence, it results:

$$W_\varphi = \beta\rho'_{\max}\sin\vartheta \quad (2.15)$$

where $\rho'_{\max}$ is the maximum transverse radius of Σ.

By employing the triangular inequality, it can be easily recognized that $R^+ - R^- \leq 2\rho'$ and, accordingly, the following sharp bound for W_φ follows:

$$W_\varphi = \frac{\beta}{2}\max_{z'}\left(R^+ - R^-\right) \leq \beta\rho'_{\max} \quad (2.16)$$

On the other hand, when the observation circumference C collapses on Σ, as shown in Appendix 2A.1, it results:

$$W_\varphi \underset{\rho\to\rho'(z)}{\approx} \beta\rho'(z) \quad (2.17)$$

As shown in Appendix 2A.2, the number N of samples at Nyquist spacing on any closed observation surface (also unbounded) surrounding the sources is [104]

$$N \cong \frac{\text{area of } \Sigma}{(\lambda/2)^2} \tag{2.18}$$

Namely, such a number is always finite and essentially coincident with the number of degrees of freedom of the EM field radiated by them [103].

Relation (2.18) clearly stresses the role that the source modelling plays in minimizing the overall number of required samples: the surface Σ must fit very well the shape of the antenna, must be analytically regular and such that the optimal parameter and phase function can be determined in a possibly simple way.

As representative example, the NR sampling representation based on the spherical modelling of the antenna is explicitly considered in the following. NR sampling representations using more effective models for quasi-planar or elongated antennas are presented in Sections 2.2 and 2.3.

By choosing as surface Σ a sphere of radius a, for any meridian curve C, it results (Figure 2.4):

$$R_1 = R_2 = \sqrt{r^2 - a^2};\ s_1' = (\vartheta - \alpha)\,a;\ s_2' = (\vartheta + \alpha)\,a;\ \alpha = \cos^{-1}(a/r) \tag{2.19}$$

Moreover, since $\ell' = 2\pi a$, it results $W_\xi = \beta a$ and

$$\gamma = \beta\sqrt{r^2 - a^2} - \beta a \cos^{-1}(a/r); \qquad \xi = \vartheta \tag{2.20}$$

For an azimuthal circumference, by putting $z' = a\cos\vartheta'$ and $\rho' = a\sin\vartheta'$, it can be shown [104] that the maximum in (2.14) is achieved in correspondence of $\vartheta' = \cos^{-1}(a\cos\vartheta/r)$ and, accordingly,

$$W_\varphi = \beta a \sin\vartheta \tag{2.21}$$

An effective modelling for flat antennas is a disk Σ (see Figure 2.5) with diameter $2a$ equal to the antenna maximum dimension. In such a case, the curve C' has

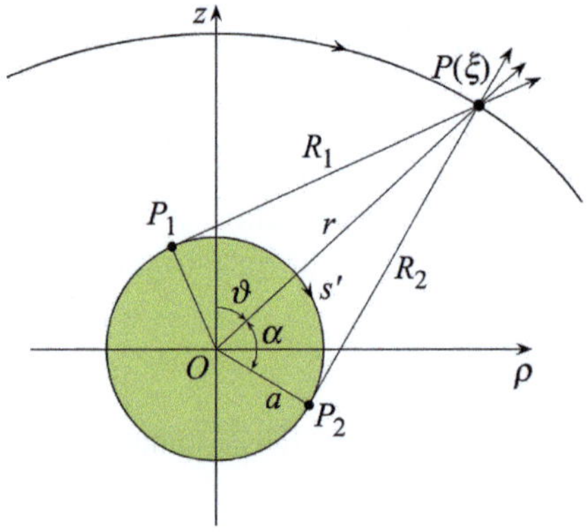

Figure 2.4 Spherical modelling of the antenna

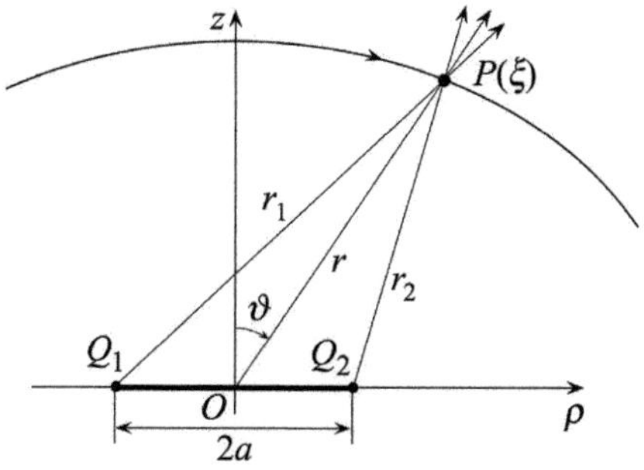

Figure 2.5 *Disk modelling of a flat antenna*

length $\ell' = 4a$ and, accordingly, $W_\xi = 4a/\lambda$. The phase function γ and the optimal parameter ξ relevant to a meridian curve C, specified by the polar coordinates $(r(\vartheta), \vartheta)$, can be obtained by substituting the values of the distances $R_{1,2}$ and curvilinear abscissae $s'_{1,2}$ in (2.8) and (2.11). It can be easily recognized that the expressions of $R_{1,2}$ and $s'_{1,2}$ change depending on the positions of the points $P_{1,2}$, which in turn depend on the angle ϑ. When $0 \le \vartheta \le \pi/2$, the point P_1 coincides with the point Q_1 and P_2 with Q_2, so that $s'_1 = -a$, $s'_2 = a$ and, hence,

$$\gamma = \frac{\beta}{2}\left[r_1 + r_2 - 2a\right]; \qquad \xi = \frac{\pi}{4a}\left[r_1 - r_2\right] \tag{2.22}$$

When $\pi/2 < \vartheta \le \pi$, the point P_1 coincides with the point Q_2 and P_2 with Q_1, so that $s'_1 = a$, $s'_2 = 3a$ and, hence,

$$\gamma = \frac{\beta}{2}\left[r_1 + r_2 - 2a\right]; \qquad \xi = \frac{\pi}{4a}\left[r_2 - r_1 + 4a\right] \tag{2.23}$$

It is noteworthy that the parameter ξ is not analytic for $\vartheta = \pi/2$. As a consequence, the bandlimitation error can significantly increase in the neighbourhoods of such a direction [104].

From relation (2.14), it can be easily realized that the bandwidth relevant to an azimuthal circumference is given by:

$$W_\varphi = \frac{\beta}{2}|r_1 - r_2| \tag{2.24}$$

2.2 Spheroidal source modellings

In this section, effective NR sampling representations, attained by modelling an elongated antenna with a prolate spheroid‡ having major and minor semi-axes equal to a and b (see Figure 2.6) and a quasi-planar one with an oblate spheroid

‡A prolate spheroid is obtained by rotating an ellipse around its major axis, whereas an oblate spheroid is got by means of a rotation around the minor axis.

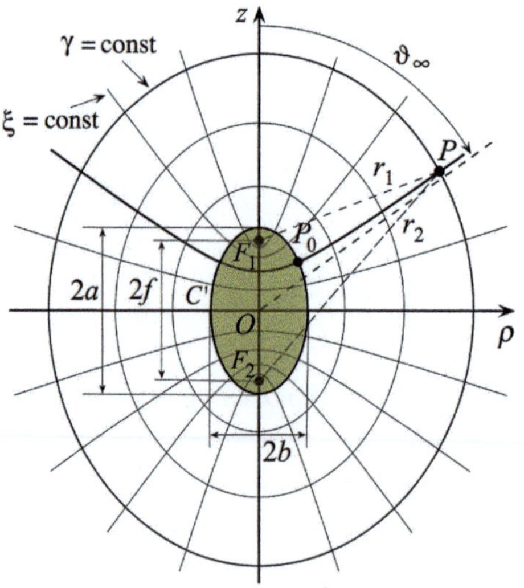

Figure 2.6 Spheroidal modelling of the antenna: the prolate case

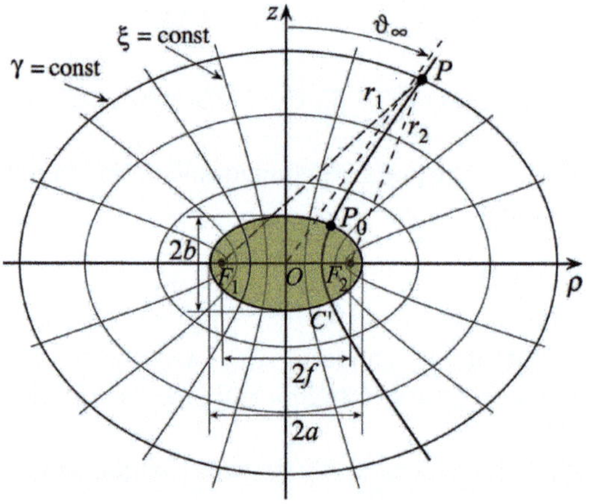

Figure 2.7 Spheroidal modelling of the antenna: the oblate case

(see Figure 2.7), are presented. These modellings are quite general and contain the spherical one as particular case. As shown in Appendix 2A.3, in both the cases, the curves ξ = constant and γ = constant in any meridian plane are hyperbolas and ellipses having the same foci of the ellipse C' [104]. Accordingly, γ and ξ are functions only of the elliptic coordinates $v = (r_1 + r_2)/2a$ and $u = (r_1 - r_2)/2f$, respectively, $2f$ being the focal distance and $r_{1,2}$ the distances from the foci to the observation point P. According to relation (2.10), since $\ell' = 4a\,\mathrm{E}(\pi/2|\varepsilon^2)$, it results:

$$W_\xi = \frac{4a}{\lambda}\mathrm{E}(\pi/2|\varepsilon^2) \tag{2.25}$$

wherein

$$\mathrm{E}\left(\alpha|\varepsilon^2\right) = \int_0^\alpha \sqrt{1-\varepsilon^2 \sin^2\alpha}\, \mathrm{d}\alpha \tag{2.26}$$

is the elliptic integral of second kind [293] and $\varepsilon = f/a$ is the eccentricity of $\mathcal{C}'$.

Since the value of γ is the same for all points lying on a confocal ellipse, the evaluation of γ at P is simplified by considering, instead of P itself, the intersection point between the z axis and the confocal ellipse passing through P. As shown in Appendix 2A.3, it results [104]:

$$\gamma = \beta a\left[v\sqrt{\frac{v^2-1}{v^2-\varepsilon^2}} - \mathrm{E}\left(\cos^{-1}\sqrt{\frac{1-\varepsilon^2}{v^2-\varepsilon^2}}\,|\,\varepsilon^2\right)\right] \tag{2.27}$$

Analogously, the evaluation of ξ at P is simplified by considering, instead of P itself, the point P_0 resulting from the intersection between the hyperbola branch through P and the ellipse $\mathcal{C}'$. As shown in Appendix 2A.3, it results [104]:

$$\xi = \frac{\pi}{2}\begin{cases} \mathrm{E}(\sin^{-1} u|\varepsilon^2)/\mathrm{E}(\pi/2|\varepsilon^2) + 1 & \text{prolate spheroid} \quad (2.28a) \\ \mathrm{E}(\sin^{-1} u|\varepsilon^2)/\mathrm{E}(\pi/2|\varepsilon^2) & \text{oblate spheroid} \quad (2.28b) \end{cases}$$

Above relations stay valid when the angle ϑ corresponding to the point P belongs to the range $[0, \pi/2]$. The case where $\pi/2 < \vartheta \leq \pi$ is easily handled by determining the value $\bar{\xi}$ of ξ corresponding to the point specified by the angle $\pi - \vartheta$ on the symmetrical hyperbola branch and then putting $\xi = \pi - \bar{\xi}$.

With reference to an azimuthal circumference, it has been be shown in [104] that the azimuthal bandwidth W_φ is the same for all the transverse circumferences on the hyperboloid of rotation fixed by a value of ξ. Therefore, it can be determined by moving the circumference to infinity along the hyperbola ξ = constant. Accordingly, by taking (2.15) into account, it results:

$$W_\varphi(\xi) = \begin{cases} \beta b \sin\vartheta_\infty(\xi) & \text{prolate spheroid} \quad (2.29a) \\ \beta a \sin\vartheta_\infty(\xi) & \text{oblate spheroid} \quad (2.29b) \end{cases}$$

where ϑ_∞ is the zenith angle of the asymptote to the hyperbola passing through P (Figures 2.6 and 2.7) and (see Appendix 2A.3) is given by:

$$\vartheta_\infty = \begin{cases} \sin^{-1} u + \pi/2 & \text{prolate spheroid} \quad (2.30a) \\ \sin^{-1} u & \text{oblate spheroid} \quad (2.30b) \end{cases}$$

As previously observed, the spherical modelling can be obtained from the spheroidal ones as particular case. In fact, it can be handled by considering a spheroid with eccentricity $\varepsilon = 0$. Accordingly, the curves γ = constant are circumferences, whereas the curves ξ = constant become radial lines. For a meridian curve, it results $W_\xi = \beta\ell'/2\pi = \beta a$, and, by taking into account that in such a case $v = r/a$, $\vartheta_\infty = \vartheta$

and E $(\vartheta/0) = \vartheta$, from (2.27) and (2.28) relations (2.20) are obtained. When C is a circumference, from (2.29) it follows (2.21).

Also the disk modelling can be obtained from the spheroidal ones as particular case. As a matter of fact, it can be easily verified [104] that such a modelling can be handled by considering an oblate spheroid with eccentricity $\varepsilon = 1$.

In order to highlight the importance of the introduction of the reduced field $\boldsymbol{F}$ for obtaining a NR representation, the real and imaginary part of the electric field and reduced electric field y-component radiated by an antenna along a line are shown in Figure 2.8a and b. The considered antenna is a uniform planar circular array with radius equal to 21 λ, lying on the plane $z = 0$. Its elements are elementary Huygens sources linearly polarized along the y axis, symmetrically placed with respect to the x and y axes, and spaced by about 0.7 λ along circumferences, radially spaced by 0.7 λ. Accordingly, it has been modelled by an oblate spheroid with $2a = 42\lambda$ and $2b = 5\lambda$.

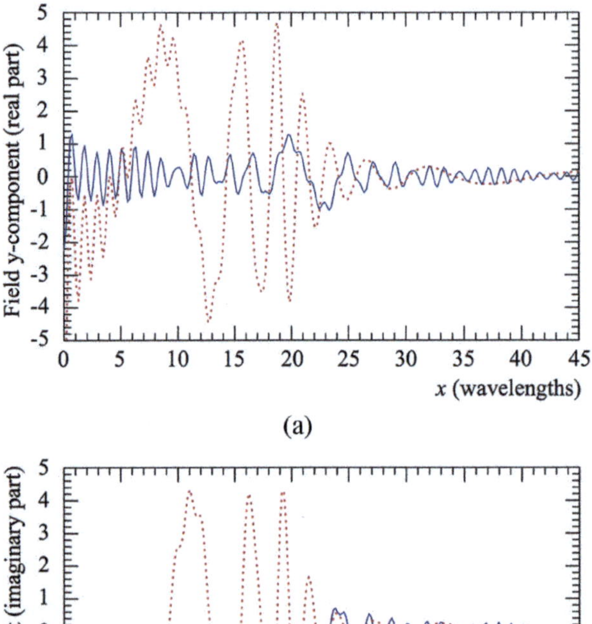

(a)

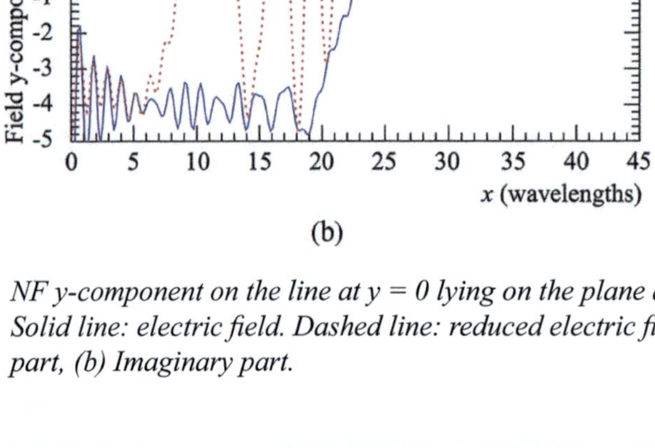

(b)

Figure 2.8 *NF y-component on the line at $y = 0$ lying on the plane at $z = 10\lambda$. Solid line: electric field. Dashed line: reduced electric field. (a) Real part, (b) Imaginary part.*

The considered straight line is the line at $y = 0$ lying on the plane at distance $d = 10\lambda$ from the antenna. As can be seen, on increasing x, the spatial variations of the real and imaginary part of the reduced field become slower and slower with respect to the those of the real and imaginary part of the electric field, since the local bandwidth $w(\xi)$ has been minimized due to the optimal choice of the phase function.

To point out the role of the optimal parameter for a convenient representation of the EM field on the observation curve, the amplitude of the electric field y-component along the same straight line is plotted as function of x and ξ in Figures 2.9 and 2.10. As can be seen, when adopting the optimal parameter, the regions where the field exhibits fast changes are displayed in an enlarged scale, whereas those characterized by slow variations are represented in a reduced scale. Thus, a uniform sampling in the optimal parameter ξ gives rise to a denser distribution of the samples where the local bandwidth $w(\xi)$ is greater and to a sparser one in the zones characterized by smaller values of $w(\xi)$ (see also Sect. 2.5).

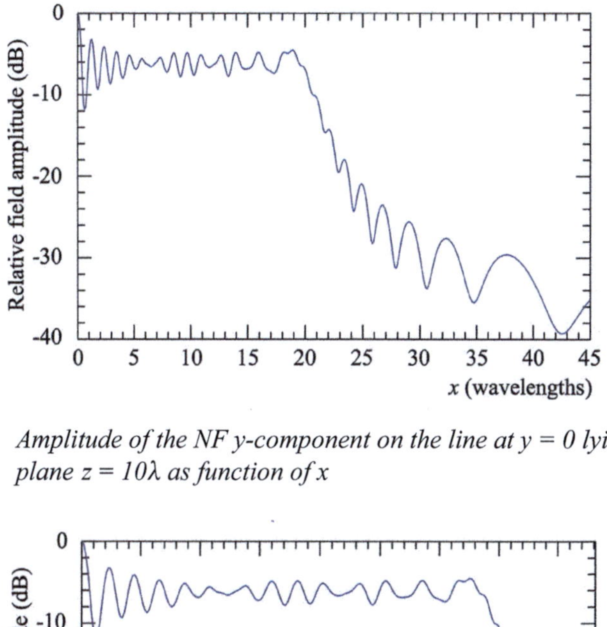

Figure 2.9 Amplitude of the NF y-component on the line at y = 0 lying on the plane z = 10λ as function of x

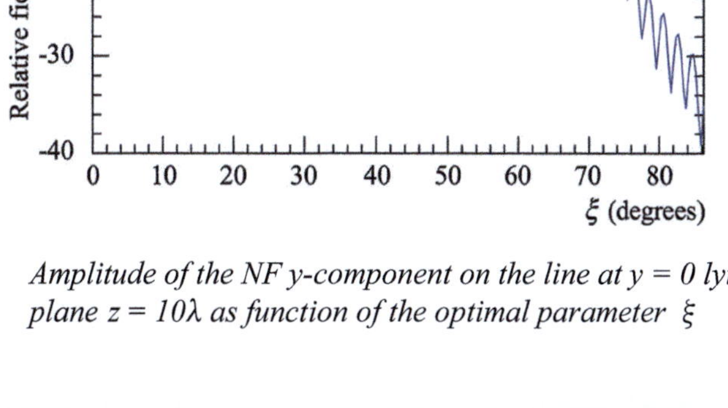

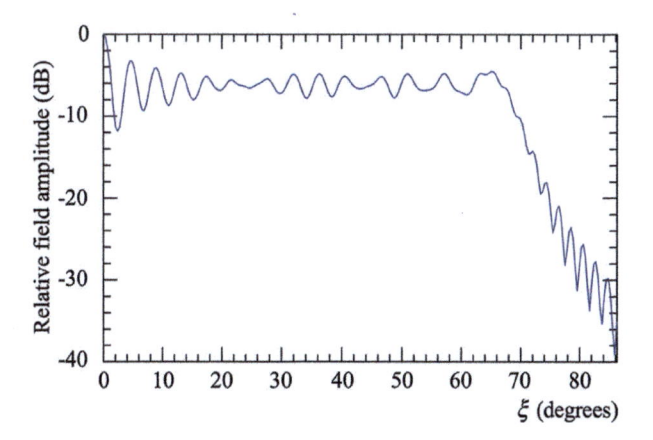

Figure 2.10 Amplitude of the NF y-component on the line at y = 0 lying on the plane z = 10λ as function of the optimal parameter ξ

2.3 Flexible source modellings

In this section, effective NR sampling representations, attained by modelling non-volumetric antennas, i.e., having one or two predominant dimensions, with flexible source modellings [95] alternative to the spheroidal ones, are described. In particular, a rounded cylinder (see Figure 2.11), i.e., a cylinder of height h' terminated by two hemi-spherical caps of radius a', is adopted to model elongated antennas,[§] whereas, when dealing with quasi-planar antennas, a double bowl (see Figure 2.12), namely a surface attained by joining together two bowls having the same circular aperture of diameter $2a$, is employed. Their lateral surfaces are determined by rotating two circular arcs, each equal to a quarter of circumference, which can have different radii

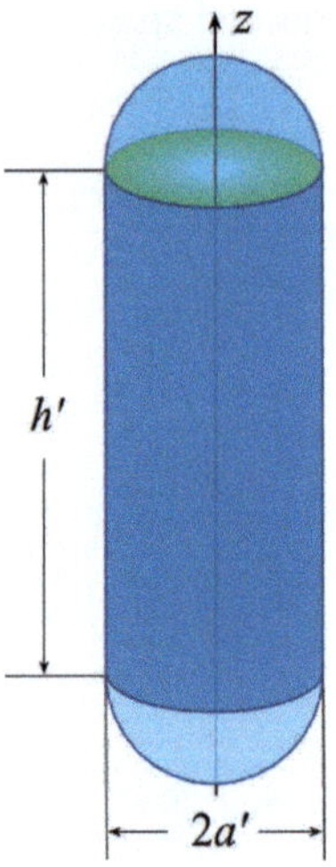

Figure 2.11 Rounded cylinder modelling of the antenna

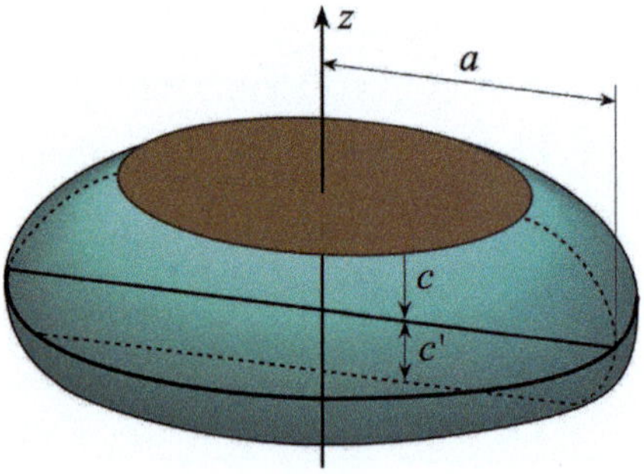

Figure 2.12 Double bowl modelling of the antenna

[§]The spherical modelling is obtained as limiting case of the rounded cylinder one, when the height h' approaches zero.

c and c' for a better fitting of the shape of the considered antenna. Accordingly, such a modelling results to be very flexible. As a matter of fact, the surface Σ becomes a sphere when $c = c' = a$ and a half-sphere if $c = 0$ and $c' = a$ (or $c = a$ and $c' = 0$). It is noteworthy that the disk modelling can be obtained from the double bowl one by putting $c = c' = 0$.

The rounded cylinder modelling is first considered. In such a case, the curve $\mathcal{C}'$ has length $\ell' = 2(h' + \pi a')$. The phase function γ and the optimal parameter ξ relevant to a meridian curve $\mathcal{C}$, specified by the polar coordinates $(r(\vartheta), \vartheta)$, can be obtained by substituting the values of the distances $R_{1,2}$ and curvilinear abscissae $s'_{1,2}$ in (2.8) and (2.11), whereas the bandwidth W_ξ is got by substituting the value of ℓ' in (2.10). It can be easily recognized that the expressions of $R_{1,2}$ and $s'_{1,2}$ change depending on the position of the tangency points $P_{1,2}$ (Figure 2.13). In particular, when $0 \leq \vartheta \leq \sin^{-1}(a'/r)$, these points are both located on the upper cap and it results:

$$R_1 = \sqrt{(r\sin\vartheta)^2 + (r\cos\vartheta - h'/2)^2 - a'^2} \tag{2.31}$$

$$s'_1 = a'\sin^{-1}\left(\frac{a'r\sin\vartheta + R_1(h'/2 - r\cos\vartheta)}{R_1^2 + a'^2}\right) \tag{2.32}$$

$$R_2 = R_1; \qquad s'_2 = a'\sin^{-1}\left(\frac{a'r\sin\vartheta - R_2(h'/2 - r\cos\vartheta)}{R_2^2 + a'^2}\right) \tag{2.33}$$

When $\sin^{-1}(a'/r) < \vartheta \leq \pi - \sin^{-1}(a'/r)$, P_1 is on the upper cap and P_2 is on the lower one, the expressions of R_1 and s'_1 are again given by (2.31) and (2.32), whereas it results:

$$R_2 = \sqrt{(r\sin\vartheta)^2 + (r\cos\vartheta + h'/2)^2 - a'^2} \tag{2.34}$$

$$s'_2 = h' + a'\left[\pi - \sin^{-1}\left(\frac{a'r\sin\vartheta + R_2(h'/2 + r\cos\vartheta)}{R_2^2 + a'^2}\right)\right] \tag{2.35}$$

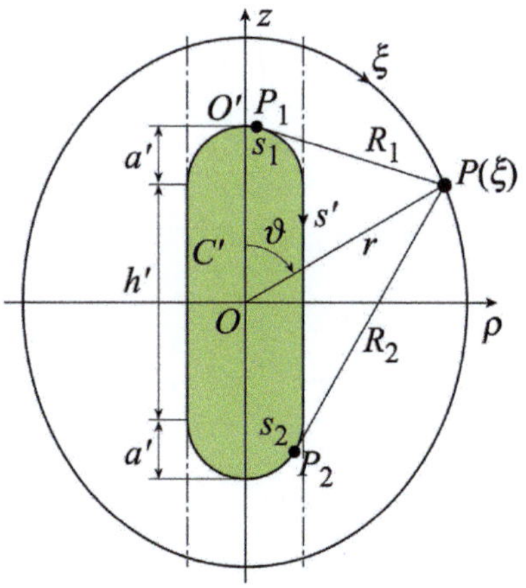

Figure 2.13 Rounded cylinder modelling: meridian curve

When $\pi - \sin^{-1}(a'/r) < \vartheta \le \pi$, the points $P_{1,2}$ are both located on the lower cap, the expressions of R_2 and s_2' are again given by (2.34) and (2.35), whereas it results:

$$R_1 = \sqrt{(r\sin\vartheta)^2 + (r\cos\vartheta + h'/2)^2 - a'^2} \tag{2.36}$$

$$s_1' = h' + a'\left[\frac{\pi}{2} - \sin^{-1}\left(\frac{R_1 r\sin\vartheta + a'(h'/2 + r\cos\vartheta)}{R_1^2 + a'^2}\right)\right] \tag{2.37}$$

The evaluation of the maximum in (2.14), allowing the determination of the azimuthal bandwidth W_φ, is now considered. It can be easily recognized that it is attained at $z' = z$, when $-h'/2 \le z \le h'/2$. In fact, for the triangular inequality $R^+ - R^- \le 2a'$, the equality being achieved in the case of a degenerate triangle. If $|z| > h'/2$, the maximum is obtained when the derivative of $R^+ - R^-$ with respect to z' is equal to zero. By summing up, the maximum in (2.14) is attained [95] at

$$z' = \begin{cases} z & |z| \le h'/2 \\ \left[\dfrac{h'}{2} + \dfrac{(|z| - h'/2)a'^2}{(r\sin\vartheta)^2 + (|z| - h'/2)^2}\right]\operatorname{sgn}(z) & |z| > h'/2 \end{cases} \tag{2.38}$$

where sgn(•) is the sign function.

The double bowl modelling is then considered. In this case, the curve C' has length $\ell' = 2[b + b' + (c + c')\pi/2]$, wherein $b = a - c$ and $b' = a - c'$. The function γ and the parameter ξ relevant to a meridian curve C can be determined by substituting the appropriate values of $R_{1,2}$ and $s_{1,2}'$ in (2.8) and (2.11), whereas the bandwidth W_ξ is got by substituting the value of ℓ' in (2.10). Also in such a case, the expressions of $R_{1,2}$ and $s_{1,2}'$ change depending on the position of the tangency points $P_{1,2}$ (Figure 2.14). In particular, when $0 \le \vartheta \le \vartheta_A = \sin^{-1}(a/r)$, these points are both located on the upper bowl and it results:

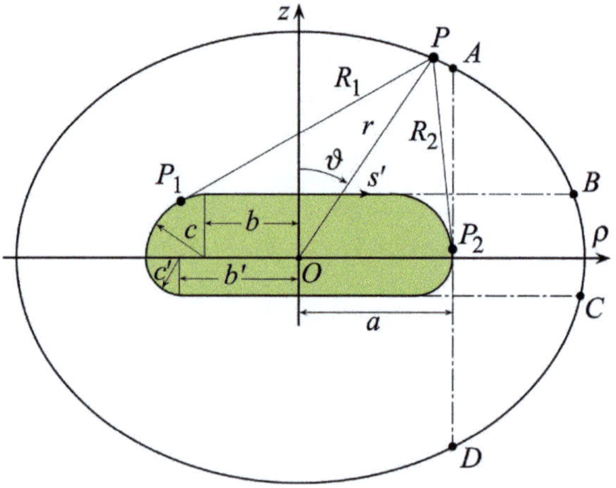

Figure 2.14 Double bowl modelling: meridian curve

$$R_1 = \sqrt{r^2 + b^2 + 2br\sin\vartheta - c^2}; \qquad s_1' = -(b + c\alpha_1) \tag{2.39}$$

$$\alpha_1 = \tan^{-1}(R_1/c) - \tan^{-1}[(b + r\sin\vartheta)/r\cos\vartheta] \tag{2.40}$$

$$R_2 = \sqrt{r^2 + b^2 - 2br\sin\vartheta - c^2}; \qquad s_2' = b + c\alpha_2 \tag{2.41}$$

$$\alpha_2 = \tan^{-1}(R_2/c) - \tan^{-1}[(b - r\sin\vartheta)/r\cos\vartheta] \tag{2.42}$$

When $\vartheta_A < \vartheta \le \vartheta_B = \cos^{-1}(c/r)$, P_1 is on the left side of the upper bowl and P_2 is on the right side of the lower one, the expressions of R_1, s_1', and α_1 are again given by (2.39) and (2.40), whereas it results:

$$R_2 = \sqrt{r^2 + b'^2 - 2b'r\sin\vartheta - c'^2}; \qquad s_2' = b + c(\pi/2) + c'\alpha_2 \tag{2.43}$$

$$\alpha_2 = \tan^{-1}(R_2/c') - \tan^{-1}[r\cos\vartheta/(r\sin\vartheta - b')] \tag{2.44}$$

When $\vartheta_B < \vartheta \le \vartheta_C = \pi - \cos^{-1}(c'/r)$, P_1 is on the right side of the upper bowl and P_2 is on the right side of the lower one, the expressions of R_2, s_2', and α_2 are again given by (2.43) and (2.44), whereas it results:

$$R_1 = \sqrt{r^2 + b^2 - 2br\sin\vartheta - c^2}; \qquad s_1' = b + c(\alpha_1 + \pi/2) \tag{2.45}$$

$$\alpha_1 = -\tan^{-1}(R_1/c) - \tan^{-1}[r\cos\vartheta/(r\sin\vartheta - b)] \tag{2.46}$$

When $\vartheta_C < \vartheta \le \vartheta_D = \pi - \sin^{-1}(a/r)$, P_1 is on the right side of the upper bowl and P_2 is on the left side of the lower one, the expressions of R_1, s_1', and α_1 are again given by (2.45) and (2.46), whereas it results:

$$R_2 = \sqrt{r^2 + b'^2 + 2b'r\sin\vartheta - c'^2}; \quad s_2' = b + 2b' + (c + c')(\pi/2) + c'\alpha_2 \tag{2.47}$$

$$\alpha_2 = \tan^{-1}(R_2/c') - \tan^{-1}[(r\sin\vartheta + b')/|r\cos\vartheta|] \tag{2.48}$$

When $\vartheta_D < \vartheta \le \pi$, the tangency points are both located on the lower bowl, the expressions of R_2, s_2', and α_2 are again given by (2.47) and (2.48), whereas it results:

$$R_1 = \sqrt{r^2 + b'^2 - 2b'r\sin\vartheta - c'^2}; \quad s_1' = b + c(\pi/2) + c'(\pi/2 - \alpha_1) \tag{2.49}$$

$$\alpha_1 = \tan^{-1}(R_1/c') - \tan^{-1}[(b' - r\sin\vartheta)/|r\cos\vartheta|] \tag{2.50}$$

In order to determine the maximum in (2.14), which allows the evaluation of the azimuthal bandwidth W_φ, it is convenient (when $\vartheta \le \pi/2$) to consider an angle η such that $z' = c\cos\eta$ and $\rho' = b + c\sin\eta$ (see Figure 2.15). It can be easily recognized [95] that such a maximum is attained in correspondence of the η value that zeroes the derivative of $R^+ - R^-$ and belongs to the range $[0, \pi/2]$. The case where $\pi/2 < \vartheta \le \pi$ can be managed in a quite similar way.

2.4 Cardinal series representations

The theoretical results presented in the previous sections can be employed to get accurate sampling interpolation formulas, which require a NR number of samples. Since the reduced electric field $\boldsymbol{F}$ can be very well approximated by bandlimited

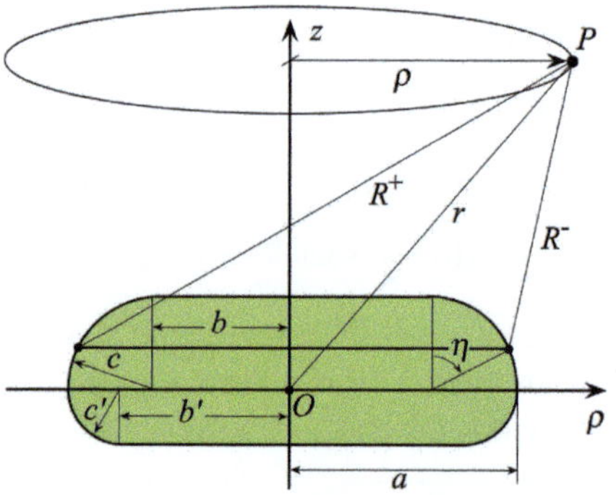

Figure 2.15 Double bowl modelling: azimuthal circumference

functions, the use of cardinal series (CS) expansions [294] springs out naturally as an appropriate tool to conveniently represent it. Accordingly, with reference to the interpolation along a meridian curve (lying on a meridian plane specified by the angle φ), it results:

$$\boldsymbol{F}\left(\xi(\vartheta), \varphi\right) = \sum_{n=-N'}^{N'} \boldsymbol{F}(\xi_n, \varphi)\, D_{N'}(\xi - \xi_n) \tag{2.51}$$

where

$$\xi_n = n\Delta\xi = 2\pi\, n/(2N' + 1); \qquad N' = \mathrm{Int}(\chi' W_\xi) + 1 \tag{2.52}$$

and

$$D_{N'}(x) = \frac{\sin((2N' + 1)\, x/2)}{(2N' + 1)\sin(x/2)} \tag{2.53}$$

is the Dirichlet function and Int (x) denotes the integer part of x.

The intermediate samples $\boldsymbol{F}(\xi_n, \varphi)$, namely, the reduced field values at the intersection points between the sampling azimuthal circumferences and the meridian curve through the observation point, can be determined by means of a similar CS expansion:

$$\boldsymbol{F}(\xi_n, \varphi) = \sum_{m=-M_n'}^{M_n'} \boldsymbol{F}(\xi_n, \varphi_{m,n})\, D_{M_n'}(\varphi - \varphi_{m,n}) \tag{2.54}$$

where

$$\varphi_{m,n} = m\Delta\varphi_n = 2m\,\pi/(2M_n' + 1); \qquad M_n' = \mathrm{Int}(\chi^* W_\varphi(\xi_n)) + 1 \tag{2.55}$$

$$\chi^* = \chi^*(\xi_n) = 1 + (\chi' - 1)[\sin\vartheta\,(\xi_n)]^{-2/3} \tag{2.56}$$

and the other symbols have the same meaning as in (2.51). The variation of the factor χ^* with ξ_n in (2.56) is required to ensure a bandlimitation error constant with respect to ξ_n (see Appendix 2A.4). It can be easily recognized that ϑ_∞ must take the place of ϑ when adopting a spheroidal source modelling.

The two-dimensional expansion over the rotational observation surface is got by properly matching the one-dimensional formulas (2.51) and (2.54). It results:

$$\boldsymbol{F}(\xi(\vartheta),\varphi) = \sum_{n=-N'}^{N'} \left\{ D_{N'}(\xi - \xi_n) \sum_{m=-M'_n}^{M'_n} \boldsymbol{F}(\xi_n, \varphi_{m,n}) D_{M'_n}(\varphi - \varphi_{m,n}) \right\} \tag{2.57}$$

If the reduced electric field $\boldsymbol{F}$ were rigorously bandlimited to W_ξ, the above expansion would be an exact representation. Unfortunately, $\boldsymbol{F}$ is only quasi-bandlimited, so that an aliasing error unavoidably arises. It can be demonstrated [103] that such an error cannot exceed $\sqrt{2}$ times the bandlimitation error and, accordingly, the aliasing error can be effectively controlled. To this end, it is worthy to remember that the bandlimitation error decreases exponentially with the bandwidth and more than exponentially with $(\chi' - 1)$ [103].

It must be stressed that the use of the CS expansion (2.57), although fully satisfactory from the accuracy viewpoint, has the shortcoming that all samples (or, at least, all the significant ones) must be considered when evaluating the field at each output point, otherwise a relatively large truncation error arises, because of the slow decay of the sampling functions. As a consequence, its use would lead to considerably large computational times, unless both the sampling and the output points are regularly spaced in ξ and φ, thus making possible to employ the fast Fourier transform (FFT) algorithm together with the null filling in evaluating the expansion (2.57). The abovementioned behaviour of the CS sampling functions gives rise to a further shortcoming when dealing with actual, i.e., non-ideal, data, as for instance measured data, which are unavoidably inaccurate. As a matter of fact, the measured data, when their values are well above the noise level, are generally affected by a nearly constant relative error and, accordingly, the absolute error in correspondence of the highest field values can be quite large. Such an error is spread out by the CS sampling functions without a significant attenuation, so that a remarkable relative error can be generated [39] in the zones where the field levels are low (see Sect. 2.6).

Both the underlined shortcomings can be overcome by resorting to an optimal sampling interpolation (OSI) expansion of central type [39, 106], wherein only the samples nearby the output point are considered in the field reconstruction and sampling functions more efficient from the truncation error viewpoint are employed. The solution to such a problem is presented in the next section.

It may be interesting to show how the samples are distributed when adopting the NR CS representation (2.51). Figure 2.16 refers to the same antenna, source modelling, and observation line considered in Sect. 2.2 and to an enlargement bandwidth factor $\chi' = 1.15$. As can be seen, the field samples are non-uniformly spaced in x, and, according to the considerations on the local bandwidth done at the end of Sect. 2.2, they are denser where this bandwidth is greater and sparser where it is smaller. The corresponding field reconstruction, obtained by considering all samples falling in the range $[-45\lambda, 45\lambda]$, is shown in Figure 2.17. As expected, the reconstruction is not very accurate in the region close to the boundary of the observation line, due to the truncation error. An extension of the zone of good reconstruction is attainable (see Sect. 2.5) by using the OSI expansions instead of the CS ones. In fact, when

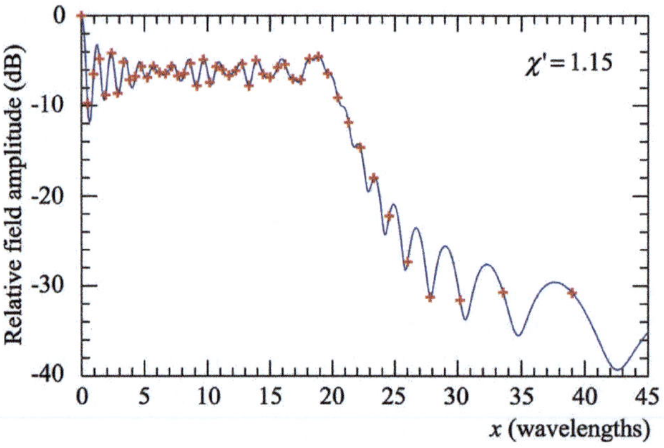

Figure 2.16 *Samples distribution on the line at y = 0 lying on the plane z = 10λ*

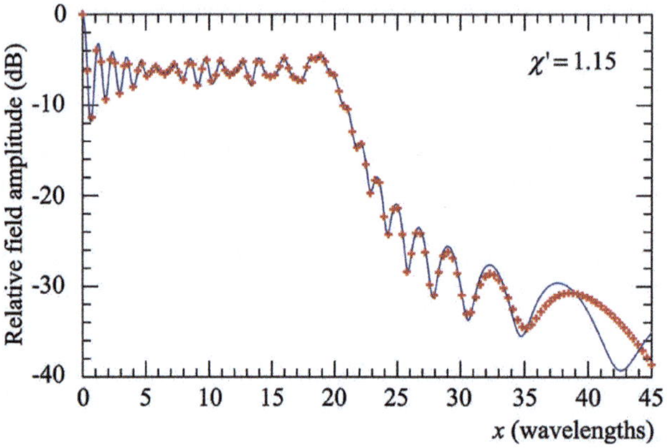

Figure 2.17 *Amplitude of the NF y-component on the line at y = 0 lying on the plane z = 10λ. Solid line: exact. Crosses: reconstructed via the CS expansion.*

evaluating the field at each output point, these last require the use of many samples to keep the truncation error low.

In order to give an objective criterion to choose the enlargement bandwidth factor χ' as function of the antenna maximum size, the maximum reconstruction error on the FF meridian at $\varphi = 90°$ has been evaluated, by considering as test source a circular uniform array of radius a, lying in the plane $y = 0$. It is formed by elementary Huygens sources linearly polarized along the z axis, symmetrically placed with respect to the x and z axes, and spaced by about 0.5 λ along circumferences, radially spaced by 0.5 λ. Since a spherical source modelling has been adopted, the corresponding bandwidth is βa. The errors have been evaluated by comparing the exact and reconstructed (via CS expansion) data on a grid with 1° step. The results of such

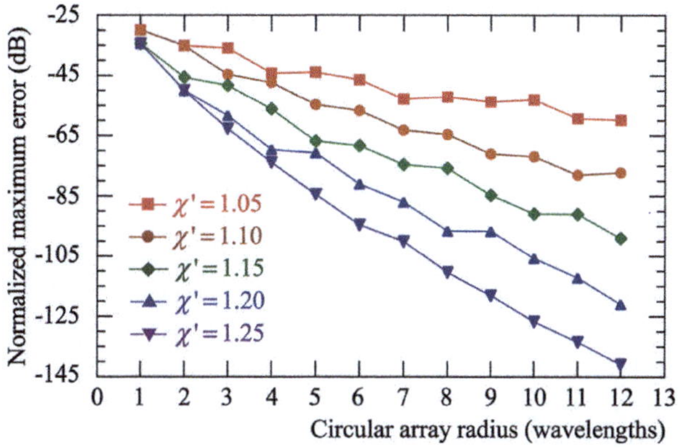

Figure 2.18 *Normalized maximum error as function of the antenna maximum size*

an evaluation are shown in Figure 2.18. It is worthy to note that, since the observation curve is closed and all samples have been considered, no truncation error occurs and the reconstruction error is only due to the aliasing. As can be seen, the error curves have the expected behaviour by decreasing exponentially with the bandwidth and more than exponentially with $(\chi' - 1)$.

2.5 Optimal sampling interpolation expansions

In light of the discussion done in the previous section, it is clear that the central interpolation algorithm to be used must minimize the truncation error for a given number of retained samples. The solution to such a problem for a closed observation curve is the OSI expansion developed in [39, 106] (see also Appendix 2A.5). Accordingly, with reference to the interpolation along a curve in a meridian plane specified by the angle φ, it results:

$$\boldsymbol{F}(\xi(\vartheta),\varphi) = \sum_{n=n_0-q+1}^{n_0+q} \boldsymbol{F}(\xi_n,\varphi)\, SF(\xi,\xi_n,\bar{\xi},N,N'') \tag{2.58}$$

where $n_0 = \mathrm{Int}(\xi/\Delta\xi)$ is the index of the intermediate sample nearest (on the left) to the output point, $2q$ is the number of the retained nearest intermediate samples,

$$\xi_n = n\Delta\xi = 2\pi\, n/(2N''+1); \qquad N'' = \mathrm{Int}(\chi N') + 1 \tag{2.59}$$

$$N' = \mathrm{Int}(\chi' W_\xi) + 1; \qquad N = N'' - N'; \qquad \bar{\xi} = q\Delta\xi \tag{2.60}$$

$\chi > 1$ being the oversampling factor, which allows the control of the truncation error, and

$$SF(\xi,\xi_n,\bar{\xi},N,N'') = \Omega_N(\xi-\xi_n,\bar{\xi})\, D_{N''}(\xi-\xi_n) \tag{2.61}$$

the OSI function. In this last,

$$\Omega_N(\xi,\bar{\xi}) = \frac{T_N[2(\cos(\xi/2)/\cos(\bar{\xi}/2))^2 - 1]}{T_N[2/\cos^2(\bar{\xi}/2) - 1]} \tag{2.62}$$

is the Tschebyscheff Sampling (TS) function, wherein $T_N(\cdot)$ is the Tschebyscheff polynomial of degree N.

It is worthy to note that the weight function $\Omega_N(\xi,\bar{\xi})$ has been obtained in [39] by paralleling the properties of the so-called sampling window (SW) function introduced by Knab [295], which (in the square norm) represents the practically optimal weight function for the case of an indefinite observation domain (where the CS sampling function is the $\sin(x)/x$ one) [294].

Apart from the aliasing error, the expansion (2.58) would be an exact representation if all samples were considered. When only $2q$ of them are retained, a truncation error $|\Delta \boldsymbol{F}|$ is introduced. By paralleling the procedure in [295], the following upper bound for the corresponding relative error has been obtained [39, 106, 296]:

$$\frac{|\Delta \boldsymbol{F}|}{\|\boldsymbol{F}\|_\infty} < \delta_q = \frac{2}{\pi\, T_N[2/\cos^2(\bar{\xi}/2) - 1]} \ln\left[\cot\left(\frac{q\pi}{4N''}\right)\right] \tag{2.63}$$

wherein $\|\boldsymbol{F}\|_\infty$ is the uniform norm for $\boldsymbol{F}$, i.e., the maximum value of the amplitude of the reduced field on the observation curve.

The intermediate samples $\boldsymbol{F}(\xi_n,\varphi)$ in (2.58) can be determined by means of a similar OSI expansion:

$$\boldsymbol{F}(\xi_n,\varphi) = \sum_{m=m_0-p+1}^{m_0+p} \boldsymbol{F}(\xi_n,\varphi_{m,n})\, SF(\varphi,\varphi_{m,n},\bar{\varphi}_n,M,M_n'') \tag{2.64}$$

where $m_0 = \mathrm{Int}\left(\varphi/\Delta\varphi_n\right)$, $2p$ is the number of retained samples along φ,

$$\varphi_{m,n} = m\Delta\varphi_n = 2m\pi/(2M_n''+1); \qquad M_n'' = \mathrm{Int}(\chi M_n') + 1 \tag{2.65}$$

$$M_n' = \mathrm{Int}(\chi^* W_\varphi(\xi_n)) + 1; \qquad M_n = M_n'' - M_n' \tag{2.66}$$

$$\chi^* = \chi^*(\xi_n) = 1 + (\chi' - 1)[\sin\vartheta(\xi_n)]^{-2/3}; \quad \bar{\varphi}_n = p\Delta\varphi_n \tag{2.67}$$

and the other symbols have the same meaning as in (2.58). Again, $\vartheta = \vartheta_\infty$ in the case of spheroidal source modellings.

The two-dimensional expansion over the observation rotational surface is obtained by properly matching the one-dimensional formulas. It results:

$$\boldsymbol{F}(\xi(\vartheta),\varphi) = \sum_{n=n_0-q+1}^{n_0+q} \left\{ SF(\xi,\xi_n,\bar{\xi},N,N'') \sum_{m=m_0-p+1}^{m_0+p} \boldsymbol{F}(\xi_n,\varphi_{m,n})\, SF(\varphi,\varphi_{m,n},\bar{\varphi}_n,M,M_n'') \right\} \tag{2.68}$$

It can be shown [106, 107] that the upper bound δ_{qp} of the truncation error relevant to interpolation on a two-dimensional domain can be expressed in terms of the one-dimensional bounds δ_q and δ_p.

$$\delta_{qp} \leq (1+\delta_q)(1+\delta_p) - 1 \cong \delta_q + \delta_p \tag{2.69}$$

The one-dimensional example shown in Figure 2.19 refers to the same antenna, source modelling, enlargement bandwidth factor, and observation curve considered in Figure 2.17. In such a case, the field reconstruction has been obtained by applying the OSI expansion with $\chi = 1.20$, $q = 7$, and again considering only the samples falling in the range [–45λ, 45λ]. As can be seen, the recovery results to be now quite accurate also in the region close to the boundary of the observation line, unlike what occurs when using the CS expansion (Figure 2.17). This confirms the validity of the OSI expansions from the truncation error viewpoint.

The efficacy of the OSI algorithms depends on the choice of the bandwidth enlargement and oversampling factors, and on the number of retained nearest samples. The choice of the first factor, which controls the aliasing error, has already been addressed in the previous section. Thus, only the choice of the other factors, which control the truncation error, will be considered in the following. To this end, the maximum and mean-square reconstruction errors (normalized to the field maximum value on the line) have been evaluated. They have been got by comparing the interpolated values of the NF y-component with those directly evaluated on a close grid in the central zone of the observation line, so that the existence of the guard samples is assured. Figures 2.20 and 2.21 show these errors for q ranging from 2 to 12, $\chi' = 1.15$, and $\chi = 1.10$, 1.15, 1.20, 1.25. As expected, the errors decrease up to very low values on increasing the oversampling factor and/or the retained samples number. In Figure 2.20, the truncation error upper bounds and an error curve relevant to the truncated CS expansion are also shown.

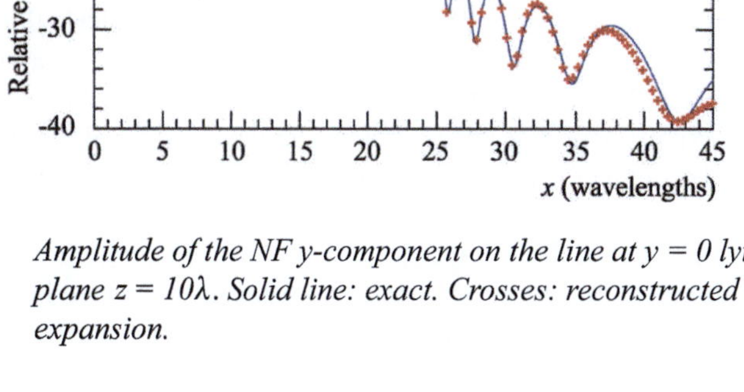

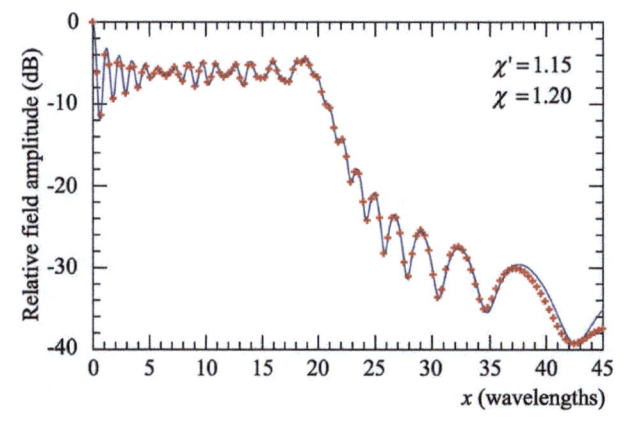

Figure 2.19 Amplitude of the NF y-component on the line at y = 0 lying on the plane z = 10λ. Solid line: exact. Crosses: reconstructed via the OSI expansion.

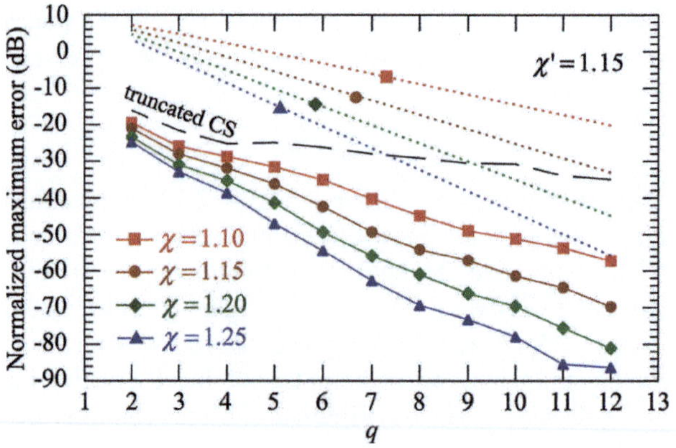

Figure 2.20 *Normalized maximum reconstruction error of the NF y-component. Solid lines: actual curves. Dashed: truncated CS ones. Dotted: theoretical upper bounds.*

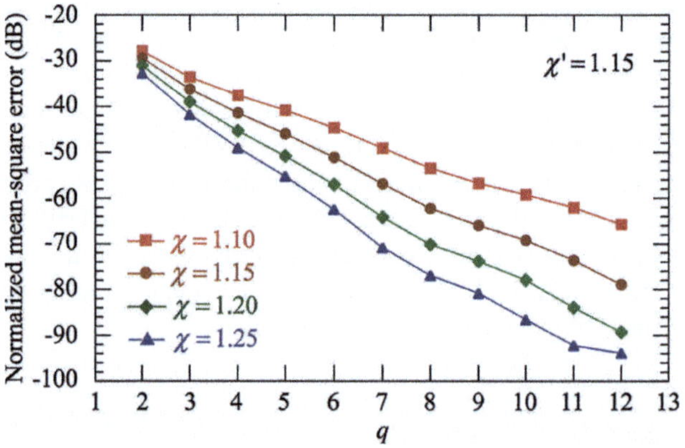

Figure 2.21 *Normalized mean-square reconstruction error of the NF y-component*

As can be seen, the evaluated error curves exhibit the same behaviour of the theoretical ones. However, the actual errors are remarkably lower than the corresponding upper bounds, resulting about 30–40 dB smaller. These theoretical upper bounds are, therefore, very pessimistic and, accordingly, can be safely lowered by about 25–30 dB in the estimation of the real truncation errors. The behaviour of the truncated CS expansion is fully unsatisfactory in the considered reasonable range of retained samples number. In fact, the related error curves show a very slow decrease with almost constant and unacceptably high error values. Moreover, it can be verified that they are practically independent from the oversampling factor [106]. The effectiveness of the OSI algorithms from the truncation error reduction viewpoint is so fully assessed. It must be stressed that quite similar results [106, 107] have been also obtained in the case of two-dimensional observation domains (see the next section).

It must be stressed that the CS and OSI expansions (2.57) and (2.68) can be applied also to represent the spherical components of the reduced EM field on a rotational surface. As a matter of fact, provided that the sign change occurring at the polar singularities of the coordinate system is taken into account, the spherical components of the reduced EM field are quasi-bandlimited functions, since they are linear combinations through simple trigonometric factors of the Cartesian ones. It is worth noting that, due to the abovementioned sign discontinuity, when performing the interpolation along the meridian curves, the sign of the intermediate samples relevant to the sampling points lying on the other side of the considered meridian curve must be changed.

It must be stressed that, as shown in [41], the voltage measured by a non-directive probe has the same effective spatial bandwidth of the field radiated by the antenna under test (AUT). Accordingly, the CS and OSI expansions (2.57) and (2.68) can be applied also to represent the probe "reduced voltage"

$$\tilde{V}(\xi) = V(\xi)\mathrm{e}^{\mathrm{j}\gamma(\xi)} \tag{2.70}$$

on a rotational surface. For reader's convenience, the related two-dimensional OSI expansion is reported in the following:

$$V(\xi(\vartheta),\varphi) = \mathrm{e}^{-\mathrm{j}\gamma(\xi)} \times \sum_{n=n_0-q+1}^{n_0+q} \left\{ SF(\xi,\xi_n,\bar{\xi},N,N'') \sum_{m=m_0-p+1}^{m_0+p} \tilde{V}(\xi_n,\varphi_{m,n})\, SF(\varphi,\varphi_{m,n},\bar{\varphi}_n,M,M_n'') \right\} \tag{2.71}$$

where all symbols have, obviously, the same meanings as in (2.68).

It is worth noting that, also in such a case, the sign change occurring at the polar singularities of the coordinate system must be properly taken into account. This implies that, when performing the interpolation along the meridian curves, the sign of the intermediate samples relevant to the sampling points lying on the other side of the considered meridian curve must be changed.

It must be stressed that the OSI expansion (2.71) will be widely applied in the next chapters to develop probe-compensated NR near-field to far-field (NF–FF) transformation techniques, where the massive NF data, needed by the corresponding traditional ones, are accurately recovered from the NR voltage samples acquired by the probe on the rotational scanning surface. It is so possible to obtain a drastic reduction of the number of the NF data to be acquired and of the related measurement time. This is a very important result, since nowadays the acquisition time is by far greater than that required to perform the NF–FF transformation. It must be stressed that the NR NF–FF transformation with spherical scanning allows one to obtain a remarkable reduction of the NF data to be acquired and related measurement time, as much as the AUT geometry departs from the spherical one. Such a reduction is even much greater in the case of NR NF–FF transformations using planar and cylindrical scannings. In fact, when applying the NR sampling representations, the number of samples required to accurately represent the measured voltage stays finite also for an unbounded scanning

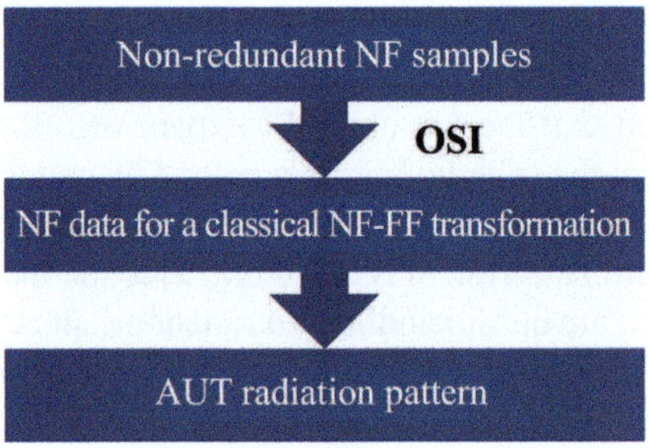

Figure 2.22 Flowchart of the non-redundant NF–FF transformations

surface. Accordingly, the reduction of the number of needed NF data becomes greater and greater on increasing the scanning area and the antenna size. Obviously, when the scanning area extension is small as compared to the AUT size, the application of the NR sampling representations becomes no longer convenient.

By summing up, the NF samples, acquired at the points prescribed by the developed NR voltage representation on the scanning surface, are interpolated through formula (2.71) to accurately reconstruct the probe and rotated probe voltages at the grid points required to obtain the antenna far field by means of the classical probe-compensated NF–FF transformation relevant to the considered scanning surface (see Figure 2.22). As regards the choice of the OSI parameters in (2.71), namely the bandwidth enlargement factor χ', the oversampling one χ, and the retained samples numbers $2p$, $2q$, it can be done by taking into account the previous considerations. In practice, their choice is done in such a way that the error related to the interpolation (aliasing plus truncation) must be smaller than the measurement one (background noise plus measurement uncertainties) and is the results of a compromise between the measurement and interpolation times. As a matter of fact, an increase of χ' and/or χ reflects in a growth of the needed NF data and, as a consequence, of the measurement time, whilst an increase of p and q lengthens the interpolation time. Moreover, as will be shown in the numerical tests reported throughout the book and relevant to reconstructions from error affected samples, the OSI algorithm results to be stable, since it does not amplify the errors present on the input samples. At last, it must stressed that the OSI kernel functions have also the capability to effectively remove the spatial harmonics of the noise exceeding the AUT spatial bandwidth, thus behaving as a low pass filter. This feature is widely testified from the experimental results reported throughout this book and, in particular, is clearly evident in the Figures 2.6 and 2.7 of ref. [109] relevant to NF reconstructions from NF samples acquired in an anechoic chamber more noisy than the one available in the Antenna Characterization Lab of the University of Salerno.

2.6 Application to far-field interpolation

In this section, two-dimensional sampling interpolation algorithms to efficiently reconstruct the far field radiated by an antenna are presented. These algorithms are surely important in the pattern recovery from measured or laboriously computed FF data. They are based on the NR sampling representations of the EM fields and adopt prolate or oblate spheroids to model the radiating antenna, thus allowing a remarkable reduction of the number of the needed samples with respect to the spherical source modelling in the case of antennas with one or two predominant dimensions.

In the light of the above consideration, an effective NR representation for the EM field radiated on the FF sphere¶ is achieved [297] by considering an elongated or quasi-planar antenna as enclosed in the smallest "usable" prolate or oblate spheroid, respectively (see Figures 2.23 and 2.24). According to the results reported in Sect. 2.2, the phase function γ to be used for obtaining the reduced field is given by (2.27), whereas the optimal parameter ξ for describing a meridian is given by (2.28a), when

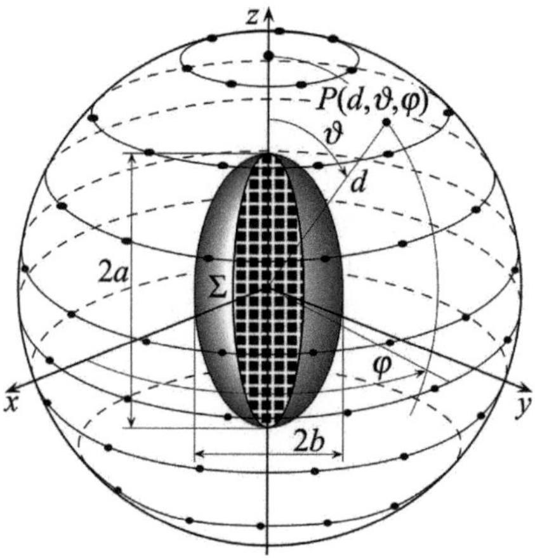

Figure 2.23 Relevant to the far-field interpolation: elongated antennas

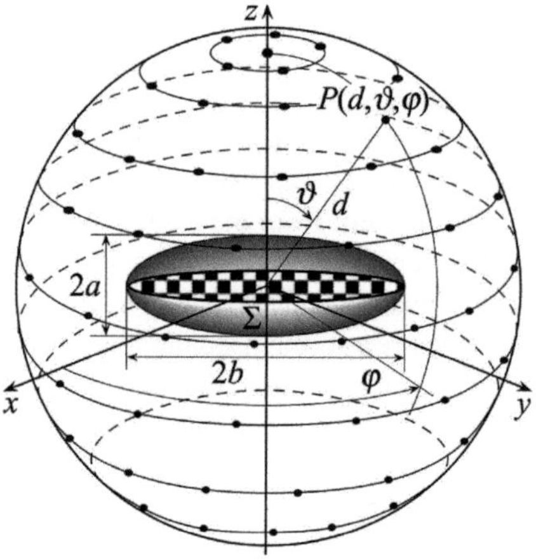

Figure 2.24 Relevant to the far-field interpolation: quasi-planar antennas

¶In the representation of the EM field radiated on the FF sphere, the factor $e^{-j\beta r}/r$ is usually omitted.

the antenna is modelled via a prolate spheroid, and by (2.28b), when the oblate spheroidal modelling is adopted. Since in the FF case $\vartheta_\infty = \vartheta$, the corresponding azimuthal bandwidth W_φ is given by:

$$W_\varphi(\xi) = \begin{cases} \beta b \sin\vartheta & \text{prolate spheroid} \quad (2.72a) \\ \beta a \sin\vartheta & \text{oblate spheroid} \quad (2.72b) \end{cases}$$

where a and b are the major and minor semi-axes (see Figures 2.6 and 2.7). Moreover, the distances $r_{1,2}$ are given by:

$$r_1 = r - f\cos\vartheta; \qquad r_2 = r + f\cos\vartheta \tag{2.73}$$

for the prolate case, and by

$$r_1 = r + f\sin\vartheta; \qquad r_2 = r - f\sin\vartheta \tag{2.74}$$

for the oblate case.

As already stressed, the spherical components of the EM field are linear combinations through trigonometric factors of the Cartesian ones and, hence, are quasi-bandlimited functions too, provided that the sign change in correspondence of the pole $\vartheta = 0$ is taken into account. Thus, by applying (2.57) to the spherical components of the electric field, it results:

$$\begin{aligned} E_{\vartheta,\varphi}(\xi(\vartheta),\varphi) &= E_{\vartheta,\varphi}(0,\varphi)D_{N'}(\xi) \\ &+ \sum_{n=1}^{N'} \Big\{ D_{N'}(\xi-\xi_n) \sum_{m=-M_n'}^{M_n'} E_{\vartheta,\varphi}(\xi_n,\varphi_{m,n}) D_{M_n'}(\varphi-\varphi_{m,n}) \\ &- D_{N'}(\xi+\xi_n) \sum_{m=-M_n'}^{M_n'} E_{\vartheta,\varphi}(\xi_n,\varphi_{m,n}) D_{M_n'}(\varphi+\pi-\varphi_{m,n}) \Big\} \end{aligned} \tag{2.75}$$

Note that there is no need to extract the phase factor from the FF components, since it is constant on the FF sphere. The minus sign in the last term of (2.75) is due to the abovementioned polar singularity. For the same reason, the value of $E_{\vartheta,\varphi}(0,\varphi)$ must be taken as the limit from the right.

As already stated in Section 2.4, the CS representation requires the use of all samples (or, at least, all the relevant ones) in the field evaluation, thus leading to large computational times, and gives rise to the propagation of errors from the high field regions to the lower field ones. To overcome these drawbacks, the two-dimensional OSI formula (2.68) can be employed. Accordingly, the proper OSI expansion for the spherical components of the electric field is

$$\begin{aligned} E_{\vartheta,\varphi}(\xi(\vartheta),\varphi) &= \sum_{n=n_0-q+1}^{n_0+q} \Big\{ SF(\xi,\xi_n,\bar{\xi},N,N'') \\ &\times \sum_{m=m_0-p+1}^{m_0+p} E_{\vartheta,\varphi}(\xi_n,\varphi_{m,n}) SF(\varphi,\varphi_{m,n},\bar{\varphi}_n,M,M_n'') \Big\} \end{aligned} \tag{2.76}$$

It must be stressed that, also in such a case, the sign change occurring at the poles $\vartheta = 0$ and $\vartheta = \pi$ must be taken into account. This implies that, when, performing the interpolation along the meridians, the sign of the intermediate samples relevant to the sampling points lying on the other side of the considered meridian must be changed.

Some numerical examples assessing the effectiveness of the described interpolation algorithms are reported in the following. For sake of brevity, only the case of the prolate spheroidal modelling will be considered, the interested reader can refer to [297] for the oblate case. The simulations refer to the far field radiated by a uniform planar array of 0.6 λ-spaced elementary Huygens sources polarized along the z axis and covering an elliptical zone in the plane $y = 0$ (see Figure 2.23), with major and minor semi-axes equal to 24 λ and 6 λ, respectively.

To select the value of the enlargement bandwidth factor χ', which allows the control of the aliasing error, the actual maximum and mean-square normalized errors have been evaluated by interpolating the field via the CS expansion. Such an evaluation has been performed, for some χ' values, by comparing the reconstructed and exact ϑ-component of the field (the most significant one) on a spherical grid with one degree step in ϑ and φ. The results are collected in Table 2.1, where the maximum and mean-square errors, relevant to various χ' values and normalized to the field maximum value, are reported together with the overall number of the corresponding samples. As expected, on increasing χ', the errors decrease quite rapidly until very low levels are reached.

Turning now to the previously considered elliptical array with semi-axes 24 λ and 6 λ, according to the data reported in Table 2.1, χ' has been chosen equal to 1.20 in the following simulations. The corresponding FF patterns reconstructed in the principal planes via the CS expansion are reported in Figure 2.25a and b. As can be seen, the exact and recovered fields are practically indistinguishable.

Following Figure 2.26a and b show the field recovering, again in the principal planes, obtained via the OSI expansion. It is worthy to note that, even with a moderate value of the oversampling factor χ and a small number of retained samples, the pattern reconstruction is very accurate in the whole angular range.

To assess in a more quantitative way the efficacy of the OSI algorithm, the maximum and mean-square reconstruction errors (normalized to the field maximum value) have been evaluated, for $\chi' = 1.20, p = q$ ranging from 3 to 12, and four different values of the oversampling factor χ, by comparing the recovered and exact

Table 2.1 Normalized aliasing errors versus χ' ($a=24\lambda$, $b=6\lambda$)

χ'	Maximum error (dB)	Mean-square error (dB)	Number of samples
1.00	−47.39	−70.67	6070
1.05	−58.59	−80.63	6792
1.10	−70.39	−90.12	7485
1.15	−83.30	−101.03	8212
1.20	−95.99	−110.41	8977
1.25	−99.87	−119.27	9762

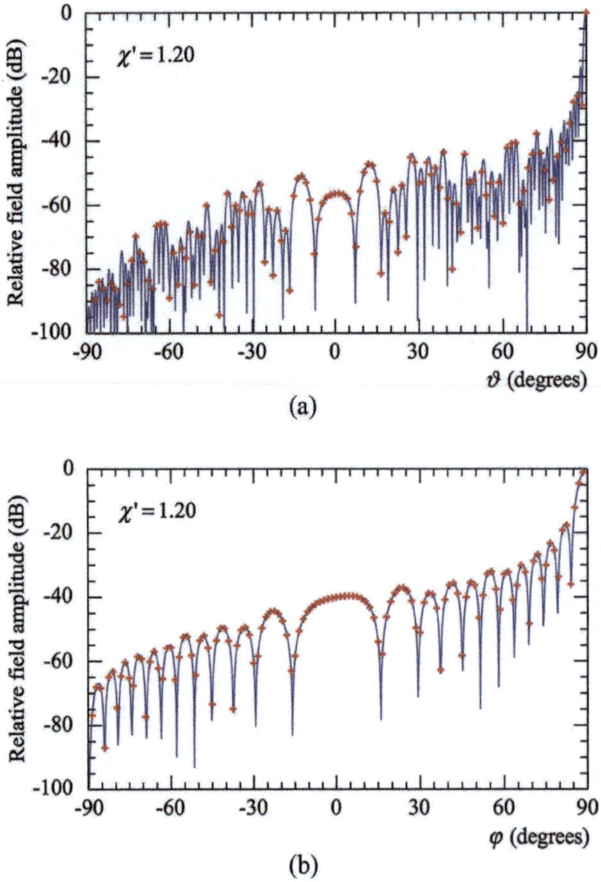

Figure 2.25 FF patterns in the principal planes. Solid line: exact. Crosses: interpolated via the CS expansion. (a) E-plane, (b) H-plane.

ϑ-component of the field on a spherical grid with one degree step in ϑ and φ. For space saving, only the plot relevant to the maximum error is shown in Figure 2.27. The curves relevant to the mean-square error exhibit a quite analogous behaviour running about 30 dB down. As can be seen, the errors decrease up to very low values on increasing the oversampling factor and/or the number of retained samples, thus assessing the effectiveness of the described sampling representation. On the same figure, the truncation errors computed by using the CS expansion truncated to the same number of samples are plotted for reference. As can be easily recognized, the behaviour of the truncated CS expansion is completely unsatisfactory in the considered reasonable range of retained samples number. In fact, the related errors decrease very slowly on increasing the retained samples number and are unacceptably high. Moreover, it can be verified that they are practically independent from the oversampling factor [106].

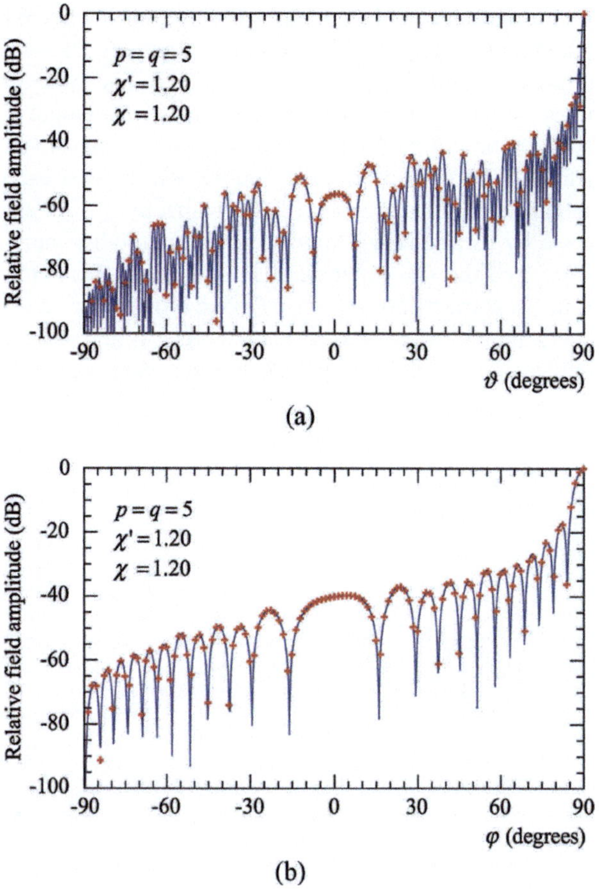

Figure 2.26 *FF patterns in the principal planes. Solid line: exact. Crosses: interpolated via the OSI expansion. (a) E-plane, (b) H-plane.*

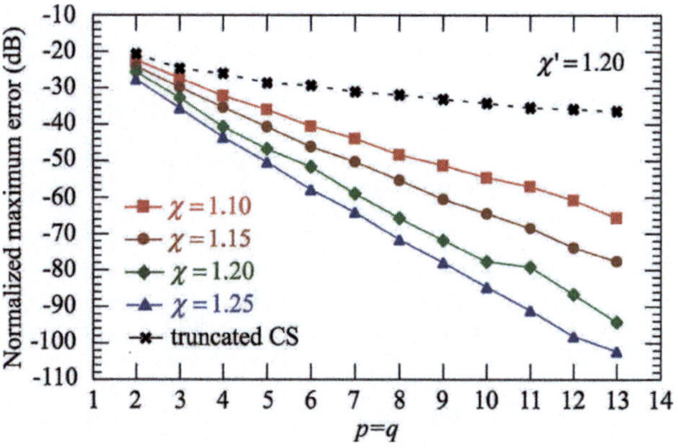

Figure 2.27 *Normalized maximum error in the reconstruction of* E_ϑ

The stability of the described algorithms has been investigated by adding uniformly distributed random errors to the exact field samples. These errors simulate a background noise (bounded to Δa dB in amplitude and with arbitrary phase) and uncertainties on the field samples (due to the measurement system inaccuracies) of $\pm\Delta a_r$ dB in amplitude and $\pm\Delta\phi$ degrees in phase. The H-plane FF pattern reconstructed from error affected samples via the OSI expansion and the CS one are shown in Figures 2.28 and 2.29, respectively. As can be seen, the pattern recovered by using the OSI algorithm is remarkably more accurate than that obtained through the CS expansion. This different behaviour can be easily explained: in the former case the error is due only to the errors affecting the nearest retained samples, whereas in the latter case, it depends by those affecting all the samples. Accordingly, in presence of a nearly constant relative error,

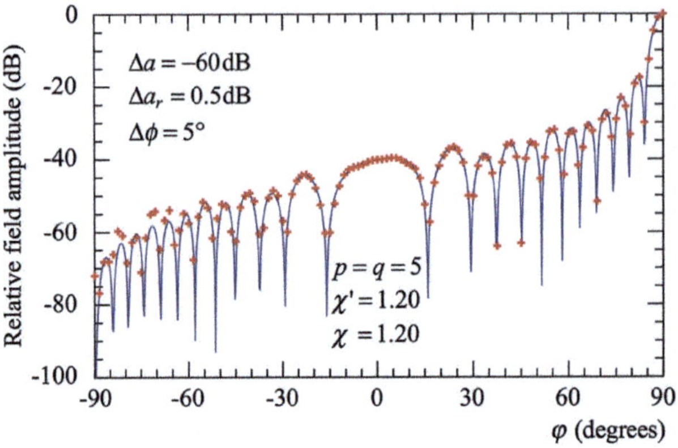

Figure 2.28 Field amplitude on the H-plane. Solid line: exact. Crosses: interpolated via OSI from error affected samples.

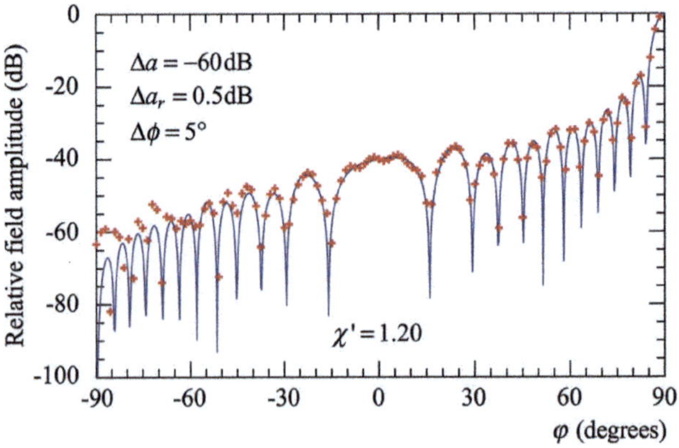

Figure 2.29 Field amplitude on the H-plane. Solid line: exact. Crosses: interpolated via CS from error affected samples.

the absolute error, relatively high in correspondence of the highest field values, propagates in the lower field ones without a significant attenuation, due to the slow decay of the CS interpolation functions, thus giving rise to a strong relative error in them.

It is worth noting that, as straightforward consequence of the frequency convolution theorem, the squared amplitude of the field is an almost bandlimited function with a bandwidth double with respect to that of the field. Accordingly, the CS and the OSI expansions can be applied to any squared amplitude field component, provided that the sampling spacings are halved.

The problem of recovering the squared amplitude of the field components at $(0, \varphi)$, whose knowledge is mandatory to apply the OSI expansion nearby the polar singularity at $\vartheta = 0$, is now considered. It can be easily shown that this is possible if the squared amplitudes of the field components are known at $(0, 0)$ and $(0, \gamma)$. As a matter of fact, the squared amplitudes of the field components at $(0, \varphi)$ are related to those at $(0, 0)$ and $(0, \gamma)$ through the relations:

$$|E_\vartheta(0,\varphi)|^2 = |E_\vartheta(0,0)|^2\cos^2\varphi + |E_\varphi(0,0)|^2\sin^2\varphi + |E_\vartheta(0,0)||E_\varphi(0,0)|\sin(2\varphi)\cos\alpha \tag{2.77}$$

$$|E_\varphi(0,\varphi)|^2 = |E_\vartheta(0,0)|^2\sin^2\varphi + |E_\varphi(0,0)|^2\cos^2\varphi - |E_\vartheta(0,0)||E_\varphi(0,0)|\sin(2\varphi)\cos\alpha \tag{2.78}$$

where

$$|E_\vartheta(0,0)||E_\varphi(0,0)|\cos\alpha = \frac{|E_\vartheta(0,0)|^2\sin^2\gamma + |E_\varphi(0,0)|^2\cos^2\gamma - |E_\varphi(0,\gamma)|^2}{\sin(2\gamma)} \tag{2.79}$$

α being the unknown phase difference between $E_\vartheta(0,0)$ and $E_\varphi(0,0)$.

The FF patterns reconstructed in the principal planes via the OSI expansion from power samples are reported in Figure 2.30a and b. As can be seen, the exact and recovered patterns are practically indistinguishable, thus demonstrating the effectiveness of the proposed sampling representation also in the power samples case.

Appendix

2A.1 Some details on the evaluation of the azimuthal bandwidth

The evaluation of the azimuthal bandwith W_φ relevant to the EM field representation on an azimuthal circumference of radius ρ and centre at the point of Cartesian coordinates $(0, 0, z)$ is now considered. In such a case, since the phase factor γ and the local bandwidth $w(\varphi)$ do not depend on the angle φ, it results from (2.2) that

$$W_\varphi = \max_{r'}\left|\beta\frac{\partial R(\varphi,\boldsymbol{r}')}{\partial\varphi}\bigg|_{\varphi=0}\right| \tag{2A.1}$$

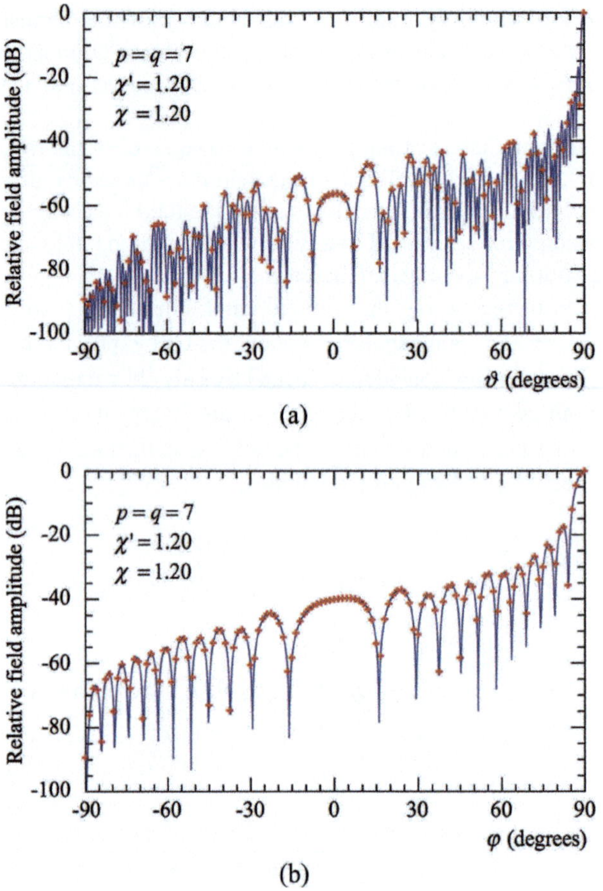

Figure 2.30 *FF patterns in the principal planes. Solid line: exact. Crosses: interpolated via the OSI expansion from power samples. (a) E-plane, (b) H-plane.*

By denoting with $(\rho' \cos\varphi', \rho' \sin\varphi', z')$ the Cartesian coordinate of the source point specified by $\boldsymbol{r}'$, it results $R = \left(\rho^2 + \rho'^2 - 2\rho\rho' \cos(\varphi - \varphi') + (z - z')^2\right)^{1/2}$ and, accordingly

$$W_\varphi = \max_{r'} \frac{\beta\rho\rho' \sin\varphi'}{\sqrt{\rho^2 + \rho'^2 - 2\rho\rho' \cos\varphi' + (z - z')^2}} \tag{2A.2}$$

Since the minimum and maximum value of $\partial R/\partial\varphi$ are opposite and attained on the surface Σ, it follows that

$$W_\varphi = \max_{z'} \left(\max_{\varphi'} \frac{\beta\rho\rho' \sin\varphi'}{\sqrt{\rho^2 + \rho'^2 - 2\rho\rho' \cos\varphi' + (z - z')^2}} \right) \tag{2A.3}$$

where $\rho'(z')$ is the equation of Σ in cylindrical coordinates.

The maximum with respect to φ' in this last relation is reached in correspondence of the φ' value such that

$$\cos\varphi'_M = \frac{[\rho^2 + \rho'^2 + (z - z')^2] - \sqrt{[\rho^2 + \rho'^2 + (z - z')^2]^2 - 4\rho^2\rho'^2}}{2\rho\rho'} \tag{2A.4}$$

and is given by:

$$\max_{\varphi'} \beta \frac{\partial R}{\partial \varphi} = \beta\sqrt{\rho\rho'\cos\varphi'_M} \tag{2A.5}$$

By substituting (2A.4) into (2A.5) and applying the formulas for the solution of double radicals, it results

$$\begin{aligned} \max_{\varphi'} \beta \frac{\partial R}{\partial \varphi} &= \frac{\beta}{2}\left(\sqrt{\rho^2 + \rho'^2 + (z - z')^2 + 2\rho\rho'} - \sqrt{\rho^2 + \rho'^2 + (z - z')^2 - 2\rho\rho'}\right) \\ &= \frac{\beta}{2}\left(\sqrt{(z - z')^2 + (\rho + \rho'(z'))^2} - \sqrt{(z - z')^2 + (\rho - \rho'(z'))^2}\right) \\ &= \frac{\beta}{2}(R^+ - R^-) \end{aligned} \tag{2A.6}$$

Relation (2.14) easily follows by taking into account (2A.6) and (2A.3).

It is now shown that the maximum in relation (2.14) is attained on the part of the surface Σ lying on the same side of the observation circumference with respect to its maximum transverse circle. To this end, let Γ_1 be a circumference of radius ρ' lying on Σ on the same side of the observation one C having radius ρ and let R_1^+ and R_1^- be the maximum and minimum distance between Γ_1 and C (see Figure 2A.1). Due to the properties of the surface Σ, it is always possible to consider a circumference Γ_2 having the same radius of Γ_1 and lying on the other side with respect to the

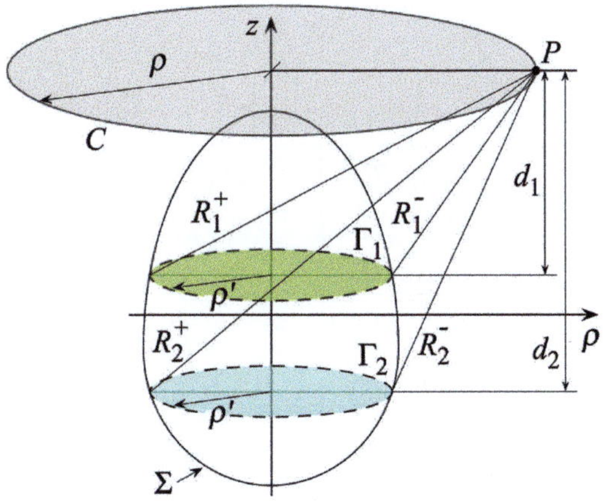

Figure 2A.1 Relevant to the evaluation of W_φ

maximum transverse circle. By denoting with R_2^+ and R_2^- the corresponding maximum and minimum distance from $\mathcal{C}$, it results:

$$R_1^+ = \sqrt{d_1^2 + (\rho + \rho')^2}; \qquad R_1^- = \sqrt{d_1^2 + (\rho - \rho')^2}$$

$$R_2^+ = \sqrt{d_2^2 + (\rho + \rho')^2}; \qquad R_2^- = \sqrt{d_2^2 + (\rho - \rho')^2}$$

Since

$$(R_1^+)^2 - (R_1^-)^2 = (\rho + \rho')^2 - (\rho - \rho')^2$$

$$(R_2^+)^2 - (R_2^-)^2 = (\rho + \rho')^2 - (\rho - \rho')^2$$

it follows

$$(R_1^+)^2 - (R_1^-)^2 = (R_2^+)^2 - (R_2^-)^2$$

which can be rewritten in the form

$$(R_1^+ + R_1^-)\,(R_1^+ - R_1^-) = (R_2^+ + R_2^-)\,(R_2^+ - R_2^-)$$

Since $(R_1^+ + R_1^-) < (R_2^+ + R_2^-)$, it results $(R_1^+ - R_1^-) > (R_2^+ - R_2^-)$.

Relation (2.17) is proved in the following. Let $\mathcal{C}$ be a circumference of radius ρ' collapsed on the surface Σ at z and Γ a generic circumference lying on Σ at $\bar{z}$ (see Figure 2A.2). According to (2.14),

$$W_\varphi = \frac{\beta}{2} \max_{\bar{z}} (R^+ - R^-)$$

Since $R^+ - R^- = BE - BD = BE - AE$, by applying the triangular inequality to the triangle ABE, it follows that $R^+ - R^- \leq 2\rho'$, the equality being attained when the points E and A coincide. Accordingly, $W_\varphi = \beta\rho'$, thus proving (2.17).

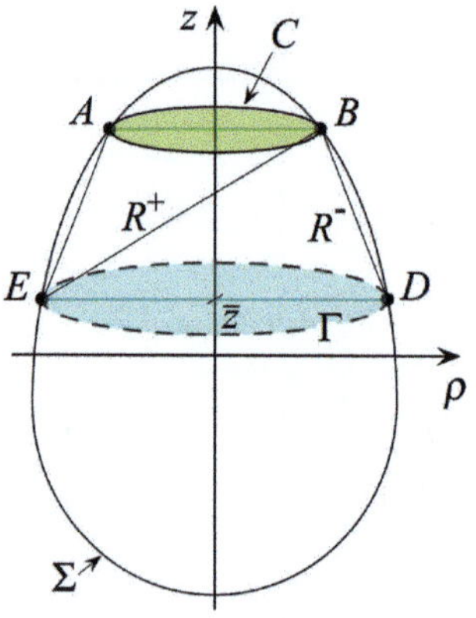

Figure 2A.2 Relevant to an observation circumference lying on Σ

2A.2 Number of samples to represent the field on a closed surface surrounding the source

It is now shown that, when using a sampling expansion along ξ and φ, the overall number N of samples needed to represent the field on any closed observation surface enclosing the radiating sources is simply related to the area of the modelling surface Σ. For electrically large sources, it is possible to choose a surface $\tilde{\Sigma}$, sufficiently far from Σ with respect to the wavelength to allow an asymptotic analysis, but enough near to it with respect to the maximum dimension of the sources so that the estimate (2.17) for W_φ is still valid. In such a case, by summing the number of samples on each of the $N_\xi/2$ sampling parallels, it results:

$$N = \sum_{i=1}^{N_\xi/2} \frac{2\pi}{\pi/W_\varphi(\xi_i)} \cong \frac{\beta}{\pi}\frac{W_\xi}{\pi} \sum_{i=1}^{N_\xi/2} 2\pi\tilde{\rho}(\xi_i)\,\Delta\xi \cong \frac{\beta}{\pi}\frac{\beta\ell'}{2\pi^2}\int_0^{2\pi} \mathrm{d}\varphi \int_0^{\pi} \tilde{\rho}(\xi)\,\mathrm{d}\xi$$
$$\cong \frac{\beta}{\pi}\frac{\beta\ell'}{2\pi^2}\int_0^{2\pi} \mathrm{d}\varphi \int_0^{\pi} \rho'(\xi)\,\mathrm{d}\xi \tag{2A.7}$$

$\tilde{\rho}$ and ρ' denoting the transverse radius of $\tilde{\Sigma}$ and Σ, respectively. By taking into account that, according to (2.11), $\xi = 2\pi s'/\ell'$ when $\tilde{C}$ coincides with C', from (2A.7) it results:

$$N \cong \frac{\beta^2}{\pi^2}\int_0^{2\pi} \mathrm{d}\varphi \int_0^{\ell'/2} \rho'(s')\,\mathrm{d}s' = \frac{\text{area of }\Sigma}{(\lambda/2)^2} \tag{2A.8}$$

2A.3 Evaluation of the optimal parameter and phase factor: spheroidal modellings

This appendix deals with the derivation of the parameter ξ and phase factor γ when considering a prolate spheroidal modelling of the antenna. Moreover, the evaluation of the angle ϑ_∞ needed to determine W_φ is also addressed. The oblate spheroidal case can be tackled in a similar way and, hence, it will not be explicitly considered in the following.

In accordance with relations (2.12) and (2.13), the curves at γ = constant and ξ = constant passing through an arbitrary point P in the meridian plane bisect the angles between the tangents from P to the curve C' (see Figure 2A.3). By taking into account that the angles between the tangents from an external point and those between the straight lines from the point to the foci have the same bisectors in a centred conical section and that the bisectors of the lines from a point P to the foci coincide with the tangent and the normal to the confocal ellipse through P, it results that the curves at γ = constant and ξ = constant are ellipses and hyperbolas confocal with C' [104].

For reader's convenience, some definitions and analytical properties of such conics are recalled in the following.

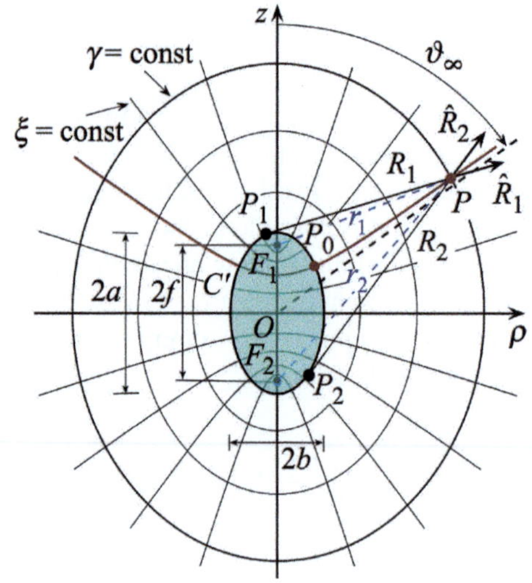

Figure 2A.3 Curves at γ = constant and ξ = constant

An ellipse having major and minor semi-axes equal to a and b (see Figure 2A.3) is the locus of the points P such that $r_1 + r_2 = 2a$, wherein $r_{1,2}$ are the distances from the foci $F_{1,2}$ to P. It can be described in Cartesian coordinates as

$$\frac{z^2}{a^2} + \frac{\rho^2}{b^2} = 1 \tag{2A.9}$$

or in parametric form as

$$\begin{cases} \rho = b\cos\alpha \\ z = a\sin\alpha \end{cases} \tag{2A.10}$$

where $\alpha = \tan^{-1}\left(b/\left(a\ \tan\vartheta\right)\right)$ is an angular like parameter. Moreover, it results

$$b = \sqrt{a^2 - f^2} = a\sqrt{1-\varepsilon^2} \tag{2A.11}$$

wherein $2f$ is the focal distance and $\varepsilon = f/a$ the eccentricity.

A hyperbola (see Figure 2A.4) is the locus of the points P such that $r_1 - r_2 = \mp 2a'$, where $r_{1,2}$ are again the distances from the foci $F_{1,2}$ to P, $2a'$ is the distance between the vertices, and the sign minus (plus) applies to the upper (lower) hyperbola branch. In Cartesian coordinates, it results

$$\frac{z^2}{a'^2} - \frac{\rho^2}{b'^2} = 1 \tag{2A.12}$$

wherein $2b'$ is the distance between the intersections of the line parallel to the ρ-axis and passing through a vertex with the asymptotes to the hyperbola. In analogy with

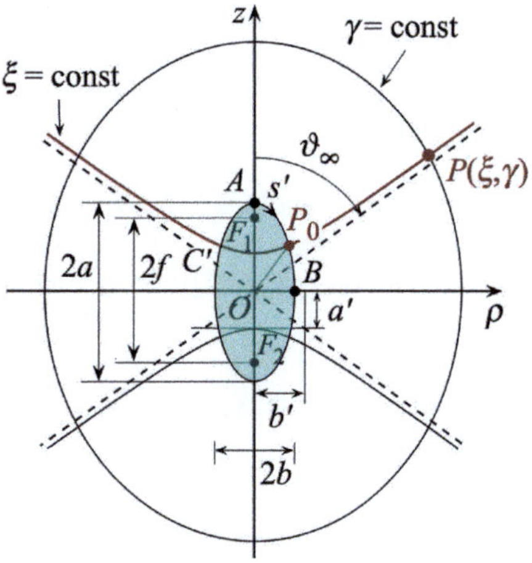

Figure 2A.4 *Relevant to the evaluation of ξ*

the ellipse, a' and b' are the major and minor semi-axes, which are related to the focal distance $2f$ as follows:

$$b' = \sqrt{f^2 - a'^2} \tag{2A.13}$$

The parameter ξ corresponding to a point P lying on a meridian curve and such that $\vartheta \in [0, \pi/2]$ is now evaluated. Since the same value corresponds to all points on the confocal hyperbola branch passing through P, it is convenient to consider, instead of P itself, the intersection point P_0 between the considered hyperbola branch and C' (see Figure 2A.4). In such a case, according to (2.11),

$$\xi = 2\pi s_0'/\ell' \tag{2A.14}$$

s_0' being the arclength between A and P_0, which can be evaluated as difference between $\ell'/4$ and the length of the ellipse arc from B to P_0. The involved integral can be computed by using the parametric equation (2A.10) of C', thus getting

$$\int_B^{P_0} \mathrm{d}s' = a \int_0^{\alpha} \sqrt{1-\varepsilon^2 \sin^2\alpha}\, \mathrm{d}\alpha = a\,\mathrm{E}\left(\alpha|\varepsilon^2\right) \tag{2A.15}$$

where $\mathrm{E}(\cdot|\cdot)$ (see (2.26)) denotes the elliptic integral of second kind [293]. To evaluate the parameter α, it is necessary to determine the intersection between the ellipse C' and the hyperbola passing through P. By imposing that such a hyperbola is confocal with the ellipse C', so that $f = \varepsilon a$, and taking into account that for this hyperbola branch is $r_1 - r_2 = -2a'$, it results

$$\begin{cases} a' = -\varepsilon a u \\ b' = \varepsilon a\sqrt{1-u^2} \end{cases} \tag{2A.16}$$

Thus, the equation (2A.12) can be rewritten as

$$\frac{z^2}{(\varepsilon\, a\, u)^2} - \frac{\rho^2}{(\varepsilon\, a)^2\,(1-u^2)} = 1 \tag{2A.17}$$

By substituting the parametric equations (2A.10) of the ellipse C' into (2A.17) and taking into account (2A.11), after some straightforward analytical manipulations, it results $\sin^2\alpha = u^2$. Therefore, for $\alpha \in [0, \pi/2]$, it results

$$\alpha = -\sin^{-1} u \tag{2A.18}$$

since in such an interval $\alpha > 0$ and $u < 0$. Accordingly,

$$s'_0 = \ell'/4 - a\,\mathrm{E}\left(-\sin^{-1} u|\varepsilon^2\right) \tag{2A.19}$$

wherein $\ell' = 4a\,\mathrm{E}\,(\pi/2|\varepsilon^2)$. By substituting (2A.19) into (2A.14) and taking into account that E is an odd function, relation (2.28a) is finally obtained.

The evaluation of the phase function γ at the point P is now tackled. Since the same value of γ corresponds to all points lying on a confocal ellipse, it can be convenient to consider, instead of P, the intersection point Q between the symmetry axis and the confocal ellipse through P (see Figure 2A.5). It can be easily verified that, in such a case, $R_1 = R_2 = R$ and $s'_2 = s'_1 + 2s'_{BP_1}$, s'_{BP_1} being the length of the ellipse arc from B to P_1. Accordingly, (2.8) becomes

$$\gamma = \beta\,(R - s'_{BP_1}) \tag{2A.20}$$

In order to determine R, we must impose the tangency condition between the ellipse C', described by the equation (2A.9), and the sheaf of straight lines passing through Q of Cartesian equations:

$$z = m\,(\rho - b_P) \tag{2A.21}$$

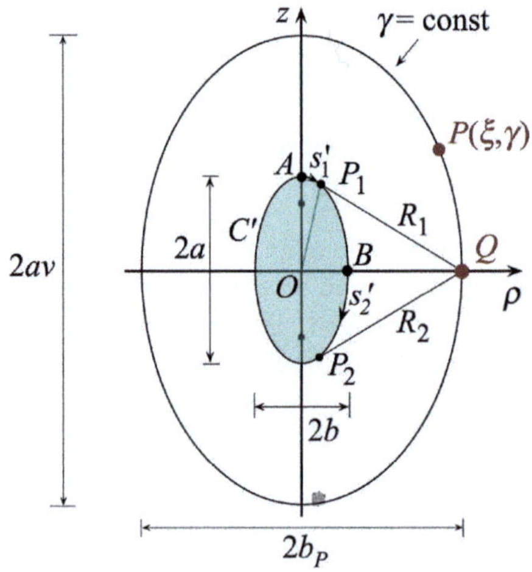

Figure 2A.5 Relevant to the evaluation of γ

wherein

$$b_P = \sqrt{(av)^2 - f^2} = a\sqrt{v^2 - \varepsilon^2} \tag{2A.22}$$

is the minor semi-axis of the confocal ellipse passing through Q, whose major semi-axis, according to the relation $v = (r_1 + r_2)/2a$, is av.

After some straightforward, but lengthy analytical manipulations, the following expressions for the Cartesian coordinates of the tangency points $P_{1,2}$ are obtained:

$$\begin{cases} \rho_1 = \rho_2 = \dfrac{a\left(1-\varepsilon^2\right)}{\sqrt{v^2-\varepsilon^2}} = \dfrac{a\left(1-v^2\right)}{\sqrt{v^2-\varepsilon^2}} + b_P \\ z_1 = -z_2 = a\sqrt{\dfrac{v^2-1}{v^2-\varepsilon^2}} \end{cases} \tag{2A.23}$$

Accordingly,

$$R = \sqrt{z_1^2 + \left(b_P - \rho_1\right)^2} = av\sqrt{\frac{v^2-1}{v^2-\varepsilon^2}} \tag{2A.24}$$

For evaluating the length s'_{BP_1} of the ellipse arc from B to P_1, it is necessary to determine the value α_{P_1} of α corresponding to P_1. By taking into account (2A.10), it results

$$\cos\alpha_{P_1} = \frac{\rho_1}{b} = \sqrt{\frac{1-\varepsilon^2}{v^2-\varepsilon^2}} \tag{2A.25}$$

Accordingly,

$$s'_{BP_1} = a\,\mathrm{E}\left(\cos^{-1}\sqrt{\frac{1-\varepsilon^2}{v^2-\varepsilon^2}}\,\middle|\,\varepsilon^2\right) \tag{2A.26}$$

By substituting (2A.24) and (2A.26) into (2A.20), relation (2.27) is obtained.

For what concerns the zenith angle ϑ_∞ of the asymptote to the hyperbola through P, by simple geometrical considerations (see Figure 2A.4), it results

$$\tan\left(\frac{\pi}{2} - \vartheta_\infty\right) = \frac{a'}{b'} \tag{2A.27}$$

which, by taking into account (2A.16), can be rewritten as:

$$\tan\left(\frac{\pi}{2} - \vartheta_\infty\right) = \frac{-u}{\sqrt{1-u^2}} \tag{2A.28}$$

Since $u < 0$ and $0 \le \vartheta_\infty \le \pi/2$, it follows:

$$\pi/2 - \vartheta_\infty = \sin^{-1}(-u) \tag{2A.29}$$

from which (2.30a) results.

2A.4 On the choice of the azimuthal enlargement bandwidth factor

As shown in [102], in the case of an azimuthal observation circumference and a spherical source modelling, the bandlimitation error decreases exponentially with $\beta a(\chi^* - 1)^{3/2} \sin\vartheta$. Accordingly, in order to ensure a bandlimitation error constant with respect to ϑ, we must enforce that $\beta a(\chi^* - 1)^{3/2} \sin\vartheta$ be constant. As a consequence, by imposing that the azimuthal bandwidth enlargement factor χ^* for $\vartheta = 90°$ coincides with that χ' relevant to a meridian curve, it results

$$\beta a(\chi^* - 1)^{3/2} \sin\vartheta = \beta a(\chi' - 1)^{3/2}$$

By solving such a relation with respect to χ^*, it results:

$$\chi^* = \chi^*(\vartheta) = 1 + (\chi' - 1)\,[\sin\vartheta]^{-2/3} \tag{2A.30}$$

Although such a relation has been rigorously derived in the case of the spherical source modelling, it can be heuristically used also when adopting a different antenna modelling.

2A.5 On the Tschebyscheff sampling function

In order to show the derivation of the Tschebyscheff sampling function employed in the OSI algorithm, some concepts concerning the use of the sampling windows for the central interpolation of finite energy bandlimited functions on rectilinear observation domains are recalled.

As shown in [295] by Knab, if $F(s)$ is a function belonging to the set $\mathcal{E}_W$ of the functions having finite energy $\mathcal{E}$ and bandlimited to W, then $F(s)$ can be rigorously represented in terms of an infinite number of its samples at the sampling rate $\Delta s = \pi/(\chi W)$ as

$$F(s) = \sum_{n=-\infty}^{\infty} F(n\Delta s)\,\Psi(s - n\Delta s)\,\mathrm{sinc}\left[\frac{\pi}{\Delta s}(s - n\Delta s)\right] \tag{2A.31}$$

where $\chi > 1$ is the oversampling factor and $\Psi(s)$ is a function belonging to the set $\mathcal{B}_\Gamma$ of the bounded functions bandlimited to $\Gamma = (\chi - 1)W$.

An interpolation scheme based on the above result makes use of a finite number of samples centred on the output point. In such a case, the relation (2A.31) can be rewritten as

$$F_t(s) = \sum_{n=-q+1}^{q} F(n\Delta s)\,\Psi(s - n\Delta s)\,\mathrm{sinc}\left[\frac{\pi}{\Delta s}(s - n\Delta s)\right] \tag{2A.32}$$

where $F_t(s)$ is the truncated version of $F(s)$, $\mathrm{sinc}(s)$ is the $\sin(s)/s$ function, $2q$ is the number of retained samples and, without any loss of generality, the output point is assumed to lie in the interval $[0, \Delta s]$. Moreover, it is convenient that the convergence factor $\Psi(s)$ satisfies the condition $\Psi(0) = 1$ so that, in correspondence of a sampling point, the expansion (2A.32) gives the corresponding sample value. Since an unavoidable truncation error is introduced when using (2A.32), the function $\Psi(s)$ must be properly chosen to minimize such an error.

According to [295], a convergence factor $\Psi(s)$, which is as low as possible outside the retained samples interval $[-q\Delta s,\ q\Delta s]$, allows one to minimize the truncation error. It has been proved that the Knab's Sampling Window (SW) function

$$\Psi(s) = \frac{\cosh\left[\pi\, v\, q\sqrt{1-(s/\bar{s})^2}\right]}{\cosh(\pi\, v\, q)} \tag{2A.33}$$

satisfies the above requirement. In (2A.33), $v = (1 - 1/\chi)$ and $\bar{s} = q\Delta s$.

The representation of a bandlimited function $F(\varphi)$ on a circular observation domain from its samples collected in correspondence of $\varphi_n = n\Delta\varphi$ at the over-sampled spacing $\Delta\varphi = 2\pi/(2N''+1)$ is now considered, wherein $N'' = \mathrm{Int}(\chi N') + 1$ and $2N' + 1$ is the number of samples at the Nyquist rate. In such a case, the Dirichlet function $D_{N''}(\varphi)$ takes the role of the sinc (s) function in (2A.32). The related convergence factor $\Omega(\varphi)$ can be found by requiring [39, 106] that it parallels the properties of $\Psi(s)$. In other words, $\Omega(\varphi)$ must be:

i a trigonometric polynomial of degree $N = N'' - N' = vN''$;
ii equal to one for $\varphi = 0$;
iii and as low as much possible outside the interval $[-q\Delta\varphi,\ q\Delta\varphi]$ of the retained samples in order to minimize the truncation error.

It can be easily realized that the above conditions are the same to be used when synthesizing an array pattern which has the lowest side lobe level for a given width of the main lobe. The solution to such a problem is given by [298]

$$\Omega(\varphi) = B\, T_N\left[a + b\cos\varphi\right] \tag{2A.34}$$

where $T_N(\cdot)$ is the Tschebyscheff polynomial of degree N. The constants B, a, and b can be found by imposing condition ii) and requiring that the argument of $T_N(\cdot)$ is equal to -1 and 1 when $\varphi = \pi$ and $\varphi = \bar{\varphi} = q\Delta\varphi$, respectively. The following expression for the convergence factor $\Omega(\varphi)$ is so obtained:

$$\Omega(\varphi) = \Omega(\varphi, \bar{\varphi}) = \frac{T_N\left[2\left(\cos(\varphi/2)/\cos(\bar{\varphi}/2)\right)^2 - 1\right]}{T_N\left[2/\cos^2(\bar{\varphi}/2) - 1\right]} \tag{2A.35}$$

Accordingly, the following interpolation formula quite analogous to (2A.32) results

$$F_t(\varphi) = \sum_{n=-q+1}^{q} F(n\Delta\varphi)\, D_{N''}(\varphi - \varphi_n)\, \Omega_N(\varphi - \varphi_n, \bar{\varphi}) \tag{2A.36}$$

wherein the output point is assumed to lie in the interval $[0, \Delta\varphi]$. When the output point doesn't lie in this interval, the above formula can easily generalized, see (2.58), wherein the nearest q samples before and the q nearest ones after the output point are considered.

Chapter 3
Theoretical foundations of spiral scannings

In the previous chapter, the non-redundant (NR) sampling representations of the radiated electromagnetic (EM) fields have been described in detail and it has been stressed that they can be conveniently exploited to get an effective sampling representation of the voltage, acquired by the probe on the scanning surface, using a NR number of its samples. This allows the development of NR near-field to far-field (NF–FF) transformation techniques (see the next chapters), requiring a number of NF data remarkably lower than that needed by the corresponding classical NF–FF transformations, so that a considerable acquisition time saving is gained.

In this chapter, efficient sampling representations of the probe voltage over a rotational scanning surface, from a NR number of its samples collected along a proper spiral wrapping such a surface, are presented. These representations allow, as will be shown in the following chapters, the development of effective NR NF–FF transformations which adopt innovative spiral scannings. As a matter of fact, the massive NF data needed by the corresponding traditional NF–FF transformation are accurately determined, using an optimal sampling interpolation (OSI) expansion, from the NR samples acquired along the spiral. A further considerable measurement time saving can be so achieved, since the acquisition of these samples is sped up by gathering them on fly and adopting continuous and synchronized motions of the antenna under test (AUT) and probe. The NR representations and related OSI expansions are obtained by adopting a suitable model of the AUT, by considering a spiral such that its step coincides with the sample spacing needed for the interpolation along the related meridian curve, and by determining the NR sampling representation along the spiral.

3.1 The unified theory of spiral scannings for quasi-spherical antennas

Let a quasi-spherical AUT be considered as enclosed in a sphere Σ of radius a and let the voltage be acquired by an electrically small non-directive probe scanning a proper spiral, which wraps an arbitrary rotational surface $\mathcal{M}$, attained by rotating a meridian curve external to the cone tangent to the sphere Σ and with the vertex at the observation point P. The Cartesian coordinates of a point Q on this spiral are given by:

$$\begin{cases} x = r(\theta)\sin\theta\cos\phi \\ y = r(\theta)\sin\theta\sin\phi \\ z = r(\theta)\cos\theta \end{cases} \tag{3.1}$$

wherein ϕ is the angular parameter which describes the spiral and the angle θ is a monotonically increasing function of ϕ. The function $r(\theta)$ is determined by the meridian curve generating the rotational surface $\mathcal{M}$. It can be easily recognized that $r(\theta) = d/\cos\theta$ in the case of a planar spiral on the plane $z=d$, $r(\theta) = d/\sin\theta$ in the case of a helix which wraps a cylinder of radius d, whereas $r(\theta) = d$ for a spherical spiral. It must be observed that θ, unlike the zenithal angle ϑ, can assume also negative values. As a matter of fact, when the spiral makes a whole round on $\mathcal{M}$, moving from the south pole to the north pole and then returning to the south one, θ varies from $-\pi$ to π. Moreover, the angular parameter ϕ is always continuous, whereas, according to relations (3.1), the azimuthal angle φ has a discontinuity jump of π when the spiral crosses the poles.

Since, as shown in [41], the voltage V measured by a non-directive probe has almost the same effective spatial bandwidth of the AUT radiated field, the NR sampling representations of EM fields [104, 105] can be usefully applied to represent the reduced voltage $\tilde{V}(\xi) = V(\xi)\, e^{j\gamma(\xi)}$, where ξ is the optimal parameter to be used for describing the observation curve and $\gamma(\xi)$ is a suitable phase function.

According to the unified theory of spiral scannings for quasi-spherical antennas [122], the two-dimensional OSI expansion for determining the voltage on the surface $\mathcal{M}$ from a NR number of its samples collected along an appropriate spiral which wraps it can be achieved: (i) by choosing a spiral whose step coincides with the sampling spacing needed to interpolate the NF data along a meridian curve; (ii) by developing a NR sampling representation of the voltage along the spiral. In this way, the voltage at a given point P on $\mathcal{M}$ can be reconstructed by evaluating the intermediate reduced voltage samples (those in correspondence of the intersection points between the spiral and the meridian curve through P), by means of an interpolation along the spiral, and then interpolating the so determined intermediate samples along the meridian curve. In the considered case of spherical AUT modelling, the optimal parameter for describing a meridian curve is the zenithal angle ϑ, the related bandwidth is βa, and the phase function is given by:

$$\gamma = \beta\sqrt{r^2 - a^2} - \beta a \cos^{-1}(a/r) \tag{3.2}$$

Accordingly, the angular step of the spiral must be constant and hence $\theta = k\phi$. For the condition (i), the angular step of the spiral $\Delta\theta = 2\pi k$ must be equal to the sampling spacing $\Delta\vartheta$ required in the interpolation along a meridian curve. As shown in the previous chapter, this last is given by $\Delta\vartheta = 2\pi/(2N'' + 1)$, where $N'' = \text{Int}(\chi N') + 1$, with $N' = \text{Int}(\chi'\beta a) + 1$. Accordingly, $k = 1/(2N'' + 1)$. The spiral can be so viewed as obtained by radially projecting on $\mathcal{M}$ the corresponding spiral which wraps the sphere Σ with the same angular step.

The development of the NR sampling representation along the spiral is now tackled. According to the results in [104], the optimal parameter η for describing the spiral and the related phase function ψ are given by (2.6) and (2.4), here reported for reader's convenience:

$$\eta = \frac{\beta}{2W_\eta} \int_0^s [\max_{r'} \hat{R} \cdot \hat{t} - \min_{r'} \hat{R} \cdot \hat{t}] \mathrm{d}s \tag{3.3}$$

$$\psi(\eta) = \frac{\beta}{2} \int_0^s [\max_{r'} \hat{R} \cdot \hat{t} + \max_{r'} \hat{R} \cdot \hat{t}] \mathrm{d}s \tag{3.4}$$

It can be easily recognized that the maximum and minimum values of $\hat{R} \cdot \hat{t}$ occur [122] in correspondence of the two tangency points $P_{1,2}$ (see Figure 3.1) of the sphere Σ with the straight lines passing through the point Q on the spiral and belonging to the plane specified by the unit vectors $\hat{t}$ (tangent to the spiral at Q) and $\hat{r}$ (pointing from the origin to the point Q).

With the aid of Figure 3.1, it is a simple task to verify that

$$(\hat{R}_1 + \hat{R}_2)/2 = \hat{r} \sin\alpha = \hat{r}\sqrt{1 - a^2/r^2} \tag{3.5}$$

$$(\hat{R}_1 - \hat{R}_2)/2 = \hat{n} \cos\alpha = \hat{n}\,(a/r) \tag{3.6}$$

wherein $\hat{R}_{1,2}$ are the unit vectors from $P_{1,2}$ to Q and $\hat{n}$ is a unit vector orthogonal to $\hat{r}$ and parallel to the plane identified by $\hat{r}$ and $\hat{t}$. Hence

$$(\hat{R}_1 - \hat{R}_2) \cdot \hat{t}/2 = (\hat{n} \cdot \hat{t})(a/r) = (a/r)\cos(\delta - \pi/2) = (a/r)\sin\delta \tag{3.7}$$

δ being the angle between $\hat{t}$ and $\hat{r}$.

By taking into account that $\mathrm{d}r = \hat{r} \cdot \hat{t}\, \mathrm{d}s$, the substitution of (3.5) into (3.4) gives

$$\psi = \beta \int_0^r \sqrt{1 - a^2/r^2}\, \mathrm{d}r = \beta\sqrt{r^2 - a^2} - \beta a \cos^{-1}(a/r) \tag{3.8}$$

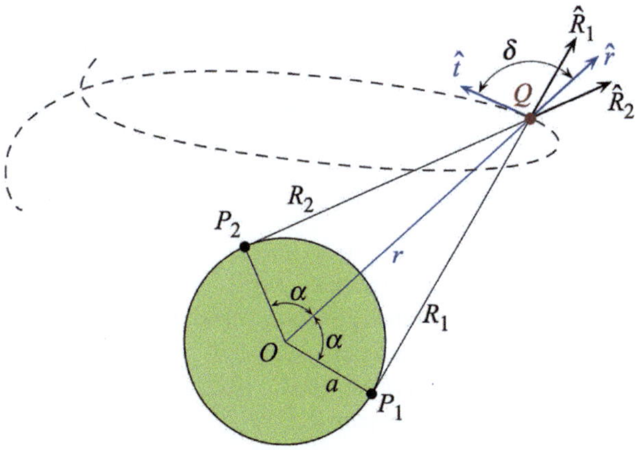

Figure 3.1 Relevant to the determination of the extreme values of $\hat{R} \cdot \hat{t}$

that is to say, when the spherical model of the antenna is adopted, the phase function ψ relevant to the representation along the spiral is the same as that γ relevant to the representation along a meridian curve.

By taking into account the spiral equations (3.1), it can be shown [122], through straightforward analytical manipulations, that

$$\mathrm{d}s = \sqrt{(\mathrm{d}x)^2 + (\mathrm{d}y)^2 + (\mathrm{d}z)^2} = \sqrt{r^2 \sin^2\theta + k^2 r^2 + k^2 \dot{r}^2}\,\mathrm{d}\phi \tag{3.9}$$

wherein $\dot{r} = \mathrm{d}r/\mathrm{d}\,\theta$. On the other hand,

$$\sin\delta = \sqrt{1 - \cos^2\delta} = \sqrt{1 - (\hat{r} \cdot \hat{t})^2} \tag{3.10}$$

where

$$\hat{r} \cdot \hat{t} = \frac{\mathrm{d}r}{\mathrm{d}s} = \frac{\mathrm{d}r}{\mathrm{d}\phi}\frac{\mathrm{d}\phi}{\mathrm{d}s} = \left[\frac{\mathrm{d}r}{\mathrm{d}\theta}\frac{\mathrm{d}\theta}{\mathrm{d}\phi}\right]\frac{\mathrm{d}\phi}{\mathrm{d}s} = k\dot{r}\,\frac{\mathrm{d}\phi}{\mathrm{d}s} = \frac{k\dot{r}}{\sqrt{r^2 \sin^2\theta + k^2 r^2 + k^2 \dot{r}^2}} \tag{3.11}$$

accordingly,

$$\sin\delta = \sqrt{\frac{r^2 \sin^2\theta + k^2 r^2}{r^2 \sin^2\theta + k^2 r^2 + k^2 \dot{r}^2}} \tag{3.12}$$

By taking into account the above relation and substituting (3.7) and (3.9) into (3.3), it results [122]:

$$\eta = \frac{\beta a}{W_\eta}\int_0^\phi \sqrt{k^2 + \sin^2 k\phi}\;\mathrm{d}\phi \tag{3.13}$$

From (3.9), it can be easily recognized that, in the case of the spiral wrapping the sphere Σ, it results:

$$\mathrm{d}s = a\sqrt{\sin^2\theta + k^2}\,\mathrm{d}\phi = a\sqrt{k^2 + \sin^2 k\phi}\,\mathrm{d}\phi \tag{3.14}$$

and, accordingly, the optimal parameter η relevant to the representation along the scanning spiral is β/W_η times the curvilinear abscissa of the corresponding point lying on the spiral wrapping Σ. It is noteworthy that this last spiral is a closed curve. This suggests to choose the bandwidth W_η in such a way that the angular-like parameter η covers a 2π range when the whole spiral on Σ is described. As a consequence:

$$W_\eta = \frac{\beta a}{\pi}\int_0^{(2N''+1)\pi} \sqrt{k^2 + \sin^2 k\phi}\;\mathrm{d}\phi \tag{3.15}$$

that is to say, when the AUT is modelled by a sphere, the bandwidth W_η is β/π times the length of the spiral which wraps Σ from the north pole to the south one.

In accordance with the above results, the reduced voltage at a given point Q on the spiral can be accurately reconstructed [122], from the NR samples gathered along the spiral, through the OSI expansion:

$$\tilde{V}(\eta) = \sum_{m=m_0-p+1}^{m_0+p} \tilde{V}(\eta_m)\, SF(\eta, \eta_m, \bar{\eta}, M, M'') \tag{3.16}$$

where $m_0 = \text{Int}\left(\eta/\Delta\eta\right)$ is the index of the sample closest (on the left) to the point Q, $2p$ is the number of the retained closest reduced voltages samples $\tilde{V}\left(\eta_m\right)$, $SF\left(\eta, \eta_m, \bar{\eta}, M, M''\right)$ is the OSI function (2.61), and

$$\eta_m = m\Delta\eta = 2\pi m/\left(2M'' + 1\right); \quad M'' = \text{Int}\left(\chi M' + 1\right) \tag{3.17}$$

$$M' = \text{Int}\left(\chi' W_\eta\right) + 1; \quad M = M'' - M'; \quad \bar{\eta} = p\Delta\eta \tag{3.18}$$

It must be underlined that, nearby the poles ($\vartheta = 0$ and $\vartheta = \pi$), small variations of η correspond to large changes of ϕ, so that the enlargement bandwidth factor χ', relevant to the representation along the spiral, must be suitably increased in these zones to avoid a significant growth of the aliasing error [122].

The OSI expansion (3.16) can be employed to evaluate the intermediate reduced voltage samples at the intersection points between the meridian curve through the observation point $P\left(\vartheta, \varphi\right)$ and the spiral. The voltage at P can be, then, reconstructed [122] by applying the following OSI expansion:

$$V\left(\vartheta, \varphi\right) = \mathrm{e}^{-\mathrm{j}\gamma(\vartheta)} \sum_{n=n_0-q+1}^{n_0+q} \tilde{V}\left(\vartheta_n\right) SF\left(\vartheta, \vartheta_n, \bar{\vartheta}, N, N''\right) \tag{3.19}$$

where $n_0 = \text{Int}\left[\left(\vartheta - \vartheta_0\right)/\Delta\vartheta\right]$ is the index of the intermediate sample closest (on the left) to P, $2q$ is the number of the retained closest intermediate samples $\tilde{V}\left(\vartheta_n\right)$, and

$$\vartheta_n = \vartheta_n\left(\varphi\right) = k\varphi + n\Delta\vartheta = \vartheta_0 + n\Delta\vartheta; \quad N = N'' - N'; \quad \bar{\vartheta} = q\Delta\vartheta \tag{3.20}$$

It is so possible to accurately reconstruct the voltage at any point of the rotational surface $\mathcal{M}$ from its NR samples acquired along the spiral and this has allowed the development of very efficient probe-compensated NF–FF transformation techniques with helicoidal [114, 115, 119, 122, 123], planar [122, 123, 127, 128], and spherical spiral [122, 123, 131, 132, 139] scannings, which require a considerably lower measurement time than those using the corresponding traditional scans. In these transformation techniques, presented in the next chapters, the NF samples acquired at the prescribed points along the spiral wrapping the scanning surface are interpolated through formulas (3.16) and (3.19) to accurately recover the probe and rotated probe voltages at the grid points required to obtain the antenna far field by means of the classical probe-compensated NF–FF transformation relevant to the considered scanning surface. As regards the choice of the OSI parameters in (3.16) and (3.19), i.e., the bandwidth enlargement factor χ', the oversampling one χ, and the retained samples numbers $2p$, $2q$, the same considerations done in Chapter 2 with reference to the OSI expansion (2.71) hold.

It is worth noting that, when dealing with quasi-planar or elongated antennas, the use of the spherical AUT modelling in the above NF–FF transformations gives rise to an ineffective increase in the number of needed NF measurements with respect to that required by the NR NF–FF transformations, which employ the corresponding traditional scanning and rely on the appropriate spheroidal or flexible modelling. Moreover, since the scanning surface must be always external to the modelling one, by denoting with D the AUT maximum size, it is not possible to consider a measurement cylinder with radius smaller than $D/2$ in the helicoidal scanning and a plane at distance smaller than $D/2$ in the planar spiral scanning. Accordingly, in the case of elongated or quasi-planar AUTs, this reflects in an increase of the error related to the truncation of the scanning surface in the helicoidal or planar spiral scanning, respectively.

It is clear that, to effectively overcome the above drawbacks, a proper extension of the unified theory of the spiral scannings to the case of non-spherical antennas is required.

3.2 The unified theory of spiral scannings for non-spherical antennas

In this section, the previously described sampling representation will be extended to the case of a non-spherical antenna, namely, a radiating structure with one or two predominant dimensions. In such a case, it is no longer opportune to consider the smallest sphere as surface containing it but, as suggested in [104], a suitable rotational surface Σ which bounds a convex domain (see Figure 2.1). Let the voltage be acquired by an electrically small probe, which scans a proper spiral wrapping an arbitrary rotational surface $\mathcal{M}$, attained by rotating a meridian curve external to the cone with the vertex at the observation point P and tangent to the modelling surface Σ (see Figure 2.2).

As in the previously considered case of quasi-spherical antennas, the OSI expansion to determine the voltage on $\mathcal{M}$ from its NR samples acquired along an opportune spiral wrapping it can be attained [124–126]: (i) by choosing a spiral whose step coincides with the sample spacing required to interpolate the NF data along a meridian curve; (ii) by developing an NR sampling representation of the voltage along such a spiral.

As shown in the previous chapter, the phase function γ and the optimal parameter ξ relevant to the NR sampling representation of the probe voltage on a meridian curve are given by (2.8) and (2.11), here reported for reader's convenience:

$$\gamma = \frac{\beta}{2}\left[R_1 + R_2 + s_1' - s_2'\right] \tag{3.21}$$

$$\xi = \frac{\pi}{\ell'}\left[R_1 - R_2 + s_1' + s_2'\right] \tag{3.22}$$

In such a case, the parametric equations of the spiral become:

$$\begin{cases} x = r\left[\theta\left(\xi\right)\right] \sin\theta\left(\xi\right) \cos\phi \\ y = r\left[\theta\left(\xi\right)\right] \sin\theta\left(\xi\right) \sin\phi \\ z = r\left[\theta\left(\xi\right)\right] \cos\theta\left(\xi\right) \end{cases} \tag{3.23}$$

where the function $r\left[\theta\left(\xi\right)\right]$ is again fixed by the meridian curve generating the rotational surface $\mathcal{M}$ and ξ is related to the angle ϕ describing the spiral by $\xi = k\,\phi$.

The constant k is determined by imposing that the spiral step $2\pi k$ must be equal to the sampling spacing $\Delta\xi$ required in the interpolation along a meridian curve, which, according to (2.59), is given by $\Delta\xi = 2\pi / (2N'' + 1)$. Accordingly, it results $k = 1 / (2N'' + 1)$.

It is noteworthy that the scanning spiral can be viewed as attained by projecting on the surface $\mathcal{M}$ the corresponding spiral which wraps with the same step the modelling surface Σ. The projection is achieved by means of the curves at ξ= constant, which have in this case the same role played by the radial lines in the spherical modelling.

As regards the parameter η to be employed for describing the spiral, the related phase function ψ and bandwidth W_η, they can be heuristically derived [124] by paralleling the corresponding rigorous procedure valid when adopting the spherical AUT modelling. As a consequence, η is β/W_η times the arclength of the projecting point lying on the spiral which wraps the surface Σ, ψ coincides with the phase function γ relevant to a meridian curve, and W_η is β/π times the length of the spiral wrapping Σ from the north to the south pole. That is to say, the spiral, η, ψ, and W_η are such that they coincide with those relevant to the spherical modelling when the surface Σ approaches a sphere.

As already stressed, the voltage at any point P on $\mathcal{M}$ can be recovered from the samples gathered along the spiral by first evaluating, through an interpolation along the spiral, the intermediate reduced voltage samples and then reconstructing the voltage at P by interpolating, along the meridian curve, the intermediate samples. The related two-dimensional OSI expansion is:

$$V\left(\xi(\vartheta), \varphi\right) = \mathrm{e}^{-\mathrm{j}\gamma(\xi)}$$

$$\times \sum_{n=n_0-q+1}^{n_0+q} \left[SF\left(\xi,\ \xi_n,\ \bar{\xi}, N, N''\right) \sum_{m=m_0-p+1}^{m_0+p} \tilde{V}\left(\eta_m\right) SF\left(\eta\left(\xi_n\right), \eta_m, \bar{\eta}, M, M''\right) \right] \quad (3.24)$$

where $n_0 = \mathrm{Int}\left[\left(\xi - \xi_0\right)/\Delta\xi\right]$ is the index of the intermediate sample closest (on the left) to P, $2q$ is the number of the retained intermediate samples $\tilde{V}\left(\xi_n\right)$,

$$\xi_n = \xi_n\left(\varphi\right) = k\varphi + n\Delta\xi = \xi_0 + n\Delta\xi; \quad N = N'' - N'; \quad \bar{\xi} = q\Delta\xi \quad (3.25)$$

$$N'' = \mathrm{Int}\left(\chi N'\right) + 1; \quad N' = \mathrm{Int}\left(\chi' W_\xi\right) + 1; \quad W_\xi = \beta\ell'/2\pi \quad (3.26)$$

and the other symbols have the same or analogous meanings as in (3.16).

It is noteworthy that, when reconstructing the intermediate samples in the neighbourhood of the poles, the enlargement bandwidth factor χ', relevant to the representation along the spiral, must be properly increased to prevent a significant growth of the bandlimitation error in those zones [124].

The application of the OSI expansion (3.24) has made possible the development of very effective probe-compensated NF–FF transformation techniques with helicoidal [116–118, 120, 121, 124–126], planar [124–126, 129, 130] and spherical spiral [124–126, 133–138, 140, 141] scannings, which, in the case of quasi-planar or elongated antennas, allow a considerable further reduction of the amount of needed NF data and of related measurement time with respect to the corresponding ones based on the spherical modelling.

3.3 The spheroidal spiral case

In this section, the previously derived results are suitably exploited to develop an efficient sampling representation of the electric field, radiated by an electrically long antenna on the surface of a prolate spheroid, using an NR number of its samples acquired on a proper spiral wrapping this spheroid. Although such a problem is here considered mainly for its generality, the related OSI expansion allows one to accurately reconstruct the NF data required by the NF–FF transformation [292] from the NR samples acquired along the spiral.

Since the considered antenna is, for hypothesis, elongated, it can be conveniently shaped by the smallest prolate spheroid Σ with major and minor semi-axes equal to a and b (Figure 3.2). In such a case, as shown in Chapter 2, the bandwidth W_ξ, the phase function γ, and the parameter ξ relevant to the NR representation along a meridian curve are given by (2.25), (2.27), and (2.28a), respectively. Moreover, in any meridian plane, the curves ξ= constant and γ= constant are hyperbolas and ellipses having the same foci of the ellipse C', intersection of Σ with the meridian plane.

The parametric equations of the prolate spheroid, around which the spiral is wrapping, are:

$$\begin{cases} x = b_S \cos\alpha \cos\varphi \\ y = b_S \cos\alpha \sin\varphi \\ z = a_S \sin\alpha \end{cases} \tag{3.27}$$

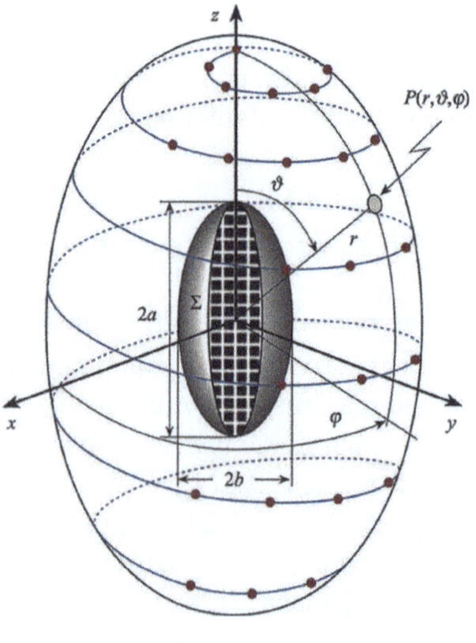

Figure 3.2 Spheroidal spiral scanning

wherein a_S and b_S are its major and minor semi-axes and the angular-like parameter $\alpha = \tan^{-1}\left(b_S/\left(a_S \tan\vartheta\right)\right)$ varies in the range $\left[-\pi/2, \pi/2\right]$. Accordingly, the Cartesian coordinates of a point of the spiral are

$$\begin{cases} x = b_S \cos\delta \cos\phi \\ y = b_S \cos\delta \sin\phi \\ z = a_S \sin\delta \end{cases} \tag{3.28}$$

As already stressed, the angular parameter ϕ, describing the spiral is always continuous, whereas, the azimuthal angle φ has a discontinuity jump of π when the spiral crosses the poles. Accordingly, the angular-like parameter δ is given by:

$$\delta = \delta\left(\xi\right) = \tan^{-1}\left(\frac{b_S}{a_S \tan\theta\left(\xi\right)}\right) + i\pi \tag{3.29}$$

where

$$i = \begin{cases} +1 \text{ if } -\pi/2 < \theta\left(\xi\right) < 0 \\ \;\;0 \text{ if } \;\;0 \;\; < \theta\left(\xi\right) < \pi \\ -1 \text{ if } -\pi \;\; < \theta\left(\xi\right) < -\pi/2 \end{cases} \tag{3.30}$$

as a consequence, δ covers the range $\left[-\pi, \pi\right]$ when the spiral describes a whole round on the surface $\mathcal{M}$.

It can be easily verified that, in the considered spheroidal spiral case, it results

$$r\left[\theta\left(\xi\right)\right] = \sqrt{b_S^2 \cos^2\delta + a_S^2 \sin^2\delta} \tag{3.31}$$

By taking into account (3.27), it can be easily shown that the unit vector $\hat{\tau}$ tangent to the meridian curve is given by:

$$\hat{\tau} = \frac{\mathrm{d}\boldsymbol{r}}{\mathrm{d}s} = \frac{-b_s \sin\alpha\left(\hat{x}\cos\varphi + \hat{y}\sin\varphi\right) + \hat{z}\, a_s \cos\alpha}{\sqrt{b_s^2 \sin^2\alpha + a_s^2 \cos^2\alpha}} \tag{3.32}$$

The numerical tests reported in the following are relevant to a uniform planar array lying on the plane $y = 0$, symmetric with respect to the plane $z = 0$. Its elements are elementary Huygens sources linearly polarized along the z axis, are spaced by $0.5\,\lambda$ (with λ being the wavelength) both along x and z, and cover an elliptical zone in the plane $y = 0$, with minor and major semi-axes equal to $12\,\lambda$ and $36\,\lambda$, respectively. The prolate spheroid, around which the spiral is wrapping, has minor and major semi-axes equal to $25\,\lambda$ and $48\,\lambda$.

In Figure 3.3a and b, the amplitude and phase of the electric field τ -component (the most significant one) on the meridian curve at $\varphi = 90°$, recovered (crosses) from the NR samples collected along the spiral, are compared with the exact ones (solid line). As can be clearly seen, a very good agreement results, thus assessing the effectiveness of the representation. It is worth noting that an enlargement bandwidth factor such that the sample spacing is reduced exactly by a factor 9 has been adopted in the zones of the spiral specified by the 80 samples around the poles to avoid a significant increase of the bandlimitation error nearby them. The performances of the representation are assessed in a more quantitative way by the values of the maximum reconstruction error (normalized to the field maximum value over the spheroid), shown in Figure 3.4 for p=q ranging from 3 to 9, $\chi' = 1.20$ (save for the zones nearby the poles), and χ=1.10, 1.15, 1.20, 1.25. The evaluation has been performed by comparing the reconstructed electric field τ-component with the exact one on a thick grid of the spheroid and, as expected, the errors decrease up to very low values on increasing the number of retained samples and/or the oversampling factor. The

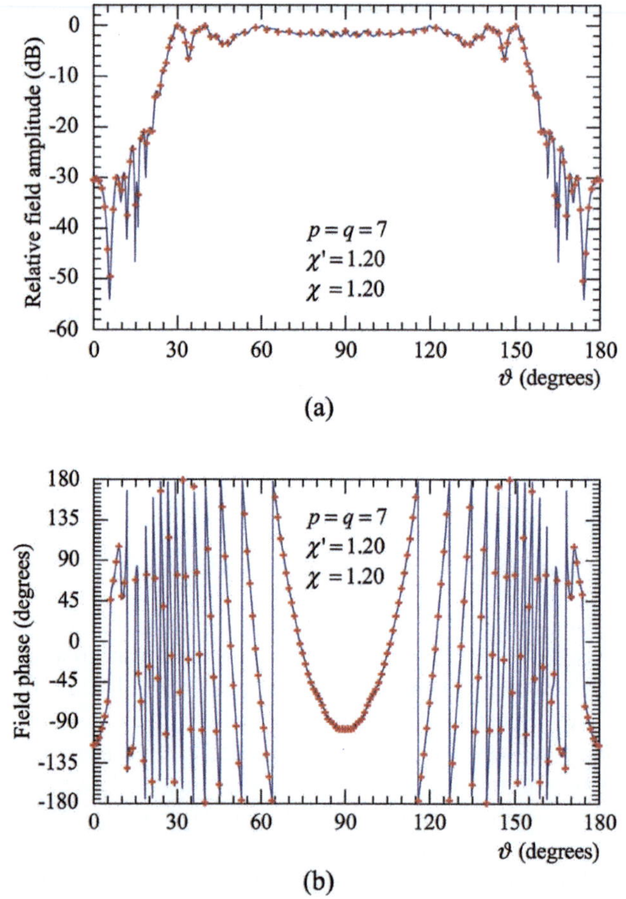

Figure 3.3 *Field τ-component on the meridian curve at $\varphi = 90°$. Solid line: exact. Crosses: interpolated from the NR samples on the spiral. (a) Amplitude, (b) Phase.*

analogous curves relevant to the mean-square error, not reported here for space saving, exhibit a quite analogous behaviour running about 20 dB down with respect to the maximum error ones. At last, the exact samples have been corrupted by random errors to verify the stability of the OSI algorithm. These errors simulate a background noise (bounded to Δa in amplitude and with arbitrary phase) and measurement uncertainties on the field samples of $\pm\Delta a_r$ in amplitude and $\pm\Delta\sigma$ in phase. As shown in Figure 3.5, the interpolation algorithm works well also when dealing with error affected data.

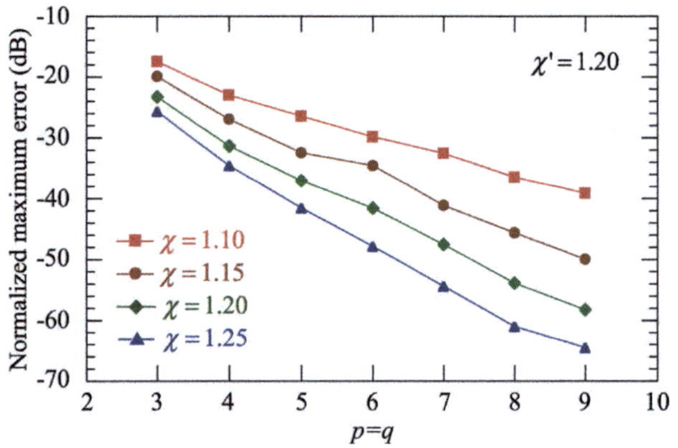

Figure 3.4 Maximum reconstruction error of the field τ-component

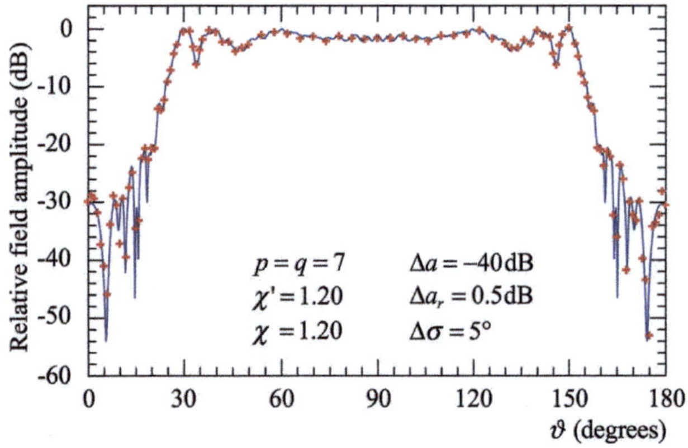

Figure 3.5 Amplitude of the field τ-component on the meridian curve at $\varphi = 90°$. Solid line: exact. Crosses: interpolated from error affected samples.

It is noteworthy that the number of samples acquired along the spiral is 38 743. In particular, the "regular samples" at spacing $\Delta\xi$ are 37 463, whereas the "extra samples" at reduced spacing are 1 280.

Chapter 4

Near-field to far-field transformations in planar scanning geometry

4.1 Introduction

The near-field to far-field (NF–FF) transformation techniques with planar scannings have received over the years large attention as testified by a lot of papers concerning them appearing in the open literature [15, 24–51, 108, 109, 112, 122–130, 143]. They are usually adopted to characterize highly directive quasi-planar antennas, which radiate a pencil beam pattern, well within the angular region specified by the edges of the antenna under test (AUT) and of the scanning area, so that the truncation (see Sect. 4.4) does not severely affect the pattern reconstruction in the region of greatest interest.

In the plane-rectangular scanning [24–31] (see Figure 4.1), the probe is mounted on a x-y positioner so that it can acquire the NF amplitude and phase data on the wanted plane-rectangular grid. From these data, measured for two different orientations of the probe (the probe is rotated by 90° around its longitudinal axis in the second set), and taking into account the probe effects, one can compute the antenna FF pattern. The NF–FF transformation with plane-rectangular scanning is the most simple and efficient one from the analytical and computational viewpoints. As a matter of fact (see Sect. 4.3), by expanding the field of the AUT in plane waves and applying the Lorentz reciprocity theorem, the AUT far field can be related [25, 26] to the two-dimensional Fourier transforms of the output voltages of the probe for both the orientations and to the far fields radiated by the probe and the rotated probe, when used as transmitting antennas. As a consequence, the AUT far field can be efficiently computed via a two-dimensional fast Fourier transform (FFT) algorithm. A different approach to compensate the directional effects introduced by an actual probe makes use of the plane wave scattering matrix formulation [27–29]. For the reader's convenience, the plane-rectangular NF–FF transformation without probe compensation is briefly summarized in Sect. 4.2.

A non-redundant (NR) effective NF–FF transformation with a non-conventional plane-rectangular scan, which allows one to significantly reduce the amount of required NF data and, as a consequence, the related measurement time with respect to that employing the traditional plane-rectangular scanning, is presented in Sect. 4.5. It is named planar wide-mesh scanning (PWMS) [108, 109], since its sample grid has meshes becoming larger and larger when moving away from the scanning

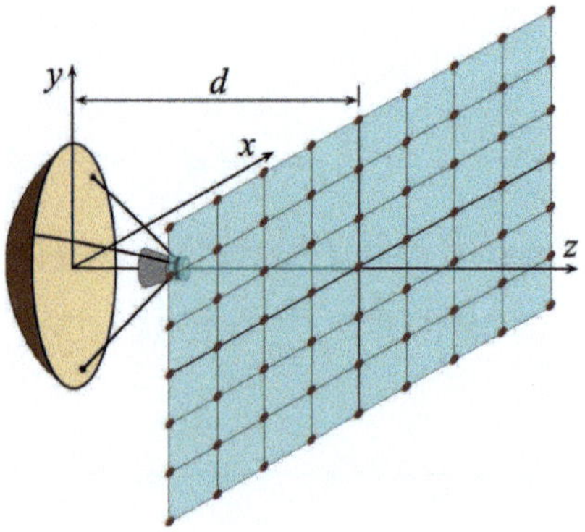

Figure 4.1 *Plane-rectangular scanning*

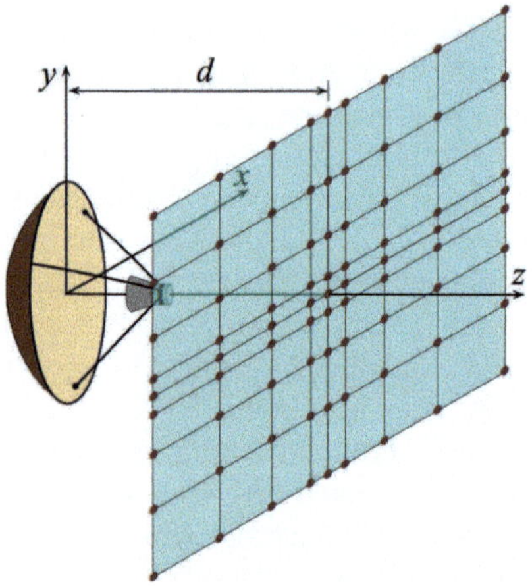

Figure 4.2 *Planar wide-mesh scanning*

plane centre (see Figure 4.2), and has been developed by applying the NR sampling representations [104, 105] to the voltage acquired by the measuring probe and modelling a quasi-planar AUT by an oblate spheroid or a double bowl. Then, the utilization of proper optimal sampling interpolation (OSI) formulas enables the reconstruction of the massive NF data required by the classical probe-compensated plane-rectangular NF–FF transformation [26] from the acquired NF samples.

In a plane-polar scanning facility [32–45], the AUT is mounted on a turntable, whereas the probe moves along a line perpendicular to the rotation axis of the turntable. In such a way, the probe can gather the NF amplitude and phase data which lie on azimuthal concentric circumferences in the scanning plane (see Figure 4.3). The plane-polar scanning has some peculiar characteristics, which make it particularly attractive. That is to say, its mechanical simplicity, the scanning area greater than in a plane-rectangular facility for a given size of the measurement chamber, and the possibility of a "fine tuning" of the anechoic chamber, since the AUT always points in a fixed direction. The large computer time, needed in the earliest approach

[32, 33] to reconstruct the antenna far field, has been drastically reduced in [34] by recovering the plane-rectangular data from the plane-polar ones via the bivariate Lagrange interpolation, thus enabling the use of the FFT algorithm. As shown in Sect. 4.6, even better results from the reconstruction accuracy and the data and measurement time reduction viewpoints can be achieved [40–45] by applying the aforementioned NR representations and the OSI expansions.

In the bi-polar NF facility [15, 46–51], proposed by Rahmat-Samii *et al.* in [15, 46, 47], the scanning is achieved by an axial rotation of the AUT and a rotation, around an axis parallel to the AUT one, of the arm at whose end is attached the probe. Accordingly, the NF data are gathered on a lattice consisting of concentric rings and radial arcs (see Figure 4.4). Such a scanning keeps all the advantages of the plane-polar one, by using a mechanically simpler and cheaper measurement system. In fact, as the arm is anchored at an end and the probe is attached at the other one, the bending is constant and this allows one to preserve the planarity. Furthermore, rotational movements are more convenient than linear ones, being the turntables more precise than linear positioners. As shown in Sect. 4.7, also in such a case, the application of the NR representations and the use of OSI expansions to recover the NF data required by the classical plane-rectangular NF–FF transformation [26] have allowed the development of NF–FF transformations with bi-polar scanning [49–51],

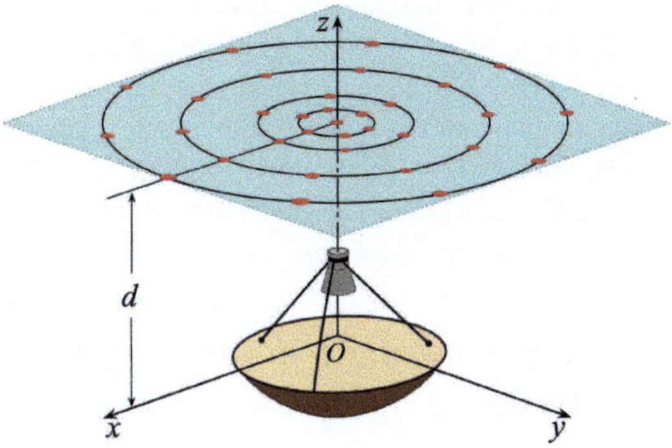

Figure 4.3 Plane-polar scanning

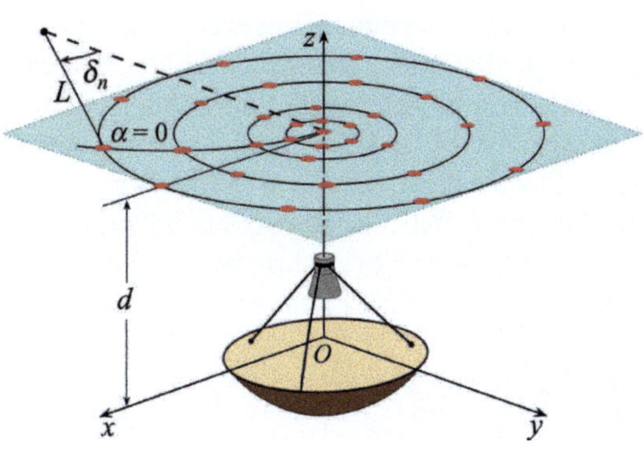

Figure 4.4 Bi-polar scanning

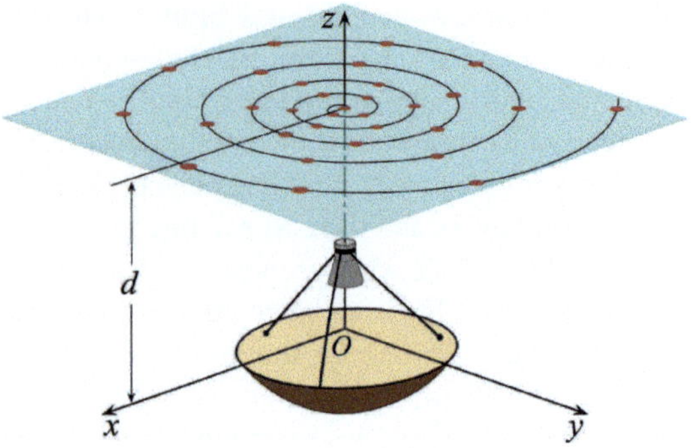

Figure 4.5 *Planar spiral scanning*

which are accurate and very efficient from the data and measurement time reduction viewpoints.

The measurement time saving can be achieved not only by reducing the number of required NF samples, but also by making faster their acquisition, as occurs in the planar spiral scanning [112, 122–130, 143] (see Figure 4.5). In fact, as suggested by Rahmat-Samii *et al.* in [112], the acquisition of the NF samples can be sped up by collecting them on fly along spirals obtained by continuous and synchronized movements of probe and AUT positioners. As shown in Sect. 4.8, the application of the unified theory of spiral scannings for volumetric AUTs [122] and its extension to the non-volumetric ones [124] have allowed the development of effective NR NF–FF transformations with planar spiral scanning [122–130] (see Figure 4.5). They allow a further measurement time saving with respect to the approaches in [112, 143], since, relying on the NR sampling representations, they require a remarkably reduced number of NF data.

4.2 Classical plane-rectangular NF–FF transformation without probe compensation

Among the NF–FF transformation techniques, that adopting the plane-rectangular (PR) scanning (Figure 4.1) is surely the most simple and effective one both from the computational and analytical viewpoints. However, as all the NF–FF transformations using a planar scan, it allows the reconstruction of the antenna FF pattern only within a cone having an aperture angle smaller than 180°. Accordingly, it is particularly suitable to characterize AUTs radiating pencil beam patterns which fall within the solid angle specified by the AUT edges and the scanning zone ones.

As shown in the Appendix 4A.1, the electric field $\boldsymbol{E}$ radiated by an antenna in a homogeneous source-free region can be represented as superposition of elementary plane waves, namely:

$$\boldsymbol{E}(x,y,z) = \int_{-\infty}^{+\infty}\int_{-\infty}^{+\infty} \hat{\boldsymbol{E}}(k_x,k_y)\,\mathrm{e}^{-\mathrm{j}(k_x x+k_y y+k_z z)}\,\mathrm{d}k_x\,\mathrm{d}k_y \tag{4.1}$$

where k_x, k_y, and k_z are the Cartesian components of the propagation vector $\boldsymbol{k}$. Accordingly, the tangential components E_x and E_y of the electric field radiated by the AUT can be expressed on the scanning plane at $z = d$ (see Figure 4.1) as [25]:

$$E_x(x, y, d) = \int_{-\infty}^{+\infty}\int_{-\infty}^{+\infty} \left[\hat{E}_x(k_x, k_y)\, \mathrm{e}^{-\mathrm{j}k_z d}\right] \mathrm{e}^{-\mathrm{j}(k_x x + k_y y)}\, \mathrm{d}k_x \mathrm{d}k_y \tag{4.2a}$$

$$E_y(x, y, d) = \int_{-\infty}^{+\infty}\int_{-\infty}^{+\infty} \left[\hat{E}_y(k_x, k_y)\, \mathrm{e}^{-\mathrm{j}k_z d}\right] \mathrm{e}^{-\mathrm{j}(k_x x + k_y y)}\, \mathrm{d}k_x \mathrm{d}k_y \tag{4.2b}$$

By Fourier inverse transforming (4.2a) and (4.2b), it results:

$$\hat{E}_x(k_x, k_y) = \frac{1}{4\pi^2} \mathrm{e}^{\mathrm{j}k_z d} \int_{-\infty}^{+\infty}\int_{-\infty}^{+\infty} E_x(x, y, d)\, \mathrm{e}^{\mathrm{j}(k_x x + k_y y)}\, \mathrm{d}x\, \mathrm{d}y \tag{4.3a}$$

$$\hat{E}_y(k_x, k_y) = \frac{1}{4\pi^2} \mathrm{e}^{\mathrm{j}k_z d} \int_{-\infty}^{+\infty}\int_{-\infty}^{+\infty} E_y(x, y, d)\, \mathrm{e}^{\mathrm{j}(k_x x + k_y y)}\, \mathrm{d}x\, \mathrm{d}y \tag{4.3b}$$

By assuming that the tangential components E_x and E_y of the electric field radiated by the AUT outside the scanning area are negligible, the Fourier transform integrals in (4.3a) and (4.3b) can be efficiently evaluated via a two-dimensional FFT algorithm, as will be shown in the following. As regards the sampling spacings of the NF data in the classical NF–FF transformation with PR scanning [25, 26], they must satisfy the following conditions:

$$\Delta x \le \pi/\beta = \lambda/2; \qquad \Delta y \le \pi/\beta = \lambda/2 \tag{4.4}$$

β and λ being the wavenumber and the wavelength, respectively. As a matter of fact, if the measurement plane is situated in the radiating NF region,* wherein the electromagnetic (EM) field does not contain evanescent waves, then the plane wave spectrum is zero for $|k_x| > \beta$ and $|k_y| > \beta$. Therefore, according to the two-dimensional Nyquist sampling theorem [299], the EM field radiated by the AUT can be accurately reconstructed on the scanning plane from the knowledge of its samples, whose spacings satisfy relations (4.4).

Once the Cartesian components $\hat{E}_x$ and $\hat{E}_y$ of the plane wave spectrum have been determined, the FF components of the electric field in the spherical coordinate system (R, Θ, Φ), adopted to specify an observation point in the antenna FF region, can be evaluated (see appendix 4A.2) by means of the relations:

$$E_R(R, \Theta, \Phi) = 0 \tag{4.5}$$

$$E_\Theta(R, \Theta, \Phi) = 2\pi\, \mathrm{j}\beta \cos\Theta\, \hat{E}_\Theta(\beta \sin\Theta \cos\Phi, \beta \sin\Theta \sin\Phi) \frac{\mathrm{e}^{-\mathrm{j}\beta R}}{R} \tag{4.6}$$

*In practice, at a distance of at least 4–5 wavelengths from the AUT.

$$E_\Phi(R,\Theta,\Phi) = 2\pi \,\mathrm{j}\, \beta \cos\Theta \hat{E}_\Phi(\beta \sin\Theta\cos\Phi,\ \beta\sin\Theta\sin\Phi)\frac{\mathrm{e}^{-\mathrm{j}\beta R}}{R} \tag{4.7}$$

wherein

$$\hat{E}_\Phi = -\hat{E}_x \sin\Phi + \hat{E}_y \cos\Phi \tag{4.8}$$

$$\hat{E}_\Theta = (\hat{E}_x \cos\Phi + \hat{E}_y \sin\Phi)/\cos\Theta \tag{4.9}$$

The use of the FFT algorithm to compute the two-dimensional Fourier transforms in (4.3a) and (4.3b) from the acquired NF data is now described. By taking explicitly into account the truncation due to the finite dimension $2L_x \times 2L_y$ of the scanning area, relations (4.3a) and (4.3b) can be rewritten in the form:

$$\hat{E}_{x,y}(k_x,k_y) = \frac{1}{4\pi^2}\mathrm{e}^{\mathrm{j}k_z d}\int_{-L_y}^{L_y}\int_{-L_x}^{L_x} E_{x,y}(x,y,d)\,\mathrm{e}^{\mathrm{j}(k_x x+k_y y)}\mathrm{d}x\,\mathrm{d}y \tag{4.10}$$

The integration over x is now considered. According to conditions (4.4), $\Delta x = 2L_x/N_x \le \lambda/2$, where N_x is the number of the NF data acquired on each of the $N_y = 2L_y/\Delta y$ measurement lines parallel to the x-axis on the scanning plane. By putting $x_n = -L_x + n\Delta x = \left(n - N_x/2\right)\Delta x$, it follows that:

$$G(k_x) = \int_{-L_x}^{L_x} g(x)\mathrm{e}^{\mathrm{j}k_x x}\,\mathrm{d}x \approx \Delta x \sum_{n=0}^{N_x-1} g(x_n)\,\mathrm{e}^{\mathrm{j}k_x(n-N_x/2)\Delta x} \tag{4.11}$$

When $G\left(k_x\right)$ is evaluated at $k_{x_i} = i\Delta k_x = 2\pi i/\left(N_x\Delta x\right)$, it results:

$$G\left(k_{x_i}\right) \approx \frac{2L_x}{N_x}\mathrm{e}^{-\mathrm{j}\pi i}\sum_{n=0}^{N_x-1} g\left(x_n\right)\,\mathrm{e}^{\mathrm{j}2\pi ni/N_x} \tag{4.12}$$

By comparing this relation with (A.91), it can easily recognized that the integration with respect to x in (4.10) can be efficiently accomplished by means of an inverse FFT algorithm [300], provided that the attained results are multiplied by the factor $2L_x\mathrm{e}^{-\mathrm{j}\pi i}$. Quite analogous results hold, obviously, also for the integration with respect to y. Accordingly, to evaluate each of the Cartesian components $\hat{E}_x$ and $\hat{E}_y$ of the plane wave spectrum are required N_y one-dimensional FFTs of size N_x and N_x FFTs of size N_y. In such a way the plane wave spectrum is determined in correspondence of the k_x and k_y values given by

$$k_{x_i} = 2\pi i/\left(N_x\Delta x\right); \qquad k_{y_p} = 2\pi p/\left(N_y\Delta y\right) \tag{4.13}$$

and, accordingly, the antenna FF components E_Θ and E_Φ are obtained at the values of Θ, Φ specified by

$$\Theta_{i,p} = \sin^{-1}\left(\lambda\sqrt{\left(i/\left(N_x\Delta x\right)\right)^2 + \left(p/\left(N_y\Delta y\right)\right)^2}\right) \tag{4.14}$$

$$\Phi_{i,p} = \tan^{-1}\left(\frac{p N_x \Delta x}{i N_y \Delta y}\right) \tag{4.15}$$

To determine the FF components with a greater resolution, it is enough "zero-filling" the NF data. That is to say, to increase the number of output values of k_x from N_x to N_X and those of k_y from N_y to N_Y, a new matrix of NF data must be considered. It differs from the original one for the first and last $(N_Y - N_y)/2$ rows and first and last $(N_X - N_x)/2$ columns, whose elements are all set to zero.

It is noteworthy that the plane waves relevant to k_x and k_y values such that $k_x^2 + k_y^2 > \beta^2$ do not contribute to the radiated far field, but only to the reactive near field. As a matter of fact, they are evanescent waves, which formally propagate in complex directions and decay exponentially on increasing z.

A two-dimensional OSI algorithm, which allows one to obtain the FF components E_Θ and E_Φ at a given direction (Θ, Φ), is described in the following. It can be conveniently used to obtain the antenna radiation pattern at any cut plane Φ as function of the zenithal angle Θ. Its mathematical justification relies on the fact that the antenna far field is related to the plane wave spectrum and, as a consequence, to the Fourier transform of the tangential components of the field on the scanning plane. Therefore, if these last are practically negligible for $|x| > L_x$ and $|y| > L_y$, then, as a consequence of the two-dimensional Nyquist sampling theorem [299], the spectrum multiplied by the factor $\mathrm{e}^{-\mathrm{j}\,k_z d}$ can be reconstructed from the knowledge of its samples at a rectangular lattice of points separated by the spacings:

$$\Delta k_x \le \pi / L_x; \qquad \Delta k_y \le \pi / L_y \tag{4.16}$$

It can be easily recognized that the bandwidth of the spectrum alone is even smaller, so that the above spacings can be used also for sampling it. Accordingly, the following two-dimensional OSI expansion (see Appendix 2A.5) can be conveniently employed to reconstruct the tangential components of the plane wave spectrum at the given values of k_x and k_y from the knowledge of their samples:

$$\begin{aligned}\hat{E}_{x,y}(k_x, k_y) = \sum_{n=n_0-q+1}^{n_0+q} \Big\{ & \Psi(k_x - k_{x_n})\,\mathrm{sinc}\big[\pi(k_x - k_{x_n})/\Delta k_x\big] \\ & \times \sum_{m=m_0-p+1}^{m_0+p} \hat{E}_{x,y}(k_{x_n}, k_{y_m})\Psi(k_y - k_{y_m})\mathrm{sinc}[\pi(k_y - k_{y_m})/\Delta k_y]\Big\}\end{aligned} \tag{4.17}$$

where $n_0 = \mathrm{Int}\,[k_x/\Delta k_x]$ and $m_0 = \mathrm{Int}\,[k_y/\Delta k_y]$ are the indexes of the sample nearest (on the left) to the output point, $2q$ and $2p$ are the retained samples numbers,

$$k_{x_n} = n\Delta k_x = 2\pi n/(N_X \Delta x); \qquad k_{y_m} = m\Delta k_y = 2\pi m/(N_Y \Delta y) \tag{4.18}$$

sinc (s) is the $\sin(s)/s$ function and

$$\Psi(s) = \frac{\cosh\left[\pi v \ell \sqrt{1 - (s/\bar{s})^2}\,\right]}{\cosh(\pi v \ell)} \tag{4.19}$$

(see Appendix 2A.5) is the Knab's Sampling Window (SW) function [295], with $v = (1 - 1/\chi)$ and $\bar{s} = \ell/\Delta s$, χ being the oversampling factor, 2ℓ the number of retained samples, and Δs the sampling spacing. In particular, $\bar{s} = \bar{k}_x = q\,\Delta k_x$ and $\chi = N_X/N_x$ in the interpolation along k_x, whereas $\bar{s} = \bar{k}_y = p\,\Delta k_y$ and $\chi = N_Y/N_y$ in the interpolation along k_y. As previously stated, the OSI expansion (4.17) can be suitably employed to obtain the antenna radiation pattern in any cut plane at $\Phi = \text{const}$ as function of the zenithal angle Θ. In fact, once the tangential components of the plane wave spectrum at $k_x = \beta \sin\Theta \cos\Phi$ and $k_y = \beta \sin\Theta \sin\Phi$ have been evaluated via (4.17), the corresponding values of the FF components are easily obtained by means of (4.8), (4.9) and (4.6), (4.7).

4.3 Classical plane-rectangular NF–FF transformation with probe compensation

It is not a difficult issue to recognize that the tangential components E_x and E_y of the electric field radiated by the AUT on a scanning plane located in the NF region cannot be acquired by means a real (not ideal) probe. As a matter of fact, during the scanning, the probe sees the AUT centre under different directions. Furthermore, also at a given position, the direction under which each portion of the AUT is viewed is different. Thus, the response (measured voltage) of an actual probe is not proportional to the tangential components of the AUT electric field. In particular, by subdividing the AUT in N smaller portions such that the probe is situated in their FF region, the open-circuit voltage V received by the probe is given by:

$$V = \sum_{n=1}^{N} \boldsymbol{E}_n^i \cdot \boldsymbol{h}_n^r \tag{4.20}$$

wherein the symbol $(\cdot)$ denotes the inner product, $\boldsymbol{E}_n^i$ is the incident electric field on the probe due to the nth portion of the AUT, and $\boldsymbol{h}_n^r$ is the receiving effective length [16, 298] of the probe relevant to the direction under which it sees this portion. Accordingly, the far field radiated by the AUT cannot be precisely reconstructed from the acquired NF data by means of the previously described PR NF–FF transformation without probe compensation.

The probe-compensated NF–FF transformation technique with PR scanning, wherein the directional effects introduced by a real probe are properly taken into account, has been developed by Kerns in [27–29] by applying the plane wave scattering matrix formulation and by Paris *et al.* in [25, 26] by applying the Lorentz reciprocity theorem. In particular, Paris *et al.* rigorously showed that the AUT far field can be obtained from the knowledge of: a) the two-dimensional Fourier transforms of the output voltages V_p and V_r of the probe for two independent sets of measurements, usually obtained rotating the probe by 90° around its longitudinal axis in the second set; b) the FF spherical components E'_{Θ_p}, E'_{Φ_p} and E'_{Θ_r}, E'_{Φ_r} of the electric field radiated by the probe and the rotated probe when employed as transmitting antennas. In the here considered reference system, the related expressions are:

$$E_\Theta(\Theta,\Phi) = \left(I_r E'_{\Phi_p}(\Theta,-\Phi) - I_p E'_{\Phi_r}(\Theta,-\Phi)\right)/\Delta \tag{4.21}$$

$$E_\Phi(\Theta,\Phi) = \left(I_r E'_{\Theta_p}(\Theta,-\Phi) - I_p E'_{\Theta_r}(\Theta,-\Phi)\right)/\Delta \tag{4.22}$$

wherein

$$\Delta = E'_{\Theta_r}(\Theta,-\Phi)\,E'_{\Phi_p}(\Theta,-\Phi) - E'_{\Theta_p}(\Theta,-\Phi)\,E'_{\Phi_r}(\Theta,-\Phi) \tag{4.23}$$

$$I_p = A\cos\Theta\,\mathrm{e}^{\mathrm{j}\beta d\cos\Theta}\int_{-\infty}^{+\infty}\int_{-\infty}^{+\infty} V_p\left(x,y\right)\mathrm{e}^{\mathrm{j}\beta x\sin\Theta\cos\Phi}\mathrm{e}^{\mathrm{j}\beta y\sin\Theta\sin\Phi}\,\mathrm{d}x\,\mathrm{d}y \tag{4.24a}$$

$$I_r = A\cos\Theta\,\mathrm{e}^{\mathrm{j}\beta d\cos\Theta}\int_{-\infty}^{+\infty}\int_{-\infty}^{+\infty} V_r\left(x,y\right)\mathrm{e}^{\mathrm{j}\beta x\sin\Theta\cos\Phi}\mathrm{e}^{\mathrm{j}\beta y\sin\Theta\sin\Phi}\,\mathrm{d}x\,\mathrm{d}y \tag{4.24b}$$

A being an appropriate constant (see (4A.43)).

A simple derivation of the above formulas for the PR NF–FF transformation with probe compensation is reported in Appendix 4A.3. In the same Appendix, the expressions of the FF spherical components E'_{Θ_p}, E'_{Φ_p} and E'_{Θ_r}, E'_{Φ_r} of the electric field radiated by an open-ended rectangular waveguide, usually employed as scanning probe, are also reported.

The two-dimensional Fourier transforms of the voltages V_p and V_r can be, obviously, again effectively computed using the FFT algorithm and the NF grid spacings must still satisfy conditions (4.4).

It is worth noting that the "zero-filling" of the NF data, described in the previous section, can be suitably exploited to obtain the FF components with a greater resolution in Θ and Φ and that the two-dimensional OSI formula (4.17) can be applied to interpolate $\hat{E}_{x,y}(k_x,k_y)$ or $I_{p,r}(k_x,k_y)$.

In the following, a numerical example, which emphasizes the need of applying a probe-compensated NF–FF transformation technique in order to get an accurate reconstruction of the AUT far field from the NF data acquired by a real probe, is shown. The simulations refer to the field radiated by a uniform planar circular array lying in the plane $z = 0$ and having diameter $2a = 34\,\lambda$. Its elements, elementary Huygens sources linearly polarized along the y axis and symmetrically disposed with respect to the x and y axes, are about $0.5\,\lambda$ spaced along circumferences, which are radially spaced by $0.5\,\lambda$. The measurement plane is $8\,\lambda$ away from the AUT and the NF samples are $0.5\,\lambda$ spaced along x and y on a $100\lambda \times 100\lambda$ sized square. An open-ended WR-90 rectangular waveguide, operating at the frequency of 10 GHz, is chosen as probe. Figure 4.6 shows the amplitude of the probe output voltage V_p (solid line) on the line at $x = 0$ lying on the scanning plane. In the same figure, the amplitude of the NF y-component E_y (crosses) is also shown. As can be seen, the plots are significantly different, thus confirming that the voltage measured by a real probe is not proportional to the tangential component of the electric field. The antenna FF patterns in the E-plane reconstructed from the voltage data by the uncompensated

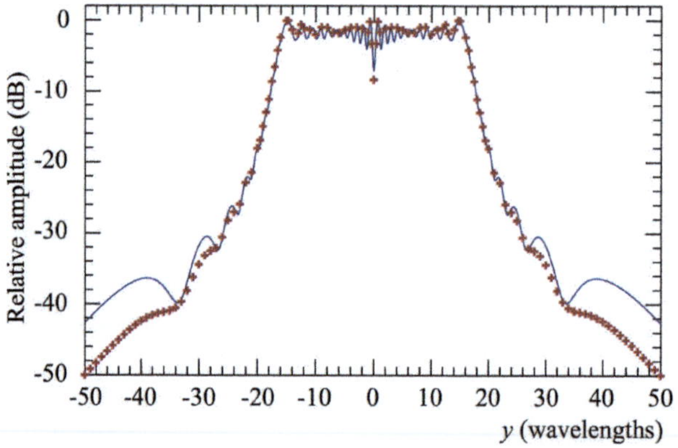

Figure 4.6 *Amplitude of the probe output voltage V_p (solid line) and of the NF y-component E_y (crosses) on the scanning line at $x = 0$.*

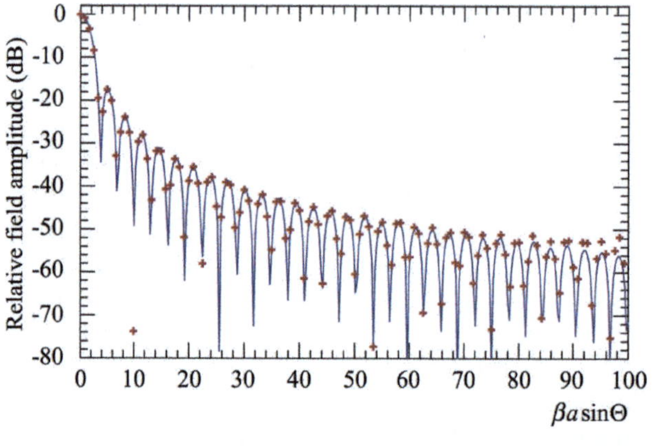

Figure 4.7 *E-plane pattern. Solid line: exact field. Crosses: reconstructed from voltage data by the uncompensated NF–FF transformation.*

and by the probe-compensated NF–FF transformation techniques are shown in Figures 4.7 and 4.8, respectively. As can be seen, some disagreements are clearly evident in the reconstruction attained by applying the NF–FF transformation without probe compensation. On the other hand, the field reconstructed via the probe-compensated NF–FF transformation and the exact one are practically indistinguishable.

The next Figures 4.9 and 4.10 are reported in order to assess the effectiveness of the OSI formula (4.17) for the accurate reconstruction of the AUT radiation pattern in any cut plane at $\Phi = \text{const}$ as function of the zenithal angle Θ. The E-plane pattern so reconstructed from the NF voltage data is compared in Figure 4.9 with the exact far field. Figure 4.10 shows the comparison between the recovered FF components

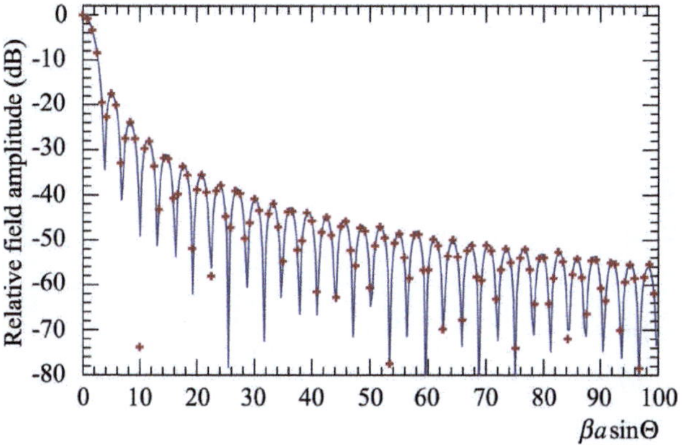

Figure 4.8 *E-plane pattern. Solid line: exact field. Crosses: reconstructed from voltage data by the probe-compensated NF–FF transformation.*

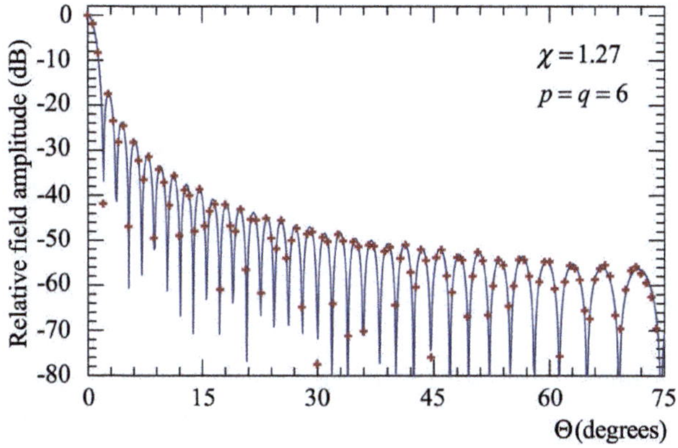

Figure 4.9 *E-plane pattern. Solid line: exact field. Crosses: reconstructed from NF data.*

and the exact ones in the cut plane at $\Phi = 30°$. As can be seen, the FF reconstruction results to be everywhere very accurate, thus assessing the effectiveness of the described technique.

4.4 Error due to the truncation of the scanning area

In the planar scannings, as well as in the cylindrical one, the measurement area is unavoidably truncated and, therefore, the reconstructed far field is affected by a truncation error, whose magnitude obviously depends on the level of the neglected NF data falling outside the scanning zone.

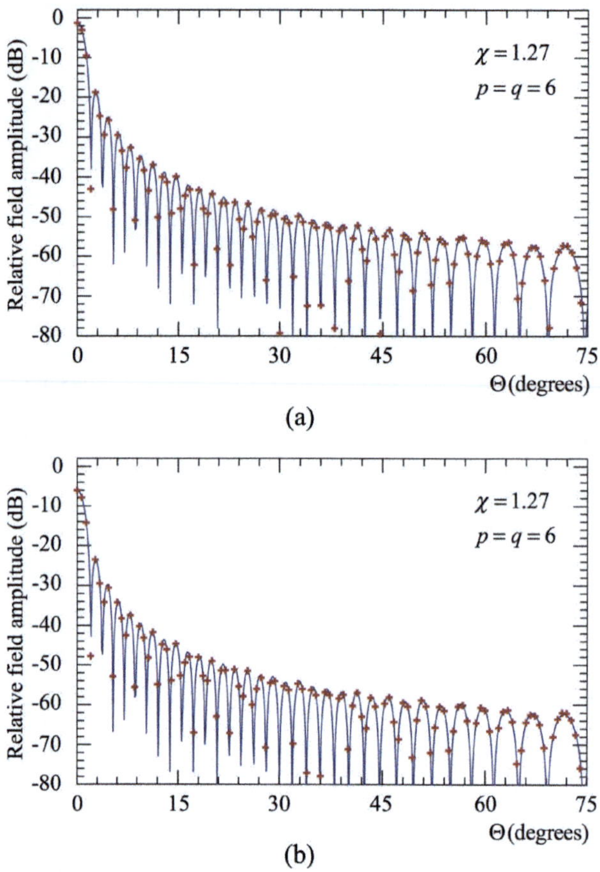

Figure 4.10 *Amplitude of the FF components in the cut plane at $\Phi = 30°$. Solid line: exact field. Crosses: reconstructed from NF data. (a) Amplitude of E_Φ, (b) Amplitude of E_Θ.*

Let a PR NF measurement facility be equipped with a scanning plane at distance d from the AUT and let $2L_x$ and $2a_x$ the dimensions along x of the scanning plane and AUT, respectively (see Figure 4.11).

A useful and practical rule [252] to predict the angular region of validity of the reconstructed FF pattern in the $x\,z$ cut plane is given by

$$-\Theta_{xc} \leq \Theta \leq \Theta_{xc} \tag{4.25}$$

wherein the validity angle Θ_{xc} is so defined:

$$\Theta_{xc} = \tan^{-1}\left(\frac{L_x - a_x}{d}\right) \tag{4.26}$$

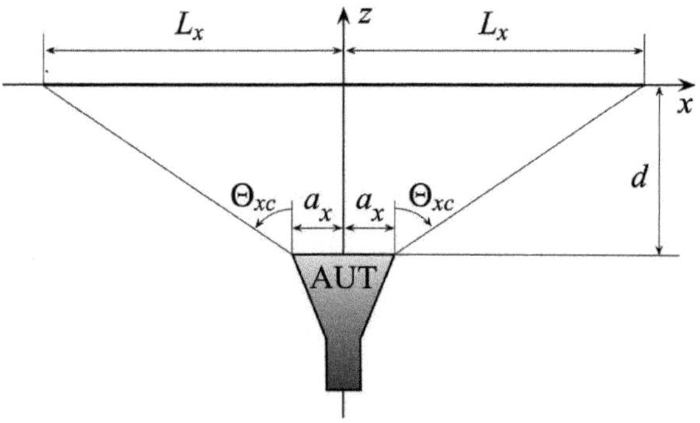

Figure 4.11 Geometry relevant to the evaluation of the validity angles

The above rule of thumb was empirically found from a large number of NF measurements performed by using different antennas and probe combinations [301] and determined by means of a theoretical analysis [302]. It must be also observed that the discontinuity of the AUT near field at the edges of the scanning area can produce a ripple in the reconstructed FF pattern, which can be present even in the validity angular region. Of course, an enlargement of such a region can be attained by decreasing the measurement plane distance. However, this distance cannot be reduced beyond certain limits, otherwise the interactions between the AUT and the probe cannot be neglected any more. Moreover, as explicitly stressed when discussing the conditions (4.4) to be satisfied by the NF data spacings, the scanning plane must be positioned in the radiating NF region, where the evanescent waves of the AUT field are negligible. In practice, a minimum distance of at least 4–5 wavelengths between the AUT and probe must be ensured during the scanning.

A rule quite analogous to (4.25) can be practically used for relating the angular region of validity of the reconstructed FF pattern in the *yz* cut plane to the dimensions $2L_y$ and $2a_y$ of the scanning plane and AUT along *y*.

Note that the case of a symmetric scanning zone has been considered for sake of simplicity. The interested reader can refer to the Chapter 5, relevant to the cylindrical scanning geometry, for the case of an asymmetrical scanning zone and for a more in depth discussion on the error due to the measurement area truncation.

The effect of the truncation on the reconstructed FF patterns in the principal planes, due to a reduced size of the scanning plane along the *y* or *x* direction, is shown in Figure 4.12 and Figure 4.13, respectively. These reconstructions are relevant to the same planar array and scanning plane distance *d* previously considered, but the sizes of the measurement zone are $2L_x = 100\ \lambda$, $2L_y = 48\ \lambda$ in the former case (corresponding to validity regions identified by $\beta a \sin\Theta_{xc} = 103.8$, $\beta a \sin\Theta_{yc} = 70.3$) and $2L_x = 48\ \lambda$, $2L_y = 100\ \lambda$ in the latter (corresponding to validity regions identified by $\beta a \sin\Theta_{xc} = 70.3$, $\beta a \sin\Theta_{yc} = 103.8$). As can be seen, the results shown in Figures 4.12 and 4.13 are in perfect agreement with the predictions based on the criterion of the validity angular region and confirm that the truncation of the

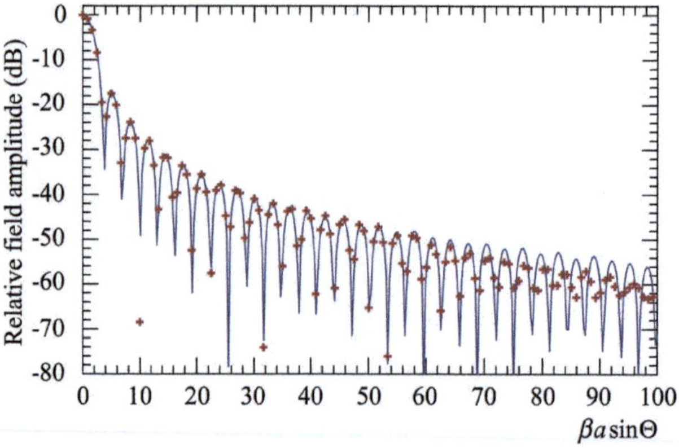

(a)

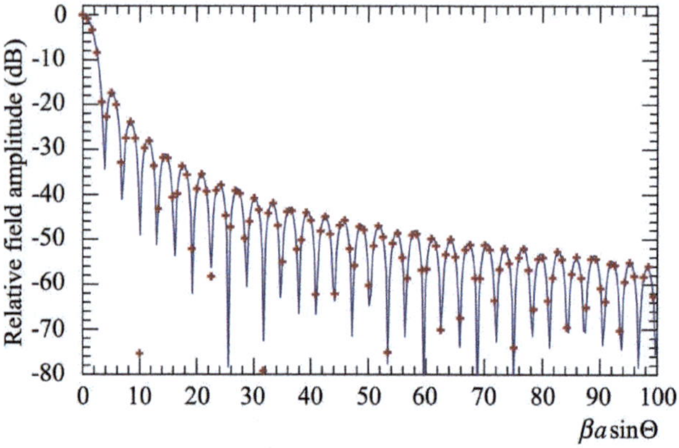

(b)

Figure 4.12 *FF patterns in the principal planes. Solid line: exact field. Crosses: reconstructed from NF data.* $2L_x = 100\,\lambda$, $2L_y = 48\,\lambda$, $\beta a \sin\Theta_{xc} = 103.8$, $\beta a \sin\Theta_{yc} = 70.3$. *(a) E-plane, (b) H-plane.*

scanning plane dimension along x mainly affects the validity angular region of the reconstructed FF pattern in the $x\,z$ cut plane, whereas that in the $y\,z$ cut plane is predominantly influenced by the scanning extension along y.

4.5 Non-redundant NF–FF transformations with non-conventional plane-rectangular scanning

Two NR NF–FF transformations [108, 109] with a non-conventional PR scan named planar wide-mesh scanning (PWMS) will be described in this section. The reason for this name is due to the sampling lattice characterized by meshes which become wider and wider when their distance from the centre of the scanning plane increases.

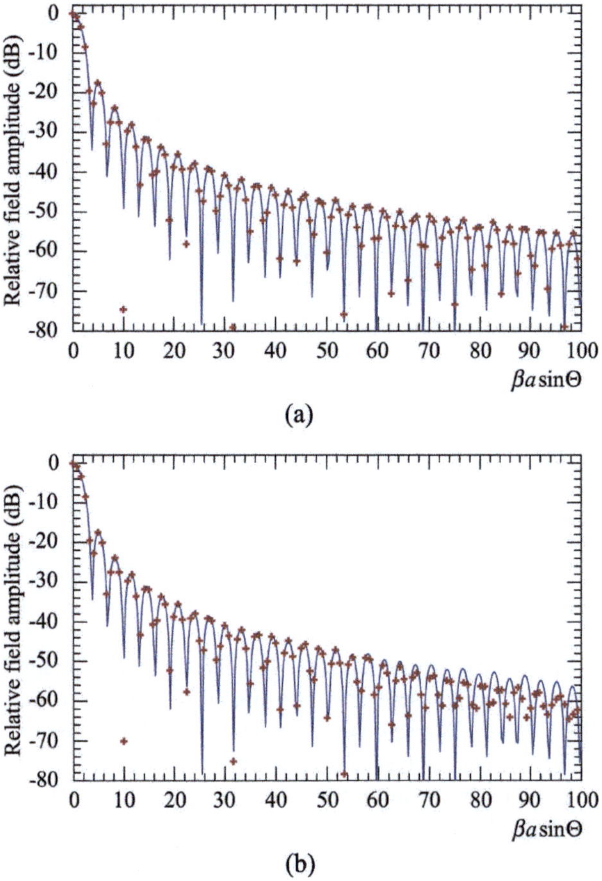

Figure 4.13 *FF patterns in the principal planes. Solid line: exact field. Crosses: reconstructed from NF data.* $2L_x = 48\,\lambda$, $2L_y = 100\,\lambda$, $\beta a \sin\Theta_{xc} = 70.3$, $\beta a \sin\Theta_{yc} = 103.8$. *(a) E-plane, (b) H-plane.*

It is so possible to get a drastic reduction of the required NF data and related acquisition time as compared to the classical PR scanning. Both these transformations are particularly tailored to quasi-planar AUTs as usually are those characterized in a planar NF facility. In the former, the AUT is considered as enclosed in an oblate spheroid, whereas in the latter, it is modelled by a double bowl.

4.5.1 NF–FF transformation with planar wide-mesh scanning: the oblate spheroidal modelling case

Let an electrically small, non-directive, probe be acquiring the NF data radiated by a quasi-planar AUT, enclosed in an oblate spheroid Σ with major and minor semi-axes equal to a and b, on a plane at distance d from the AUT aperture. A Cartesian

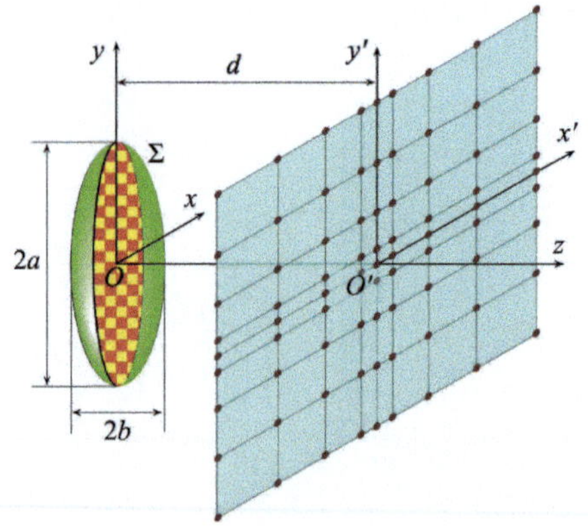

Figure 4.14 Planar wide-mesh scanning: oblate spheroidal modelling

coordinate system (x, y, z), having the origin O at the AUT aperture centre, and the associated spherical one (r, ϑ, φ) are employed to specify a given observation point. An additional Cartesian reference system (x', y', z'), whose origin O' coincides with the scanning plane centre, is also introduced to denote a point P on the plane (see Figure 4.14).

As shown in [41] and already stressed in Chapter 2, the voltage measured by a non-directive probe has the same effective spatial bandwidth of the AUT radiated field, so that the NR sampling representations [104, 105] and the related OSI expansions can be applied to the probe reduced voltage $\tilde{V}(\xi) = V(\xi)\, e^{j\gamma(\xi)}$, ξ being an optimal parameter to be used for describing the observation curve and $\gamma(\xi)$ a suitable phase function. According to the results reported in Sect. 2.2, the bandwidth W_ξ, the phase function γ and the optimal parameter ξ to describe a meridian curve are given by the expressions (2.25), (2.27) and (2.28b). Since the axes x' and y' are meridian curves, the parameters ξ and η, both determinable by means of expression (2.28b), can be used to describe them.

To make possible the factorization of the two-dimensional interpolation algorithm into one-dimensional OSI expansions [108, 109], all lines parallel to the x' axis must be described by the parameter ξ relevant to such an axis and the parameter η has to be used for those parallel to the y' axis. Hence, the sample spacings on the lines parallel to the axes coincide with those relevant to the corresponding axes and the resulting grid has meshes whose sizes are larger and larger as their distance from the measurement plane centre grows (see Figure 4.14). As concerns the phase function γ, it can be determined [108, 109] by means of (2.27) and, accordingly, it depends only on the distances $r_{1,2}$ of the considered point $P(x', y')$ from the foci of the modelling spheroid Σ.

According to (2.58), the probe and rotated probe voltages at any point P on the scanning plane can be reconstructed [108, 109] in a fast and accurate way via the OSI expansion:

$$V(\xi(x'), \eta(y')) = e^{-j\gamma(x', y')} \sum_{m=m_0-p+1}^{m_0+p} \tilde{V}(\xi, \eta_m)\, SF(\eta, \eta_m, \bar{\eta}, M, M'') \tag{4.27}$$

where $m_0 = \text{Int}(\eta/\Delta\eta)$, $2p$ is the number of the retained intermediate samples $\tilde{V}(\xi, \eta_m)$, i.e., the reduced voltages at the intersection points of the y'-directed line in P with the sampling lines parallel to the x' axis, $SF(\eta, \eta_m, \bar{\eta}, M, M'')$ is the OSI function (2.61), and

$$\eta_m = m\Delta\eta = 2\pi\, m/(2M'' + 1); \qquad M'' = \text{Int}(\chi M') + 1 \tag{4.28}$$

$$M' = \text{Int}(\chi' W_\eta) + 1; \qquad M = M'' - M'; \qquad \bar{\eta} = p\Delta\eta \tag{4.29}$$

with the bandwidth W_η given by (2.25).

The intermediate samples can be determined [108, 109] from the acquired PWMS samples by means of a similar OSI expansion:

$$\tilde{V}(\xi, \eta_m) = \sum_{n=n_0-q+1}^{n_0+q} V(\xi_n, \eta_m)\, e^{j\gamma(x'_n, y'_m)} SF(\xi, \xi_n, \bar{\xi}, M, M'') \tag{4.30}$$

where $n_0 = \text{Int}(\xi/\Delta\xi)$, $2q$ is the number of the retained samples $V(\xi_n, \eta_m)$, and

$$\xi_n = n\Delta\xi = n\Delta\eta; \qquad \bar{\xi} = q\Delta\xi \tag{4.31}$$

The two-dimensional OSI expansion, obtained by properly matching the one-dimensional ones (4.27) and (4.30), can be employed to accurately recover the values of the probe voltages V_p and V_r at the points needed by the classical probe-compensated NF–FF transformation with PR scanning [26]. It is so possible to get a probe-compensated NF–FF transformation with non-conventional PR scanning which makes use of a NR number of data.

Some numerical examples, which assess the effectiveness of the technique, are shown in the following. The simulations refer to the previously considered planar circular array and the chosen probe is again an open-ended WR-90 rectangular waveguide working at 10 GHz, which scans a square of side 100 λ on a plane 8 λ away from the AUT. The AUT has been modelled by an oblate spheroid with major and minor semi-axes equal to 17.4 λ and 2.2 λ. Figure 4.15a and b show a representative reconstruction example of the amplitude and phase of the probe voltage V_p (the most significant one) on the scanning plane line $y' = 15\,\lambda$. As can be seen, there is a very good agreement between the reconstructed voltage and the exact one. The stability of the algorithm, i.e., its robustness with respect to errors affecting the samples, has been investigated by corrupting the exact samples with random errors. These errors simulate a background noise (bounded to Δa dB in amplitude and with arbitrary phase) and uncertainties on the data of $\pm\Delta a_r$ dB in amplitude and $\pm\Delta\,\phi$ degrees in phase. As shown in Figure 4.16, the algorithm is stable. The accuracy of the OSI algorithm is assessed in a more quantitative way by the maximum and mean-square reconstruction error values shown in Figure 4.17a and b. They are normalized to the voltage maximum value over the scanning plane and have been attained by comparing the interpolated values of V_p with those directly evaluated on a close grid in the central zone of the plane, so that the existence of the required guard samples is guaranteed. As can be seen, they decrease up to very low values on increasing the number of retained samples and/or the oversampling factor. At

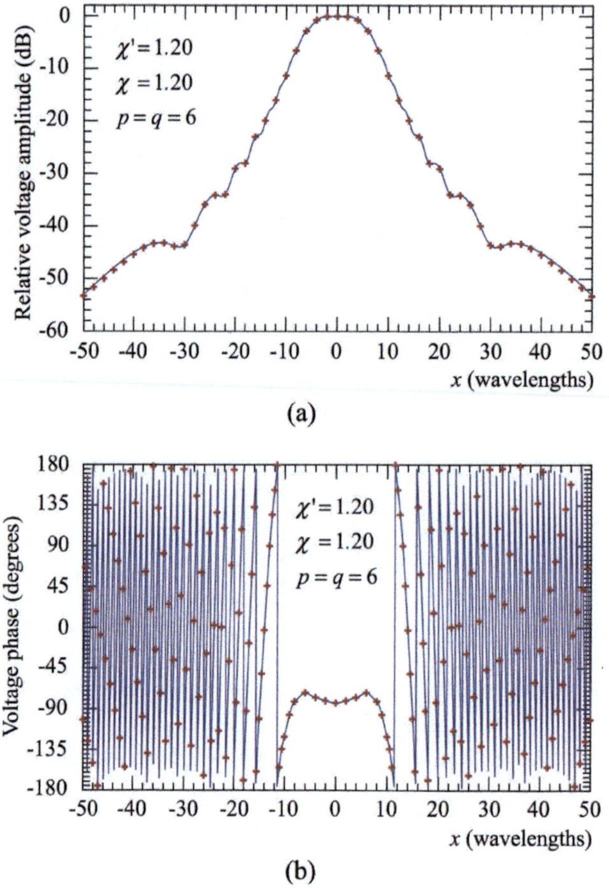

Figure 4.15 *Probe voltage V_p on the scanning plane line $y' = 15\,\lambda$. Solid line: exact. Crosses: interpolated. (a) Amplitude, (b) Phase.*

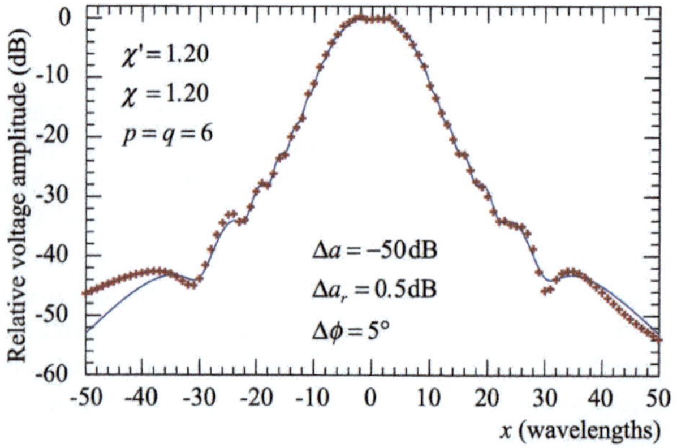

Figure 4.16 *Amplitude of the probe voltage V_p on the scanning plane line $y' = 15\,\lambda$. Solid line: exact. Crosses: interpolated from error affected samples.*

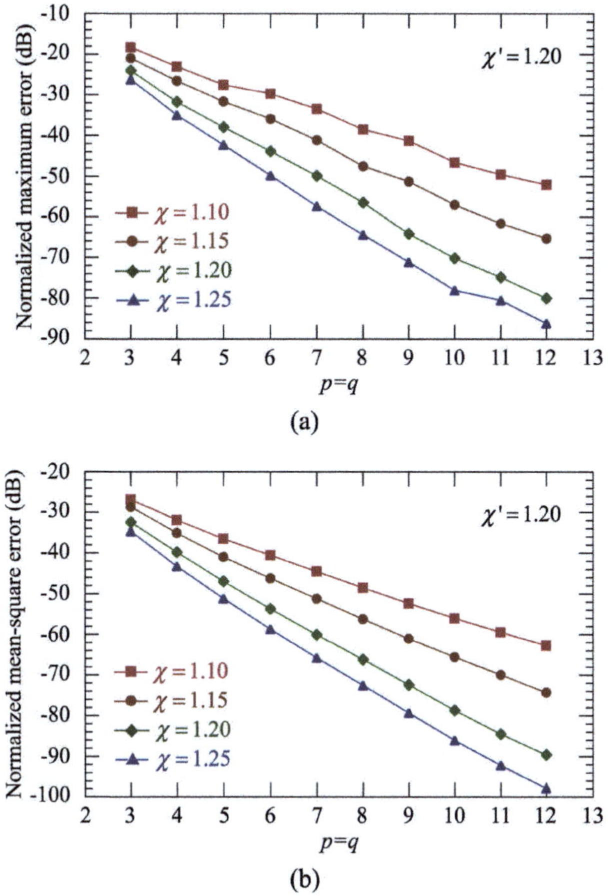

Figure 4.17 *Normalized errors in the reconstruction of the probe voltage V_p. (a) Maximum error, (b) Mean-square error.*

last, the described two-dimensional OSI algorithm has been applied to recover the NF data needed for the classical PR NF–FF transformation [26]. The reconstruction of the antenna E-plane FF pattern is shown in Figure 4.18. As can be seen, the exact pattern and the recovered one are practically indistinguishable, thus further confirming the effectiveness of the technique.

It can be interesting to compare the number of used NF samples (10201) with that (40401) which are required when applying the classical PR NF–FF transformation [26]. It is noteworthy that the number of NR NF samples, which would be needed to cover an unbounded scanning plane, are in such a case 11025. In conclusion, the described NF–FF transformation with PWMS allows a remarkable reduction of the required number of NF data and related measurement time without any loss of accuracy in the recovery of the AUT far field.

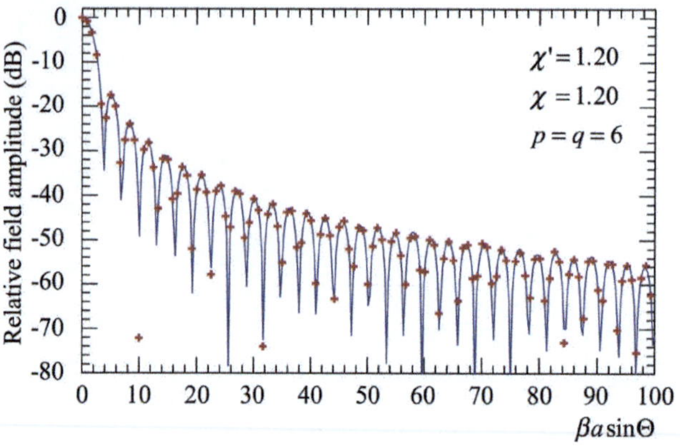

Figure 4.18 *E-plane pattern. Solid line: exact field. Crosses: reconstructed from the acquired PWMS samples.*

The interested reader can find in [109] experimental results assessing the effectiveness of such a NF–FF transformation also from a practical viewpoint.

4.5.2 NF–FF transformation with planar wide-mesh scanning: the double bowl modelling case

Let an electrically small, non-directive, probe be acquiring the NF data radiated by a quasi-planar AUT enclosed in a double bowl Σ, namely a surface attained by joining together two bowls having the same circular aperture of diameter $2a$ and lateral surfaces determined by rotating two circular arcs, each equal to a quarter of circumference, which can have different radii c and c' to fit better the shape of the considered AUT (see Figures 4.19 and 4.20). Also in such a case, it is convenient to adopt the same coordinate systems already considered in the previous subsection and to apply the NR sampling representations [104] to the probe reduced voltage $\tilde{V}(\xi) = V(\xi)\, e^{j\gamma(\xi)}$.

According to the results presented in Sect. 2.3, the bandwidth W_ξ, the phase function γ and the optimal parameter ξ for describing a meridian curve, such as the axes x' and y' of the scanning plane, are given by the expressions (2.10), (2.8) and (2.11), here reported for reader's convenience.

$$W_\xi = \beta\ell'/2\pi; \qquad \gamma = \frac{\beta}{2}\,[R_1 + R_2 + s_1' - s_2']; \qquad \xi = \frac{\pi}{\ell'}\,[R_1 - R_2 + s_1' + s_2'] \tag{4.32}$$

where $\ell' = 2\,[b + b' + (c + c')\,\pi/2]$, with $b = a - c$ and $b' = a - c'$, is the length of the curve C' intersection of Σ with a meridian plane. The explicit expressions of γ and ξ can be obtained by substituting the values of the distances $R_{1,2}$ and curvilinear abscissae $s_{1,2}'$ in (4.32). In such a case (see Figure 4.20), for $\rho < a$, the tangency points $P_{1,2}$ are both situated on the upper bowl. Accordingly, R_1 and s_1' are given by (2.39),

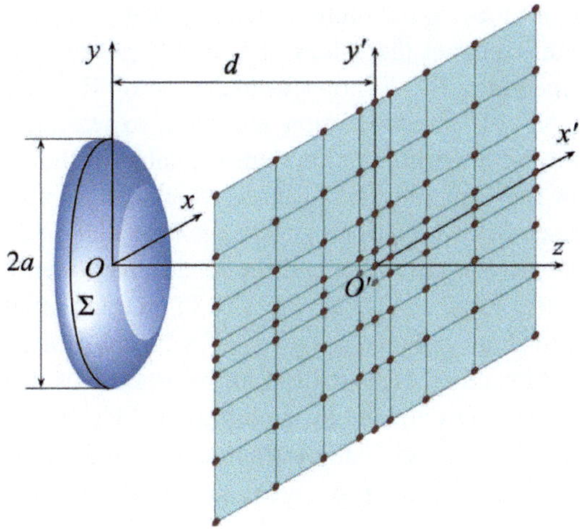

Figure 4.19 *Planar wide-mesh scanning: double bowl modelling*

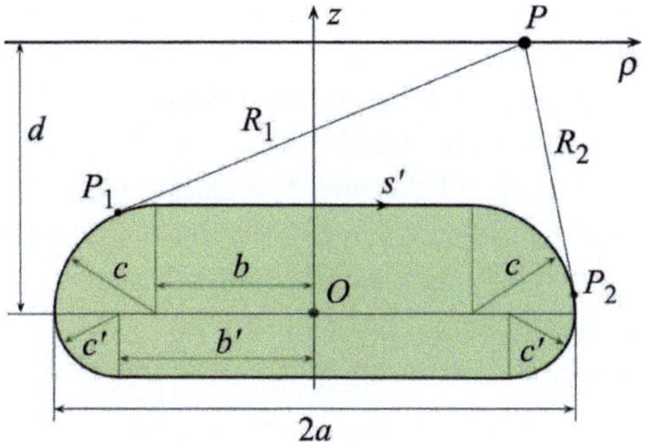

Figure 4.20 *Double bowl modelling: relevant to the evaluation of γ and ξ*

whereas R_2 and s_2' are given by (2.41). For $\rho > a$, P_1 is on the left side of the upper bowl and P_2 is on the right side of the lower one. Accordingly, the expressions of R_1 and s_1' are again given by (2.39) and those of R_2 and s_2' by (2.43).

The factorization of the two-dimensional interpolation algorithm into one-dimensional OSI expansions is again made possible [108, 109] by describing all lines parallel to the x' axis using the parameter ξ relevant to the x' axis and the lines parallel to the y' axis with that η relevant to this axis. Therefore, the resulting sampling lattice has meshes which become larger and larger on increasing their distance from the scanning plane centre. As regards the phase function γ, it is the same which would be used when interpolating along the radial line passing through the considered point and, hence, it can be evaluated [108, 109] by applying (4.32).

The two-dimensional OSI expansion, got by matching the one-dimensional ones (4.27) and (4.30), allows again the efficient reconstruction of the probe and rotated probe voltages at any point on the scanning plane and, accordingly, can be employed to accurately recover the values of V_p and V_r at the points needed by the classical probe-compensated NF–FF transformation with PR scanning [26].

The described NF–FF transformation has been experimentally assessed by means of the flexible NF measurement system available in the antenna characterization laboratory of the University of Salerno, which can be arranged to perform a plane-polar scan (see Sect. 1.2). Such an arrangement, due to the presence of an additional turntable between the probe and the linear positioner (see Figure 1.15), allows also the classical PR scan and the PWMS. In fact, the points where the corresponding NF data must be gathered are reached by means of the plane-polar scanning arrangement and the correct orientations of the probe axes, which must be kept parallel to the AUT ones, are obtained by exploiting this last turntable.

An open-ended WR90 rectangular waveguide is used as probe. The AUT is a very simple E-plane monopulse antenna, operating at 10 GHz in the sum mode. It has been realized (see Figure 4.21) by using two pyramidal horns (8.9 cm × 6.8 cm sized) of Lectronic Research Labs at a distance of 26.5 cm (between centres) and a hybrid Tee. This AUT is located in the plane $z = 0$ and the greater sides of the horn apertures are parallel to the x axis. According to the described sampling representation, it has been considered as enclosed in a double bowl having $a = 18.6$ cm and $c = c' =$ 2.7 cm. The PWMS samples have been acquired within a circle with radius 110 cm on a plane at distance $d = 16.5$ cm from the AUT.

To assess the effectiveness of the two-dimensional OSI algorithm, the amplitude and phase of the reconstructed voltage V_p (the most significant one), relevant to the scanning plane line $x' = 3.6$ cm, are compared in Figure 4.22a and b with those directly measured on the same line. The reconstruction of the amplitude of the voltage V_r (the lest significant one) relevant to the same line is shown in Figure 4.23.

Figure 4.21 Photo of the E-plane monopulse antenna. The photo was taken before covering the AUT support with the absorbers.

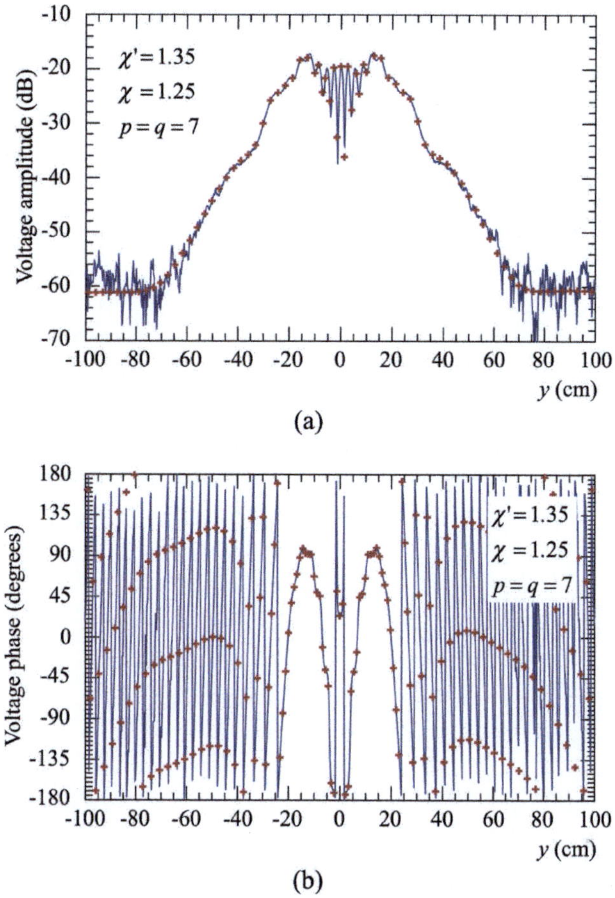

Figure 4.22 *Probe voltage V_p on the scanning plane line x' = 3.6 cm. Solid line: measured. Crosses: reconstructed from the PWMS samples. (a) Amplitude, (b) Phase.*

At last, the reconstruction of the amplitude and phase of V_p relevant to the line x' = 10.0 cm is shown in Figure 4.24 and b. As can be seen, there is an excellent agreement between the reconstructed voltages (crosses) and the measured ones (solid line), except where the voltage values are very low (below about –55 dB), where the error is caused both by the scanning zone truncation and the environmental reflections. It is worthy to note that, due to the filtering properties of the interpolation functions, the spatial harmonics relevant to the noise sources outside the AUT spatial bandwidth are cut away. This results in a smoother behaviour of the reconstructed voltage with respect to the measured one.

At last, the overall effectiveness of the described NF–FF transformation technique is assessed by comparing (see Figure 4.25) the FF patterns in the principal

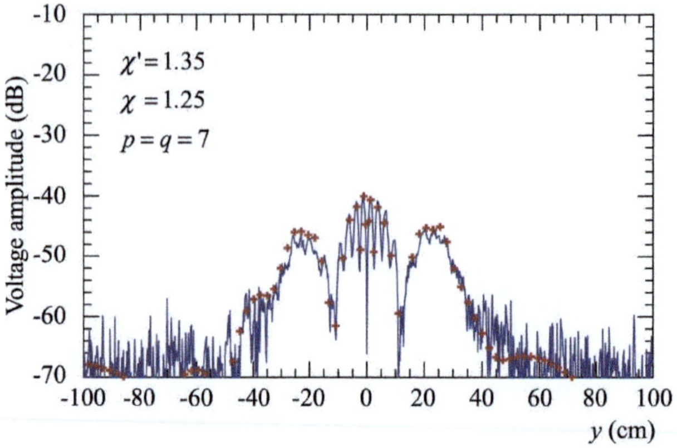

Figure 4.23 *Amplitude of the probe voltage V_r on the scanning plane line $x' = 3.6$ cm. Solid line: measured. Crosses: reconstructed from the PWMS samples.*

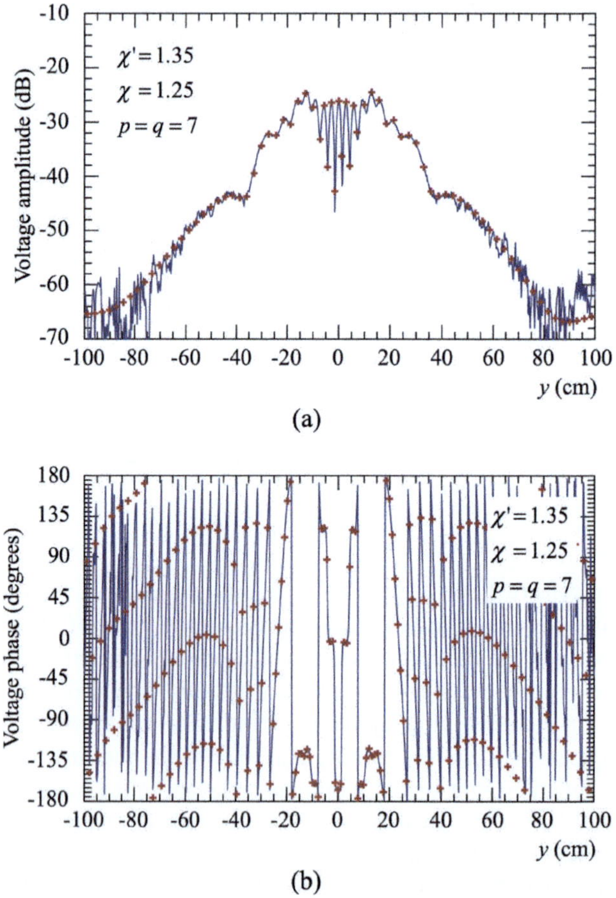

Figure 4.24 *Probe voltage V_p on the scanning plane line x'=10.0 cm. Solid line: measured. Crosses: reconstructed from the PWMS samples. (a) Amplitude, (b) Phase.*

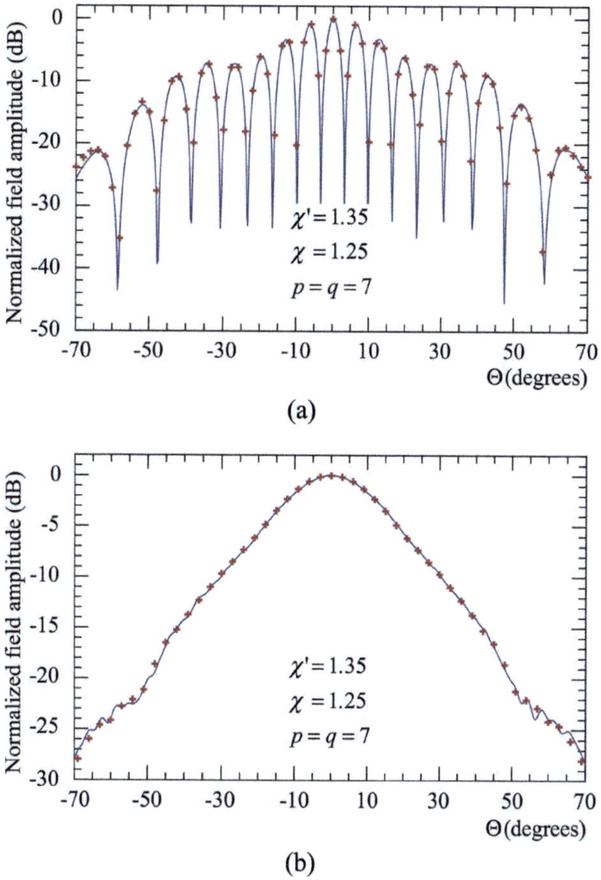

Figure 4.25 *FF patterns in the principal planes. Solid line: obtained from the NF data directly measured on the classical PR grid. Crosses: reconstructed from the PWMS samples. (a) E-plane, (b) H-plane.*

planes E and H reconstructed from the collected 2209 PWMS samples with those obtained from the 11025 PR NF data directly acquired at 0.45 λ spacing on a square of side 140 cm contained within the measurement circle. As can be seen, there is a very good agreement, thus confirming the effectiveness of such a NF–FF transformation technique.

4.6 Non-redundant NF–FF transformations with plane-polar scanning

In this section, two NR NF–FF transformations with plane-polar (PP) scanning will be presented. Both these transformations are particularly suitable to quasi-planar AUTs as usually are those characterized in a planar NF measurement system. In the former [43, 44], the AUT is modelled by an oblate spheroid, whereas in the latter [42, 45], it is considered as enclosed in a double bowl.

4.6.1 Non-redundant NF–FF transformation with plane-polar scanning: the oblate spheroidal modelling case

Let an electrically small, non-directive, probe be acquiring, by means of a PP NF measurement system, the NF data radiated by a quasi-planar AUT, enclosed in an oblate spheroid Σ with major and minor semi-axes equal to a and b, on a plane at distance d from the AUT aperture. In such a case, it is convenient to introduce also the PP coordinates system (ρ, φ) for denoting a point P on the scanning plane (see Figure 4.26).

Since the voltage measured by a non-directive probe has the same effective spatial bandwidth of the field radiated by the AUT [41], the NR representations [104, 105] and the related OSI expansions can be applied to the voltage detected by such a kind of probe. Therefore, it is convenient to introduce the probe reduced voltage $\tilde{V}(\xi) = V(\xi)\,\mathrm{e}^{\mathrm{j}\gamma(\xi)}$, wherein ξ is an optimal parameter to be adopted for describing the curves (radial lines and rings), which define the scanning plane in the PP reference frame, and $\gamma(\xi)$ a suitable phase function.

As shown in Chapter 2, the optimal parameter for describing a ring is the azimuthal angle φ and, when the AUT is modelled by an oblate spheroid, the bandwidth W_ξ, the phase function γ and the optimal parameter ξ relevant to a radial line are given by the relations (2.25), (2.27) and (2.28b), respectively, whereas the bandwidth W_φ is fixed by the expression (2.29b).

The two-dimensional OSI expansion (2.71), allowing the fast and precise recovery of the probe and rotated probe voltages at any point $P(\rho, \varphi)$ on the scanning plane, is reported in the following for reader's convenience:

$$V(\xi(\rho), \varphi) = \mathrm{e}^{-\mathrm{j}\gamma(\xi)} \times \sum_{n=n_0-q+1}^{n_0+q} \left\{ SF(\xi, \xi_n, \bar{\xi}, N, N'') \sum_{m=m_0-p+1}^{m_0+p} \tilde{V}(\xi_n, \varphi_{m,n})\, SF\left(\varphi, \varphi_{m,n}, \bar{\varphi}_n, M, M_n''\right) \right\} \tag{4.33}$$

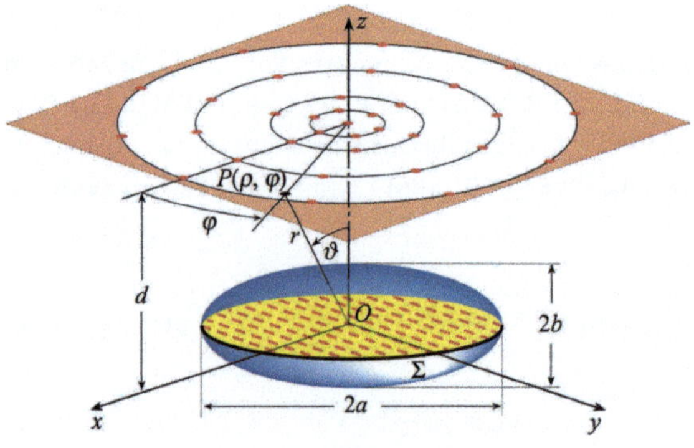

Figure 4.26 Plane-polar scanning: oblate spheroidal modelling

wherein all symbols have the same meanings as in (2.68). Note that $\vartheta_\infty(\xi_n)$, given by (2.30b), takes the place of $\vartheta(\xi_n)$ in the expression (2.67) of the azimuthal enlargement bandwidth factor χ^*. Such an expansion can be, obviously, applied to accurately reconstruct the values of the probe and rotated probe voltages at the points wherein the NF data to perform the classical NF–FF transformation with PR scanning [26] are required. It is so possible to obtain a probe-compensated NF–FF transformation with PP scanning requiring a NR number of NF data.

It must be observed that the AUT rotates in a PP scanning system and that the probe-compensated NF–FF transformation formulas (4.21) and (4.22) are valid only if the probe maintains its orientation with respect to the AUT. As a consequence, to perform the probe compensation, the measurement probe must generally co-rotate to maintain the AUT orientation (see Figure 4.27). Obviously, the positioning system is simplified when the probe co-rotation is avoided (see Figure 4.28). Probes exhibiting only a first-order azimuthal dependence in their radiated far field can be used without co-rotation, since the voltages V_p and V_r (measured by the probe and the rotated probe with co-rotation) can be evaluated

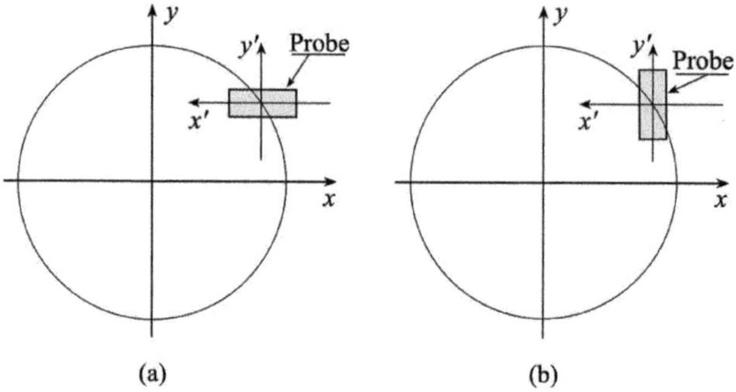

Figure 4.27 *Orientation of probe axes with co-rotation. (a) Measurement of V_p, (b) Measurement of V_r.*

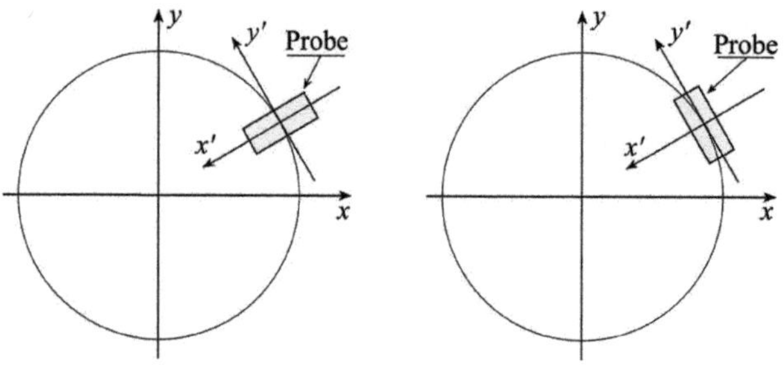

Figure 4.28 *Orientation of probe axes without co-rotation. (a) Measurement of V_φ, (b) Measurement of V_ρ.*

(see Appendix 4A.4) from the knowledge of the voltages V_φ and V_ρ (measured without co-rotation) through the relations:

$$V_p = V_\varphi \cos\varphi - V_\rho \sin\varphi; \qquad V_r = V_\varphi \sin\varphi + V_\rho \cos\varphi \tag{4.34}$$

As well known, a probe, whose radiated far field has a first-order azimuthal dependence, is an open-ended circular cross-section waveguide excited by a TE_{11} mode [127], but such a dependence is exhibited [303] in a very good approximation also by the far field radiated by an open-ended rectangular waveguide where a TE_{10} mode is propagating (see Appendix 4A.3).

Some numerical results assessing the effectiveness of the technique are shown in the following. The simulations are relevant to the previously considered circular array and the chosen probe is again an open-ended WR-90 rectangular waveguide working at 10 GHz, which scans a circular zone of radius 71 λ on a plane 8 λ away from the AUT. Such an array has been modelled by an oblate spheroid with major and minor axes equal to 34.8 λ and 4.4 λ. Figure 4.29a and b show a representative reconstruction example of the amplitude and phase of the voltage V_ρ on the radial line at $\varphi = 90°$. As can be seen, there is a very good agreement between the reconstructed voltage and the exact one. The stability of the algorithm, namely its robustness with respect to errors affecting the samples, has been investigated by altering the exact samples with random errors. These simulated errors are a background noise (bounded to Δa dB in amplitude and with arbitrary phase) and uncertainties on the data of $\pm\Delta a_r$ dB in amplitude and $\pm\Delta\phi$ degrees in phase. As testified by Figure 4.30, the algorithm is stable. The precision of the OSI algorithm is assessed in a more quantitative way by the maximum and mean-square reconstruction error values shown in Figure 4.31a and b. They are normalized to the voltage maximum value over the scanning area and have been got by comparing the interpolated values of V_ρ with those directly evaluated on a close grid in the central zone of the scanning circle in order to guarantee the availability of the necessary guard samples. As can be observed, they decrease up to very low values on increasing the oversampling factor and/or the number of retained nearest samples.

At last, the two-dimensional OSI algorithm (4.33) and the relations (4.34) have been employed to recover the massive input NF data for the classical PR NF–FF transformation [26] from the NR PP samples. The reconstruction of the antenna FF pattern in the E-plane is shown in Figure 4.32. As can be seen, the exact and recovered patterns are practically indistinguishable, thus further confirming the effectiveness of the described PP NF–FF transformation technique.

It can be interesting to compare the number of employed NR plane-polar NF samples (8570) with those 126665 and 40401 needed by the PP NF–FF transformation [32–34] and by classical PR one [26], respectively. It is worth noting that the number of NR NF samples, which would be required to cover an unbounded scanning plane, are in such a case 8889. As a conclusion, the described NR NF–FF transformation with PP scanning allows a remarkable reduction of the required number of NF data and related measurement time without any loss of precision in the reconstruction of the AUT far field pattern.

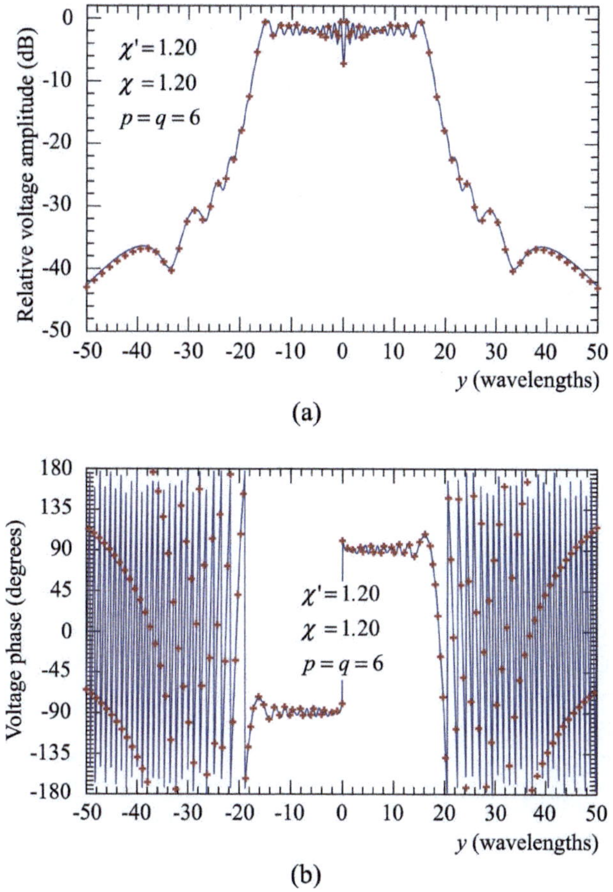

Figure 4.29 *Probe voltage V_ρ on the radial line at $\varphi = 90°$. Solid line: exact. Crosses: obtained by interpolating the NR plane-polar NF samples. (a) Amplitude, (b) Phase.*

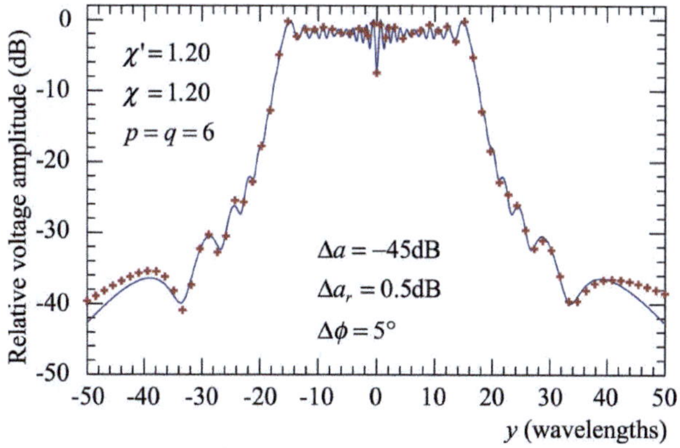

Figure 4.30 *Amplitude of the probe voltage V_ρ on the radial line at $\varphi = 90°$. Solid line: exact. Crosses: interpolated from error affected samples.*

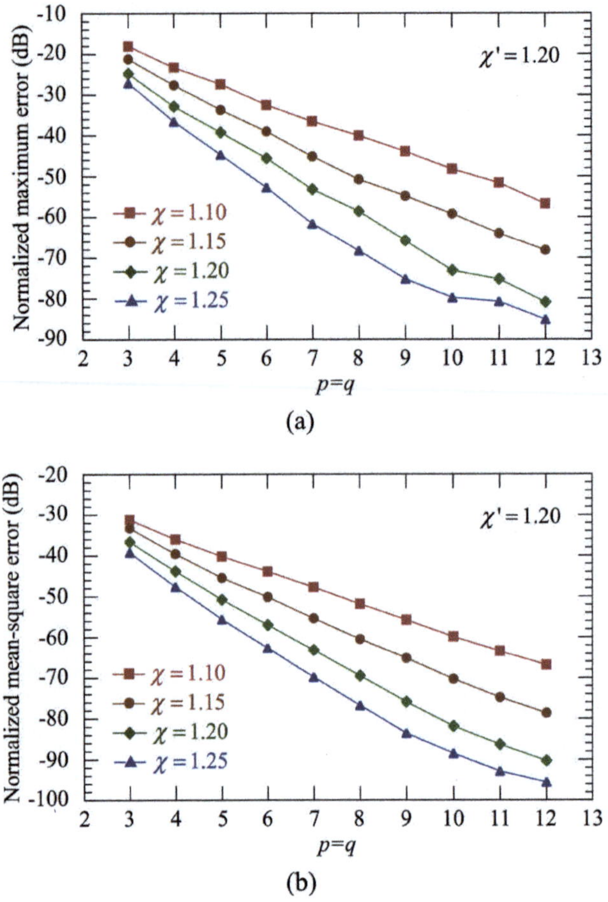

Figure 4.31 *Normalized errors in the reconstruction of the probe voltage V_ρ. (a) Maximum error, (b) Mean-square error.*

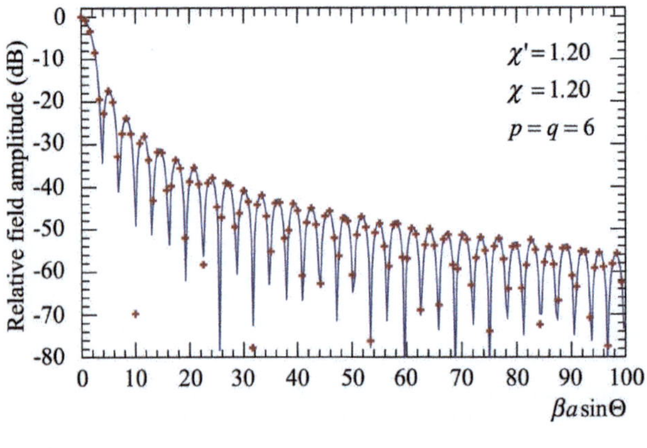

Figure 4.32 *E-plane pattern. Solid line: exact field. Crosses: reconstructed from the NR plane-polar NF samples.*

Experimental results assessing the effectiveness of such a NF–FF transformation also from a practical viewpoint can be found in [44].

4.6.2 Non-redundant NF–FF transformation with plane-polar scanning: the double bowl modelling case

Let an electrically small, non-directive, probe be acquiring, by means of a PP scanning facility, the NF data radiated by a quasi-planar AUT, enclosed in a double bowl Σ, on a plane at distance d from the antenna aperture (see Figure 4.33). As in the previous subsection, the NR sampling representations [104] can be applied to the probe reduced voltage $\tilde{V}(\xi) = V(\xi)e^{j\gamma(\xi)}$.

The bandwidth W_ξ, the phase function γ and the optimal parameter ξ relevant to a radial line are in such a case [42, 45] given by the expressions (4.32), where the related parameters can be determined as described in Sect. 4.5.2. The optimal parameter for describing a ring is again the azimuthal angle φ and the maximum in the relation (2.14), which allows the evaluation of the related bandwidth W_φ, can be determined [42, 45] as described in Sect. 2.3.

The OSI expansion (4.33), wherein the azimuthal enlargement bandwidth factor χ^* is given by (2.67), allows again the efficient reconstruction of the probe and rotated probe voltages at any point on the scanning plane and, therefore, can be employed to accurately determine their values at the points needed by the classical probe-compensated NF–FF transformation with PR scanning [26].

The described NR NF–FF transformation has been experimentally validated by means of the flexible NF measurement facility available in the Antenna Characterization Lab of the University of Salerno, which can be arranged to perform a PP scanning (see Sect. 1.2). Such a facility, due to the presence of an additional turntable between the probe and the linear positioner (see Figure 1.15), allows also the hardware co-rotation, the classical PR scanning, and the non-conventional PWMS.

The AUT is again the previously considered E-plane monopulse antenna operating at 10 GHz in the sum mode and the PP samples have been collected by the probe, an open-ended WR90 rectangular waveguide, within a circle of radius

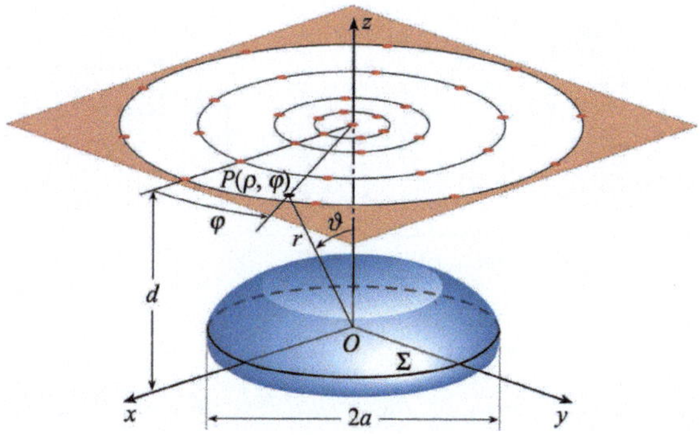

Figure 4.33 Plane-polar scanning: double bowl modelling

110 cm on a plane at distance $d = 16.5$ cm from the horns apertures. Such an AUT has been modelled by a double bowl having $a = 18.0$ cm and $c = c' = 3.0$ cm. In order to assess the effectiveness of the two-dimensional OSI expansion, the amplitude and phase of the reconstructed probe voltage V_ρ on the radial line at $\varphi = 90°$ are compared in Figure 4.34a and b with those directly measured on the same line. For completeness, the reconstructions of the amplitude of the voltage V_φ relevant to the radial lines at $\varphi = 30°$ and $\varphi = 0°$ are shown in Figures 4.35 and 4.36. All these reconstructions have been got by using $\chi' = 1.30$, $\chi = 1.20$, and $p = q = 7$. As can be seen, the agreement between the recovered voltage (crosses) and the measured one (solid line) results to be very good save for the peripheral zone, characterized by very low voltage values, wherein the error is caused both by the environmental reflections and by the truncation of the

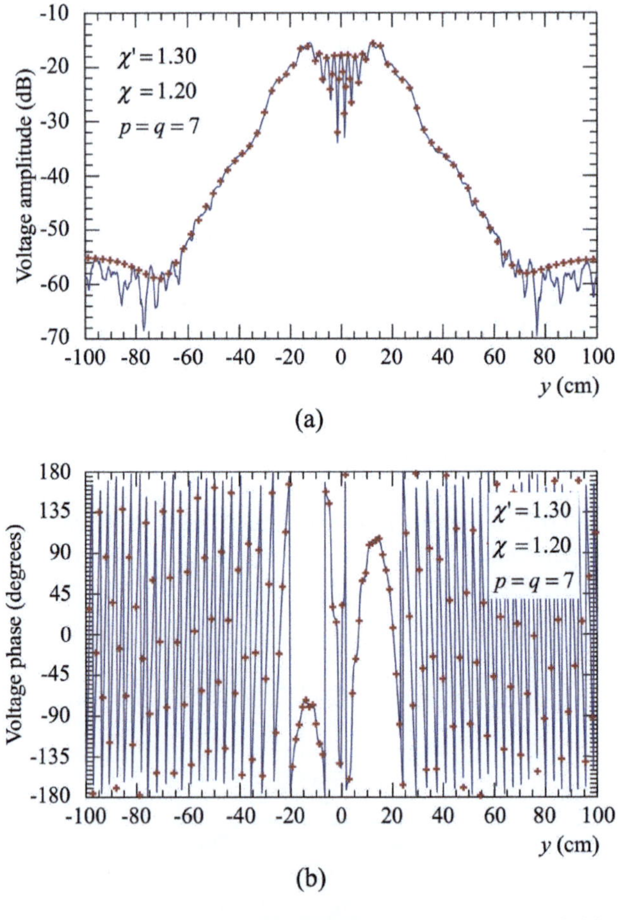

Figure 4.34 *Probe voltage V_ρ on the radial line at $\varphi = 90°$. Solid line: measured. Crosses: obtained by interpolating the NR plane-polar NF samples. (a) Amplitude, (b) Phase.*

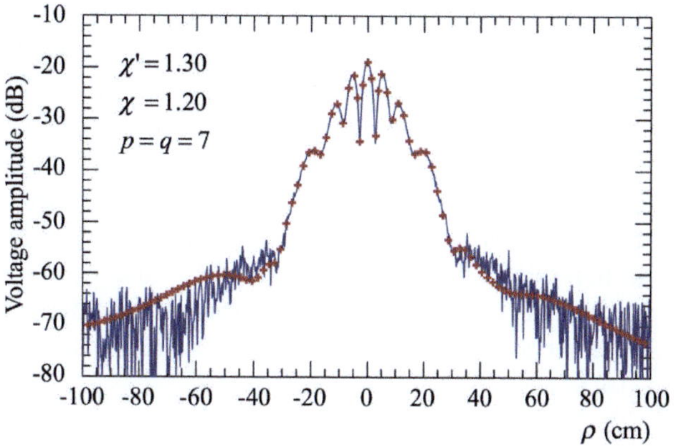

Figure 4.35 *Amplitude of the probe voltage V_φ on the radial line at $\varphi = 30°$. Solid line: measured. Crosses: obtained by interpolating the NR plane-polar NF samples.*

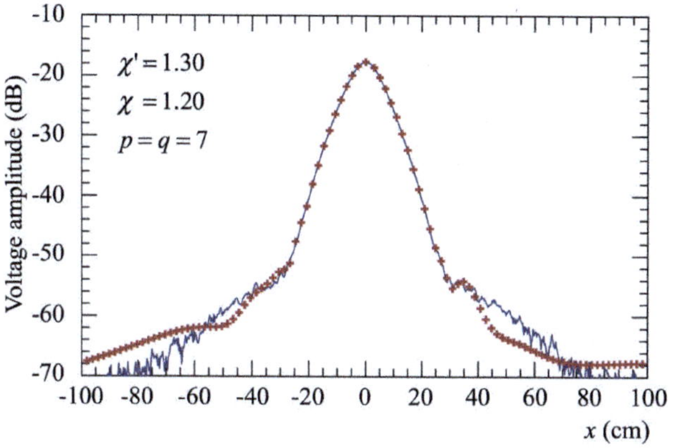

Figure 4.36 *Amplitude of the probe voltage V_φ on the radial line at $\varphi = 0°$. Solid line: measured. Crosses: obtained by interpolating the NR plane-polar NF samples.*

scanning area. Moreover, it is noteworthy that the reconstructed voltage exhibits always a smoother behaviour with respect to the directly measured one and this is due to the filtering properties of the OSI functions, which are able to cut away the spatial harmonics relevant to the noise sources which exceed the spatial bandwidth of the AUT.

At last, the OSI algorithm (4.33) and the relations (4.34) have been employed to recover the 11025 PR NF data spaced by 0.45λ on a 140 cm × 140 cm sized square from the acquired PP NF samples. The E- and H-planes FF patterns reconstructed from these data by applying the classical PR NF–FF transformation [26] are compared in Figure 4.37a and b with those obtained from the PR NF data directly measured on the same grid. As can be noticed, a very good agreement is found in the E-plane, whereas a less accurate reconstruction occurs in the H-plane. This is due to the fact that the far field radiated by an open-ended rectangular waveguide, wherein a TE_{10} mode is propagating, has only approximately a first-order azimuthal dependence [303]. The reconstruction of the H-plane pattern, attained from the PP NF samples acquired when using the hardware co-rotation, is shown in Figure 4.38 and, as can be seen, results to be very accurate in such a case.

It is worthy to point out that the number of the acquired PP NF samples is 1476 and, accordingly, remarkably less than that (33581) required by the PP NF–FF transformations [32–34] to cover the same scanning area.

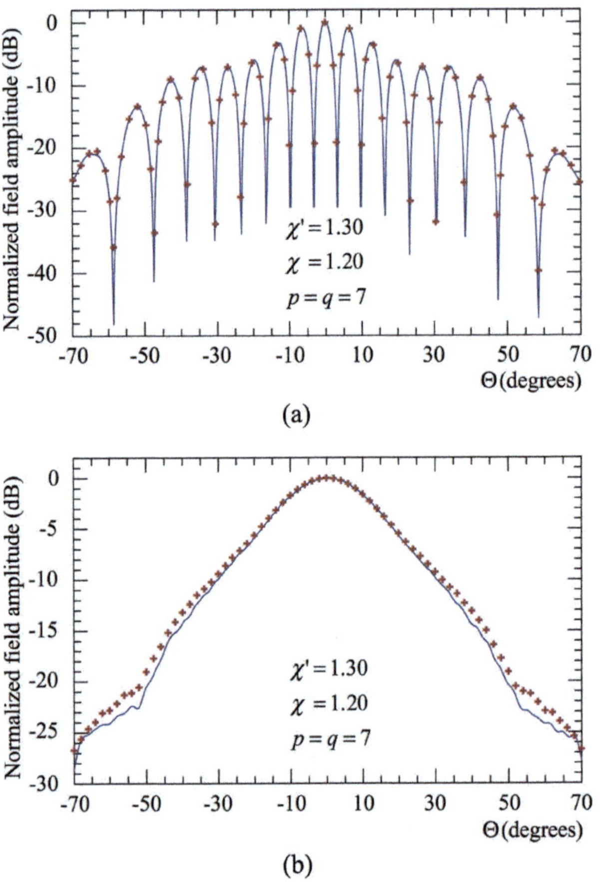

Figure 4.37 FF patterns in the principal planes. Solid line: reconstructed from the plane-rectangular NF data. Crosses: reconstructed from the NR plane-polar NF samples with software co-rotation. (a) E-plane, (b) H-plane.

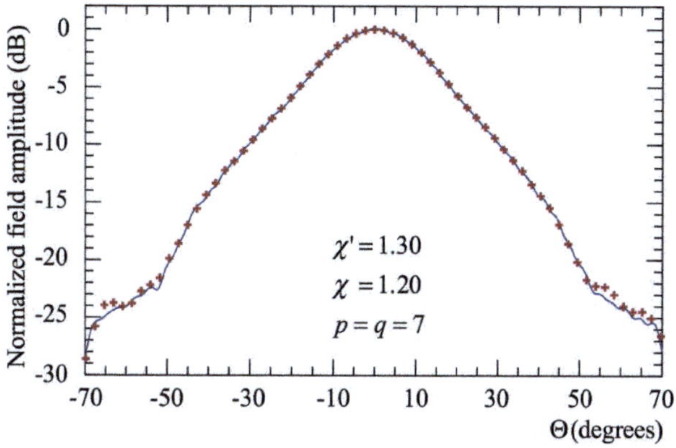

Figure 4.38 *FF pattern in the H-plane. Solid line: reconstructed from the plane-rectangular NF data. Crosses: reconstructed from the NR plane-polar NF samples with hardware co-rotation.*

Another set of experimental results assessing the effectiveness of such a PP NF–FF transformation and relevant to a different AUT is reported in [45].

4.7 Non-redundant NF–FF transformations with bi-polar scanning

Two NF–FF transformation techniques with bi-polar (BP) scanning, which require a NR number of NF samples, will be described in this section. As all other transformations employing a planar scanning, they too are particularly tailored to characterize quasi-planar AUTs. In the former [49, 50], the AUT is considered as enclosed in an oblate spheroid, whereas in the latter [51], it is modelled by a double bowl.

4.7.1 Non-redundant NF–FF transformation with bi-polar scanning: the oblate spheroidal modelling case

Let an electrically small, non-directive, probe be acquiring, through a BP NF facility, the NF data radiated by a quasi-planar AUT, enclosed in an oblate spheroid Σ with major and minor semi-axes equal to a and b, on a plane d away from the antenna aperture. In such a case, it is convenient to identify a point P on the scanning plane by means of the BP coordinates L, δ, and α, where L is the length of the bi-polar arm, and δ and α are the rotation angles of the arm and AUT, respectively (see Figure 4.39). The polar coordinates (ρ, φ) are linked to them by the relations:

$$\rho = 2L \sin(\delta/2); \qquad \varphi = \alpha - \delta/2 \tag{4.35}$$

It is noteworthy that a convenient way to get a NR representation of the voltage measured by a non-directive probe over a plane from its BP

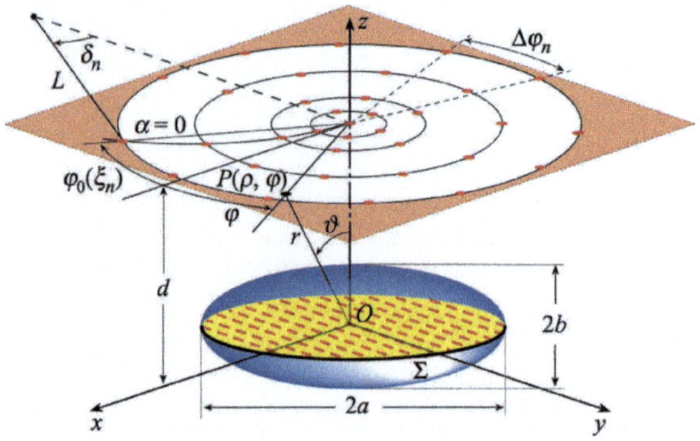

Figure 4.39 Bi-polar scanning: oblate spheroidal modelling

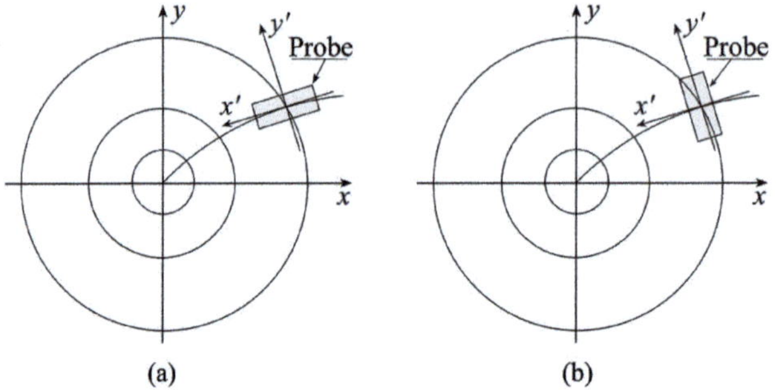

Figure 4.40 Orientation of probe axes without co-rotation. (a) Measurement of V_α, (b) Measurement of V_δ.

samples is to describe the scanning plane by means of radial lines and rings, as in the PP scanning case, since the arcs described by the arm, unlike the radial lines, are not meridian curves as required by the NR representations [104]. Accordingly, it is convenient to introduce the probe reduced voltage $\tilde{V}(\xi) = V(\xi)e^{j\gamma(\xi)}$, where ξ is the optimal parameter for describing the radial lines or the rings, $\gamma(\xi)$ a suitable phase function, and V is the voltage V_α or V_δ measured by the probe or by the rotated probe, respectively (see Figure 4.40). The optimal parameter to describe a ring is, as shown in Chapter 2, the azimuthal angle φ and, since the AUT is modelled by an oblate spheroid, the bandwidth W_ξ, the phase function γ and the optimal parameter ξ relevant to a radial line are given by the expressions (2.25), (2.27) and (2.28b), whereas the bandwidth W_φ is fixed by the relation (2.29b).

It must be taken into account that, due to the way of collecting the data in a BP scanning facility (see Figure 4.39), the positions of the samples on the n-th ring are shifted by $\varphi_0(\xi_n) = -\delta_n/2$ with respect to the corresponding ones in the PP grid. As a consequence, the OSI expansion (4.33) can be again employed to efficiently recover the probe and rotated probe voltages at the points wherein the input NF data for the classical PR NF–FF transformation [26] are necessary, but the expressions of m_0 and $\varphi_{m,n}$ must be modified as follows:

$$m_0 = \text{Int}\left[(\varphi - \varphi_0(\xi_n))/\Delta\varphi_n\right]; \qquad \varphi_{m,n} = \varphi_0(\xi_n) + m\Delta\varphi_n \tag{4.36}$$

Moreover $\vartheta_\infty(\xi_n)$, given by (2.30b), takes the place of $\vartheta(\xi_n)$ in the expression (2.67) of the azimuthal enlargement bandwidth factor χ^*.

It should be emphasized that even in a BP scanning system the AUT rotates and, therefore, its axes do not remain parallel to the probe ones, as requested for the validity of the probe-compensated NF–FF transformation formulas (4.21) and (4.22), unless the probe is suitably co-rotated. The use of a probe, whose far field has a first-order azimuthal dependence, allows one to avoid the probe co-rotation, thus simplifying the positioning system. In fact, by parallelizing the demonstration reported in Appendix 4A.4, it is possible to prove that, when this type of probe is used, the voltages V_p and V_r (measured by the probe and the rotated probe with co-rotation) can be determined from the voltages V_α and V_δ (measured without co-rotation) by means of the following relations:

$$V_p = V_\alpha \cos(\varphi - \delta/2) - V_\delta \sin(\varphi - \delta/2) \tag{4.37a}$$

$$V_r = V_\alpha \sin(\varphi - \delta/2) + V_\delta \cos(\varphi - \delta/2) \tag{4.37b}$$

thus enabling a software co-rotation.

Some numerical simulations assessing the effectiveness of the described technique are shown in the following. They are relevant to the previously considered circular array and again an open-ended WR-90 rectangular waveguide working at 10 GHz is chosen as probe. Such an array has been considered as enclosed in an oblate spheroid having major and minor axes equal to 34.8 λ and 4.4 λ, respectively. The scanning plane is 8 λ away from the AUT and the BP measurement system is characterized by arm length L = 75 λ and maximum rotation angle of the arm $\delta_{\max} \approx 56.3°$, so that the NF data lie in a circle of radius $\rho_{\max} \approx 71\lambda$. A representative reconstruction example of the amplitude and phase of the voltage V_δ on the radial line at $\varphi = 90°$ is shown in Figure 4.41a and b. As can be observed, the agreement between the reconstructed voltage and the exact one is very good. The stability of the algorithm with respect to errors affecting the samples has been tested by altering the exact samples with random errors. These simulated errors are a background noise (bounded to Δa dB in amplitude and with arbitrary phase) and uncertainties on the data of $\pm\Delta a_r$ dB in amplitude and $\pm\Delta\,\phi$ degrees in phase. As shown in Figure 4.42, the algorithm is stable. The performances of the OSI algorithm have been assessed in a more quantitative way by evaluating the maximum and mean-square errors in the reconstruction of V_δ, normalized to the voltage maximum value over the plane. They are shown in Figure 4.43a and b and have been obtained by comparing the interpolated values of V_δ with those directly evaluated on a close grid in the central zone of the scanning circle so that the existence of the needed guard samples is assured. As expected, they decrease up

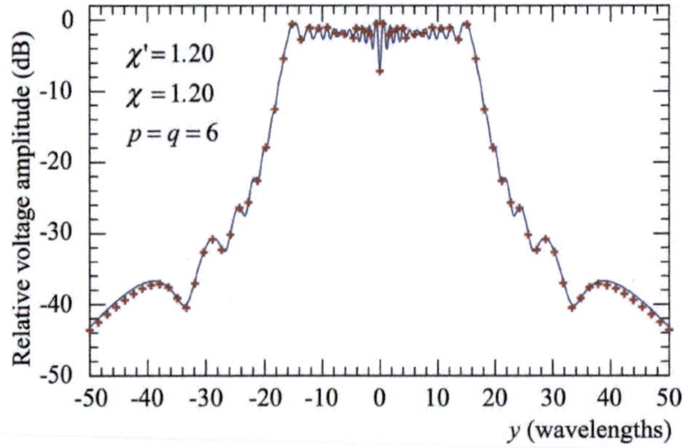

(a)

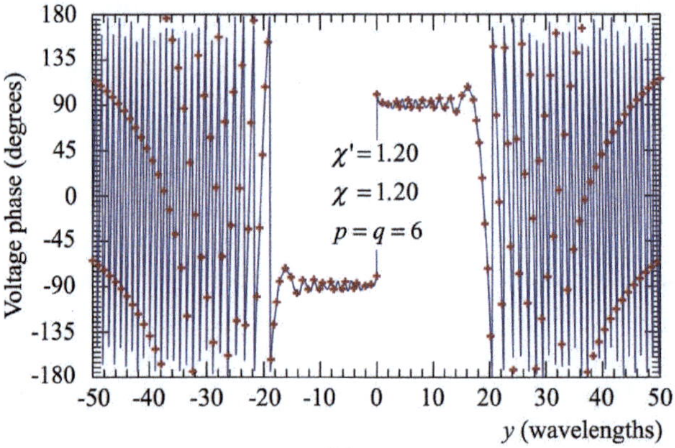

(b)

Figure 4.41 *Probe voltage V_δ on the radial line at $\varphi = 90°$. Solid line: exact. Crosses: obtained by interpolating the NR bi-polar NF samples. (a) Amplitude, (b) Phase.*

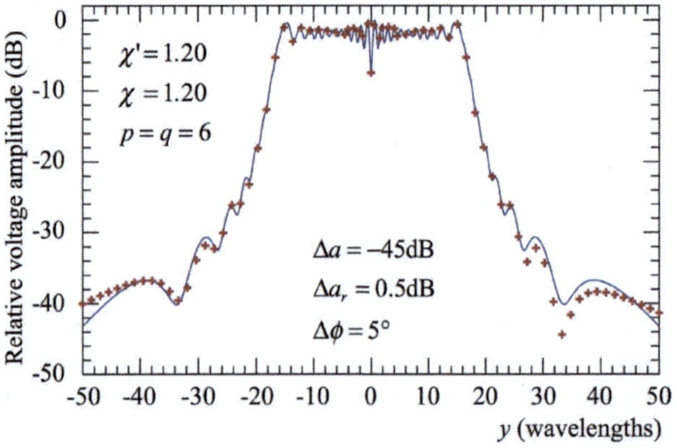

Figure 4.42 *Amplitude of the probe voltage V_δ on the radial line at $\varphi = 90°$. Solid line: exact. Crosses: interpolated from the error affected bi-polar NF samples.*

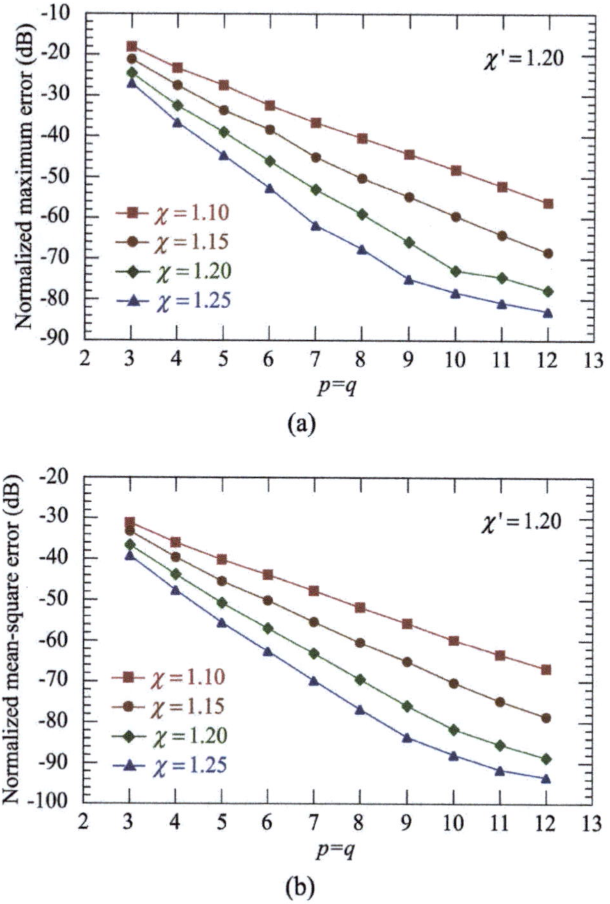

Figure 4.43 *Normalized errors in the reconstruction of the probe voltage V_δ. (a) Maximum error, (b) Mean-square error.*

to very low values on increasing the number of considered closest samples and/or the oversampling factor. The OSI algorithm and the relations (4.37) have been finally employed to reconstruct the massive input NF data for the standard PR NF–FF transformation [26] from the NR BP samples. The antenna E-plane FF pattern reconstructed from the NR BP samples is compared in Figure 4.44 with the exact one. As can be seen, the exact and recovered patterns are practically indistinguishable, thus further assessing the effectiveness of the described NF–FF transformation with BP scanning.

It must be stressed that the number of used bi-polar NF samples is 8570 and, accordingly, much smaller than those 65818 and 40401 needed, to cover the same scanning area, by the BP NF–FF transformation [15, 46, 47] and by the classical PR one [26], respectively.

The interested reader can find in [50] some experimental results assessing the effectiveness of this NF–FF transformation also from the practical viewpoint.

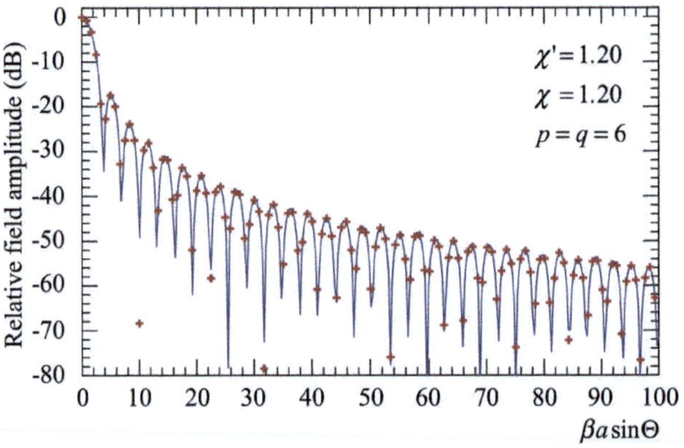

Figure 4.44 *E-plane pattern. Solid line: exact field. Crosses: reconstructed from the NR bi-polar NF samples.*

4.7.2 Non-redundant NF–FF transformation with bi-polar scanning: the double bowl modelling case

Let a non-directive probe be collecting, through a BP acquisition system, the NF data radiated by a quasi-planar AUT on a plane at distance d from its aperture (see Figure 4.45). As already stressed, the voltage V detected by this kind of probe has the same spatial bandwidth of the AUT radiated field, thus the NR sampling representations [104] can be applied to the reduced voltage $\tilde{V}(\xi) = V(\xi)\,e^{j\gamma(\xi)}$.

When the AUT is modelled by a double bowl, the bandwidth W_ξ, the phase function γ and the optimal parameter ξ relevant to a radial line are given [51] by the expressions (4.32), wherein the involved parameters can be evaluated as described in Sect. 4.5.2. The optimal parameter to describe a ring is still the azimuthal angle φ

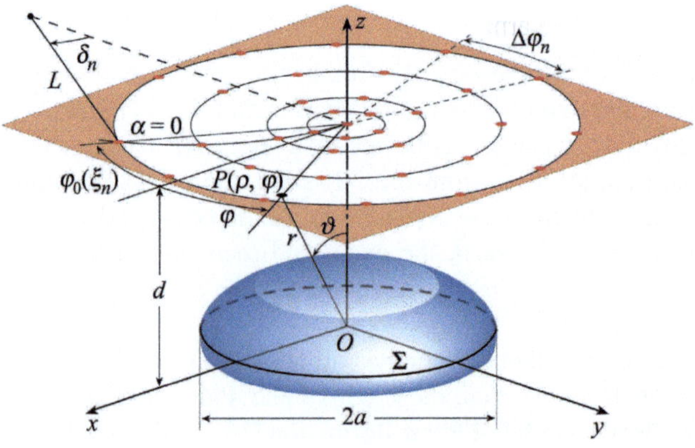

Figure 4.45 *Bi-polar scanning: double bowl modelling*

and the maximum in (2.14), which allows the evaluation of the bandwidth W_φ, can be found [51] as described in Sect. 2.3.

The OSI expansion (4.33), where m_0 and $\varphi_{m,n}$ are given by (4.36) and χ^* by (2.67), allows the efficient reconstruction of the probe and rotated probe voltages at any point on the scanning plane and, hence, can be applied to accurately recover their values at the points required by the classical PR NF–FF transformation [26].

The described NF–FF transformation with BP scanning has been experimentally validated through the flexible NF measurement facility available in the UNISA Antenna Characterization Lab, arranged in the PP scan configuration (see Sect. 1.2). In fact, in such a configuration, it is possible, due to the presence of a further turntable between the linear positioner and the probe (see Figure 1.15), to acquire the NF data which would be measured by a BP NF measurement system and by a PR one. Moreover, it is also possible to measure the NF data which would be acquired in a BP NF facility with hardware co-rotation.

The experimental tests are relevant to the already considered E-plane monopulse antenna working at 10 GHz in the sum mode and again an open-ended WR90 rectangular waveguide has been employed as scanning probe. Such an AUT has been modelled by a double bowl having now $a = 18.6$ cm and $c = c' = 2.7$ cm. The BP samples have been acquired by the probe in a circle of radius 105 cm on the plane 16.5 cm away from the AUT at the points specified by the described BP sampling representation when the arm length L is 120 cm, $\chi' = 1.35$, and $\chi = 1.25$.

The amplitude and phase of the voltage V_α on the radial line at $\varphi = 0°$ interpolated from the collected BP samples are compared in Figure 4.46a and b with those directly measured on the same line. For completeness, the reconstructions of the amplitude and phase of the voltage V_δ relevant to the radial line at $\varphi = 45°$ are shown in Figure 4.47a and b. Also in such a case, a very good agreement between the recovered and measured voltages can be observed save for the peripheral zone characterized by very low voltages, wherein the reconstructed voltages exhibit a smoother behaviour with respect to the directly measured ones. This is due to the filtering properties of the OSI functions, which are able to cut away the spatial harmonics of the noise sources higher than the AUT spatial bandwidth.

To further assess the effectiveness of the presented NF–FF transformation, the E- and H-planes FF patterns, obtained by applying the NF–FF transformation [26] to the 11025 PR NF data directly acquired at 0.45 λ spacing on a square of side 140 cm contained within the measurement circle, are compared in Figure 4.48a and b with those got from the PR NF data recovered in the same grid, by using the described OSI algorithm and the relations (4.37), from the acquired BP samples. As can be seen, the agreement in the E-plane is very good, whereas a less accurate reconstruction can be observed in the H-plane. This, as already stressed, is imputable to the use as probe of an open-ended rectangular waveguide, whose radiated far field has only approximately a first-order azimuthal dependence [303]. As can be clearly seen from Figure 4.49, when the BP samples are acquired by using the hardware co-rotation, the reconstruction is very good also in the H-plane.

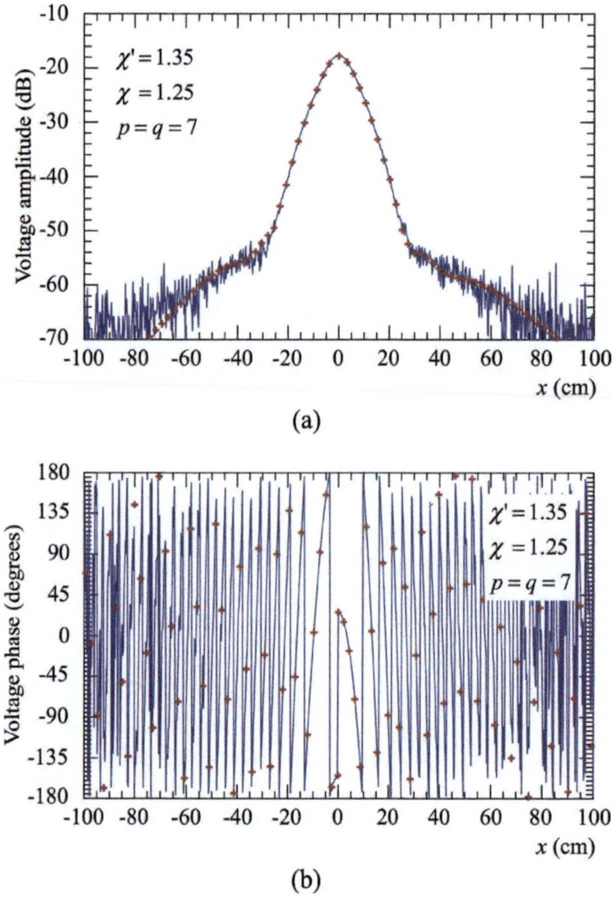

Figure 4.46 *Probe voltage V_α on the radial line at $\varphi = 0°$. Solid line: measured. Crosses: obtained by interpolating the NR bi-polar NF samples. (a) Amplitude, (b) Phase.*

It is noteworthy that the number of the acquired BP samples is 1810 and, therefore, remarkably less than that (15251) needed by the BP scanning technique [15, 46, 47] to cover the same measurement area.

4.8 Non-redundant NF–FF transformations with planar spiral scanning

In this section, NR NF–FF transformations with planar spiral scanning will be presented. Since the NR samples are acquired on fly and the scanning is performed through continuous and synchronized movements of the probe linear positioner and AUT turntable, a further remarkable saving of measurement time with respect to the

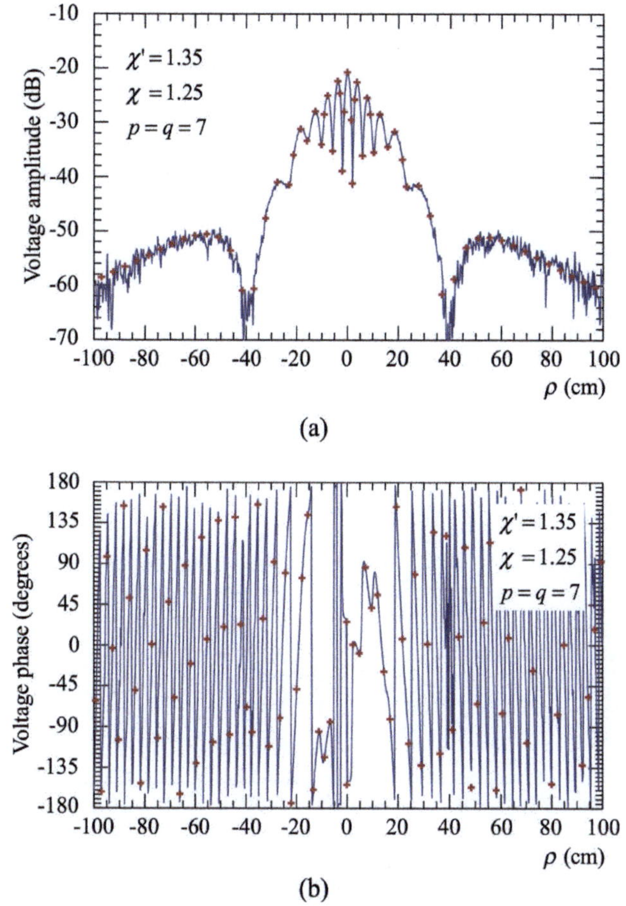

Figure 4.47 *Probe voltage V_δ on the radial line at $\varphi = 45°$. Solid line: measured. Crosses: obtained by interpolating the NR bi-polar NF samples. (a) Amplitude, (b) Phase.*

NR NF–FF transformations with PP scanning is obtained. In particular, a NF–FF transformation for volumetric AUTs adopting a uniform planar spiral scanning [304] is described in subsection 4.8.1, whereas two NR NF–FF transformations with planar spiral scanning, suitable for quasi-planar antennas modelled by a double bowl [130] or by an oblate spheroid [129, 130], are presented in the subsections 4.8.2 and 4.8.3, respectively.

It is noteworthy that, although the NF–FF transformation with uniform planar spiral scanning requires a number of NF samples greater than those needed by the other two, its hardware implementation is simpler, since, being the spiral step constant, the velocity of the linear positioner must not vary during the acquisition.

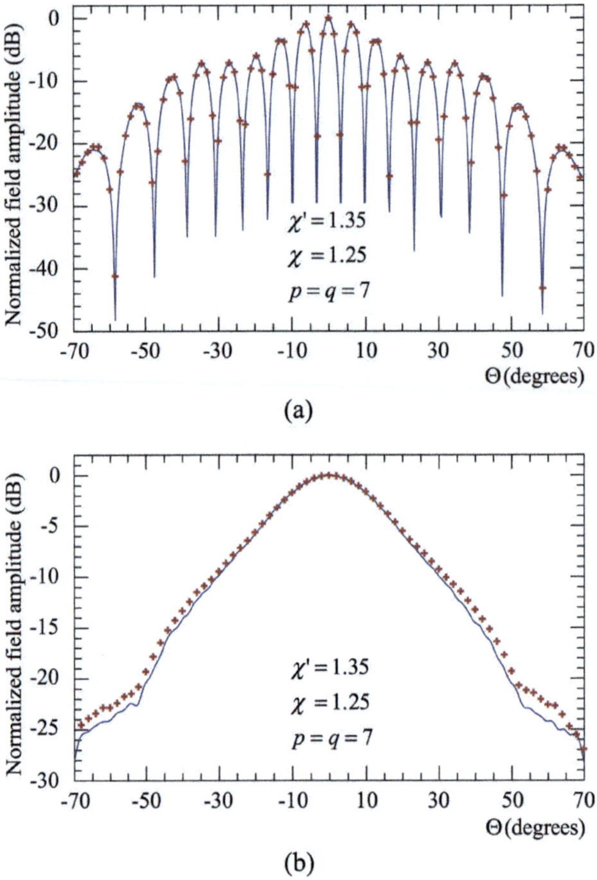

Figure 4.48 *FF patterns in the principal planes. Solid line: reconstructed from the plane-rectangular NF data. Crosses: reconstructed from the NR bi-polar NF samples with software co-rotation. (a) E-plane, (b) H-plane.*

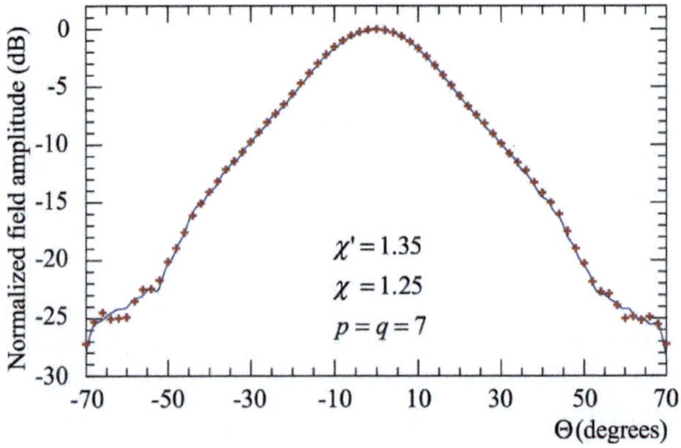

Figure 4.49 *H-plane pattern. Solid line: reconstructed from the plane-rectangular NF data. Crosses: reconstructed from the NR bi-polar NF samples with hardware co-rotation.*

4.8.1 NF–FF transformation with uniform planar spiral scanning

Let a quasi-spherical AUT be considered as enclosed in the smallest sphere Σ of radius a able to contain it and let the voltage be acquired by an electrically small non-directive probe scanning a uniform planar spiral (i.e., having constant radial step) lying on a plane at distance d from the AUT centre. As in the PP scanning case, it is convenient, besides the Cartesian coordinate system (x, y, z) with the origin O at the AUT centre and the associated spherical one (r, ϑ, φ), to introduce also the PP coordinates system (ρ, φ) to identify a point P on the scanning plane (see Figure 4.50). The Cartesian coordinates of a point Q on the spiral are given by the parametric equations:

$$\begin{cases} x = \bar{\rho}\cos\phi \\ y = \bar{\rho}\sin\phi \\ z = d \end{cases} \tag{4.38}$$

where ϕ is the angular parameter which describes the spiral and $\bar{\rho} = k\phi$. It must be pointed out that $\bar{\rho}$ becomes negative when ϕ is less than zero, whereas the radial distance ρ is always a positive quantity. Moreover, the angular parameter ϕ is always continuous, whereas the angle φ exhibits a discontinuity jump of π in correspondence of the origin, as can be easily verified from (4.38). The radial step $\Delta\rho$ of the spiral is determined by two its consecutive intersections at $Q(\phi)$ and $Q(\phi + 2\pi)$ with a radial line. Accordingly, $\Delta\rho = 2\pi k$ and, hence, $k = \Delta\rho/2\pi$.

A convenient way to determine the antenna radiated far field from a NR number of NF samples acquired along a proper planar spiral with constant radial step is the development of an efficient interpolation algorithm for the voltage reconstruction over the plane from these NR samples. In such a way, it is possible to accurately reconstruct the NF data required by the classical NF–FF transformation with PR scan [26]. As outlined in Chapter 3, the two-dimensional OSI expansion for

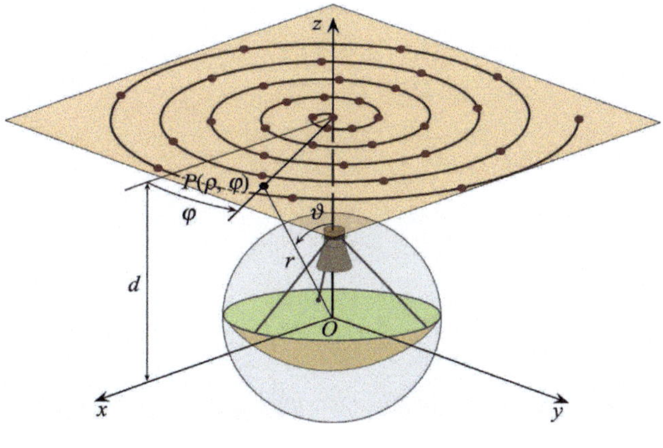

Figure 4.50 Uniform planar spiral scanning

determining the probe voltage on the plane from its samples acquired along the spiral can be obtained:

i. by choosing the radial step of the spiral coincident with the sampling spacing necessary for the interpolation of the NF data along a radial line according to the spatial bandwidth properties of EM fields [102], i.e.,

$$\Delta\rho = \frac{\pi d}{\chi\chi'\beta a} \tag{4.39}$$

ii. by developing a NR sampling voltage representation along the spiral.

Accordingly, the voltage at any point P on the plane can be recovered by evaluating the intermediate samples (those in correspondence of the intersections between the radial line through P and the spiral) via an interpolation along the spiral and then interpolating, along the radial line, the intermediate samples.

The NR sampling representation of the voltage along the spiral can be obtained [104] by choosing the phase function ψ and the optimal parameter η for describing the spiral according to the expressions (2.4) and (2.6), here reported for reader's convenience:

$$\psi(\eta) = \frac{\beta}{2}\int_0^s \left[\max_{r'} \hat{R}\cdot\hat{t} + \min_{r'} \hat{R}\cdot\hat{t}\right] ds \tag{4.40a}$$

$$\eta = \frac{\beta}{2W_\eta}\int_0^s \left[\max_{r'} \hat{R}\cdot\hat{t} - \min_{r'} \hat{R}\cdot\hat{t}\right] ds \tag{4.40b}$$

With the aid of Figure 4.51, it is a simple task to recognize that the maximum and minimum values of $\hat{R}\cdot\hat{t}$ occur [304] in correspondence of the tangency points

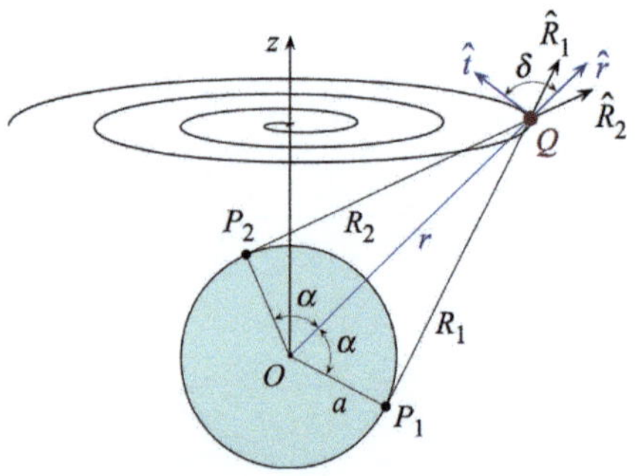

Figure 4.51 Relevant to the determination of the extreme values of $\hat{R}\cdot\hat{t}$

$P_{1,2}$ of the modelling sphere with the straight lines passing through the point Q on the spiral and lying on the plane specified by the unit vectors $\hat{r}$ (pointing from the origin to Q) and $\hat{t}$ (tangent to the spiral at Q), and that:

$$(\hat{R}_1 + \hat{R}_2)/2 = \hat{r}\,\sin\alpha = \hat{r}\sqrt{1 - a^2/r^2} \tag{4.41a}$$

$$(\hat{R}_1 - \hat{R}_2)/2 = \hat{n}\,\cos\alpha = \hat{n}(a/r) \tag{4.41b}$$

wherein $\hat{R}_{1,2}$ are the unit vectors pointing from $P_{1,2}$ to Q and $\hat{n}$ is the unit vector orthogonal to $\hat{r}$ and parallel to the plane identified by $\hat{r}$ and $\hat{t}$. By denoting with δ the angle between $\hat{r}$ and $\hat{t}$, it results:

$$(\hat{R}_1 - \hat{R}_2)\cdot\hat{t}/2 = (\hat{n}\cdot\hat{t})\,(a/r) = (a/r)\cos(\delta - \pi/2) = (a/r)\sin\delta \tag{4.42}$$

By substituting (4.41a) into (4.40a) and taking into account that $\mathrm{d}r = \hat{r}\cdot\hat{t}\,\mathrm{d}s$, it follows that:

$$\psi = \beta\int_0^r \sqrt{1 - a^2/r^2}\,\mathrm{d}r = \beta\sqrt{r^2 - a^2} - \beta a\cos^{-1}(a/r) \tag{4.43}$$

By taking into account the spiral equations (4.38), it results:

$$\mathrm{d}s = \sqrt{(\mathrm{d}x)^2 + (\mathrm{d}y)^2} = k\sqrt{1+\phi^2}\,\mathrm{d}\phi \tag{4.44}$$

On the other hand,

$$a/r = \frac{a}{\sqrt{d^2 + k^2\phi^2}} \tag{4.45}$$

and

$$\sin\delta = \sqrt{1 - \cos^2\delta} = \sqrt{1 - (\hat{r}\cdot\hat{t})^2} \tag{4.46}$$

where

$$\hat{r}\cdot\hat{t} = \frac{\mathrm{d}r}{\mathrm{d}s} = \frac{\mathrm{d}r}{\mathrm{d}\phi}\frac{\mathrm{d}\phi}{\mathrm{d}s} = \frac{1}{k\sqrt{1+\phi^2}}\frac{\mathrm{d}r}{\mathrm{d}\phi} = \frac{k\phi}{\sqrt{1+\phi^2}\sqrt{d^2 + k^2\phi^2}} \tag{4.47}$$

accordingly,

$$\sin\delta = \frac{\sqrt{d^2 + \phi^2 d^2 + k^2\phi^4}}{\sqrt{1+\phi^2}\sqrt{d^2 + k^2\phi^2}} \tag{4.48}$$

By substituting (4.42) and (4.44) into (4.40b) and choosing $W_\eta = \beta a$, its results:

$$\eta = k\int_0^{\phi(Q)} \frac{\sqrt{d^2 + \phi^2 d^2 + k^2\phi^4}}{d^2 + k^2\phi^2}\,\mathrm{d}\phi \tag{4.49}$$

It is worth noting that such an integral cannot be solved in closed form and, accordingly, has to be numerically evaluated.

The voltage value at any point Q on the spiral can be accurately reconstructed [304] from the NR spiral samples by means of the OSI expansion (see (2A.32)):

$$V(\eta) = e^{-j\psi(\eta)} \sum_{m=m_0-p+1}^{m_0+p} \tilde{V}(\eta_m)\,\Psi(\eta - \eta_m)\,\text{sinc}\,(\pi(\eta - \eta_m)/\Delta\eta) \tag{4.50}$$

wherein $m_0 = \text{Int}\,(\eta/\Delta\eta)$ is the index of the sample closest (on the left) to the point Q, $2p$ is the number of the retained closest reduced voltages samples $\tilde{V}(\eta_m) = V(\eta_m)\,e^{j\psi(\eta_m)}$, and

$$\eta_m = m\Delta\eta = \frac{m\pi}{\chi\chi'\beta a} \tag{4.51}$$

In relation (4.50), $\Psi(\eta)$ is the Knab's Sampling Window function (2A.33), wherein $\bar{\eta} = p\Delta\eta$ and $v = 1 - 1/\chi$.

The OSI expansion (4.50) can be employed to evaluate the intermediate voltages samples at the intersection points between the spiral and the radial line through the observation point P. The voltage at P can be, then, determined [304] by applying the OSI expansion:

$$V(s) = e^{-j\beta r} \sum_{n=n_0-q+1}^{n_0+q} \tilde{V}(s_n)\,\Psi(s - s_n)\,\text{sinc}\,(\pi(s - s_n)/\Delta s) \tag{4.52}$$

where $n_0 = \text{Int}\,[(s - s_0)/\Delta s]$, $s = \rho/d$, $\tilde{V}(s_n) = V(s_n)\,e^{j\beta r(s_n)}$ are the intermediate reduced voltage samples, $s_n = s_n(\varphi) = k\varphi/d + n\Delta s = s_0 + n\Delta s$ are their normalized abscissas, $\bar{s} = q\Delta s$, and the other symbols have the same meanings as in (4.50). It is so possible to accurately reconstruct the NF data required by the classical NF–FF transformation with PR scanning [26] from the NR samples gathered on the spiral.

The interested reader can find numerical and experimental tests assessing the effectiveness of the described NF–FF transformation in [304].

4.8.2 Non-redundant NF–FF transformation with planar spiral scanning: the double bowl modelling case

Let a quasi-planar AUT be considered as contained in a double bowl Σ and let the voltage be acquired by an electrically small, non-directive, probe scanning a planar spiral lying on a plane at distance d from the AUT centre (see Figure 4.52). The unified theory of spiral scans for non-spherical antennas [124] (see Sect. 3.2) can be properly exploited to develop a NR sampling representation, which allows the efficient recovery of the NF data needed by the classical probe-compensated NF–FF transformation with PR scanning [26] from the NR voltage samples acquired along the spiral [130].

The spiral equations (3.23), by taking into account that $r(\theta) = d/\cos\theta$ in the case of a spiral on the plane $z = d$, become:

$$\begin{cases} x = d\tan\theta(\xi)\cos\phi = \bar{\rho}(\xi)\cos\phi \\ y = d\tan\theta(\xi)\sin\phi = \bar{\rho}(\xi)\sin\phi \\ z = d \end{cases} \tag{4.53}$$

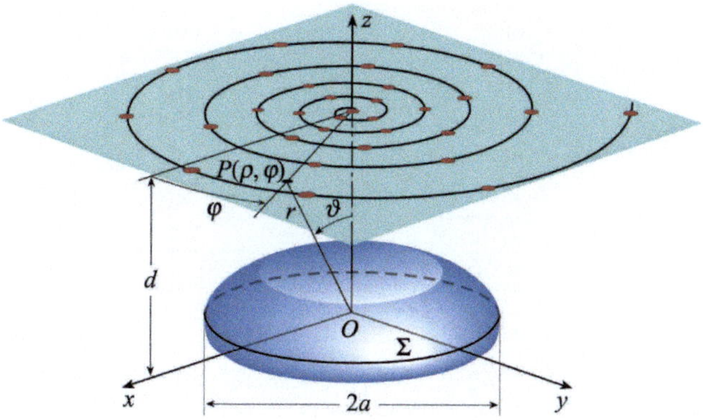

Figure 4.52 *Planar spiral scanning: double bowl modelling*

where ϕ is the angular parameter describing the spiral and $\xi = k\phi$. It must be observed that $\bar{\rho}$ and θ become negative when ϕ is less than zero, whereas the radial distance ρ and the zenithal angle ϑ are always positive quantities. Moreover, the angular parameter ϕ is always continuous, whereas the angle φ exhibits a discontinuity jump of π in correspondence of the origin, as can be easily verified from (4.53). The constant k is determined by imposing that the spiral step $2\pi k$ be equal to the sample spacing $\Delta\xi$ required for interpolating along a radial line, which, according to (2.59), is given by $\Delta\xi = 2\pi/(2N'' + 1)$. Hence, it follows that $k = 1/(2N'' + 1)$.

It must be stressed that the use of the double bowl modelling, instead of the spherical one [127, 128], allows one to reduce the distance of the scanning plane, thus giving rise to a significant mitigation of the truncation error for a given size of the scanning area. Moreover, the reduced volumetric redundancy of this model allows a considerable reduction of the number of needed NF measurements and related acquisition time.

It is noteworthy that the spiral can be viewed as attained by projecting on the scanning plane, through the curves at ξ = constant (see Figure 4.53), the spiral wrapping, with the same step, the double bowl Σ.

According to the unified theory of spiral scans for non-spherical antennas [124] (see Sect. 3.2), the NR sampling representation of the probe voltage from its samples acquired on the planar spiral, whose step is equal to the sample spacing required for the interpolation along a radial line, can be achieved by developing a NR sampling representation along this spiral.

The bandwidth W_ξ, the phase function γ and the optimal parameter ξ for describing a radial line are given by the expressions (4.32), where the related parameters can be determined as described in Sect. 4.5.2.

The NR sampling representation along the spiral can be obtained by applying the heuristic procedure [124] described in Sect. 3.2, which parallelizes the rigorous procedure valid in the spherical AUT modelling case [122]. Accordingly, the parameter η for describing the spiral is β/W_η times the arclength of the projecting point lying on the spiral which wraps the double bowl Σ, the related phase function ψ coincides with the phase function γ relevant to a radial line, and W_η is β/π times the length of the spiral wrapping Σ from the north to the south pole.

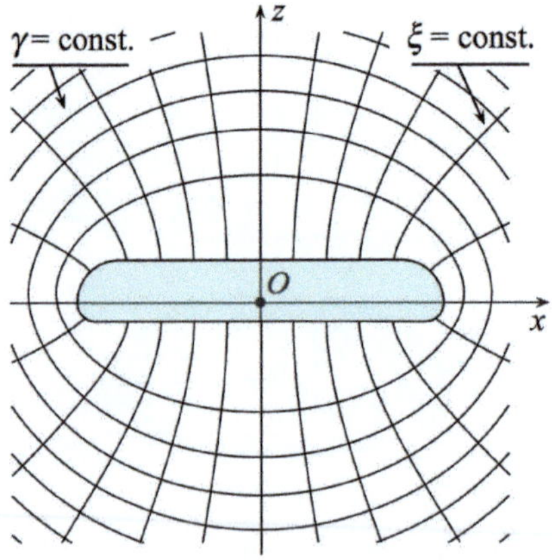

Figure 4.53 *Double bowl modelling: curves at ξ = constant and γ = constant.*

The probe and rotated probe voltages at the points lying in the scanning area and, accordingly, at the points wherein the NF data to perform the classical NF–FF transformation with PR scan [26] are required, can be efficiently reconstructed from the acquired samples by applying the two-dimensional OSI expansion (3.24), here reported for reader's convenience:

$$V(\xi(\vartheta), \varphi) = e^{-j\gamma(\xi)}$$

$$\times \sum_{n=n_0-q+1}^{n_0+q} \left[SF(\xi, \xi_n, \bar{\xi}, N, N'') \sum_{m=m_0-p+1}^{m_0+p} \tilde{V}(\eta_m)\, SF(\eta(\xi_n), \eta_m, \bar{\eta}, M, M'') \right] \tag{4.54}$$

where all symbols have the same meanings as in (3.24). It is noteworthy that, in the neighbourhood of the pole, the enlargement bandwidth factor χ' relevant to the representation along the spiral must be suitably increased to avoid a significant growth of the bandlimitation error in such a zone [124, 130].

It must be stressed that, in order to perform the probe compensation, the probe must generally co-rotate to maintain the AUT orientation, as requested for the validity of the probe-compensated NF–FF transformation formulas (4.21) and (4.22). As in the PP scanning case, a probe exhibiting only a first-order azimuthal dependence in its radiated far field can be used without co-rotation, since the voltages V_p and V_r can be evaluated from the knowledge of the voltages V_φ and V_ρ through the relations (4.34).

The effectiveness of the described technique is assessed by some numerical results shown in the following. The simulations are relevant to the previously considered circular array and the chosen probe is again an open-ended WR-90 rectangular waveguide working at 10 GHz, which scans a planar spiral covering a circular zone of radius 71 λ on a plane 8 λ away from the AUT. Such an array has been modelled by a double bowl with $a = 17\,\lambda$, $c = 2\,\lambda$, and $c' = 1.5\,\lambda$. A representative reconstruction example of the amplitude and phase of the voltage V_ρ on the radial line at $\varphi = 90°$ is shown in Figure 4.54a and b. As can be observed, an excellent agreement between the reconstructed voltage and the exact one occurs. It is noteworthy that, to control the aliasing error, the enlargement bandwidth factor χ' has been suitably increased in the zone of the spiral corresponding to the 32 samples nearby the pole in such a way that the sample spacing is reduced by a factor 9. The algorithm stability with respect to errors affecting the samples has been investigated by corrupting the exact samples with random errors. The simulated errors are a background noise (bounded to Δa dB in amplitude and with arbitrary phase) and uncertainties on the data of $\pm\Delta a_r$ dB in amplitude and $\pm\Delta\,\phi$ degrees in phase. As shown in Figure 4.55, the algorithm is stable. The accuracy of the two-dimensional OSI formula (4.54) is quantitatively assessed by the plots of the maximum and mean-square errors in the reconstruction of the voltage V_ρ

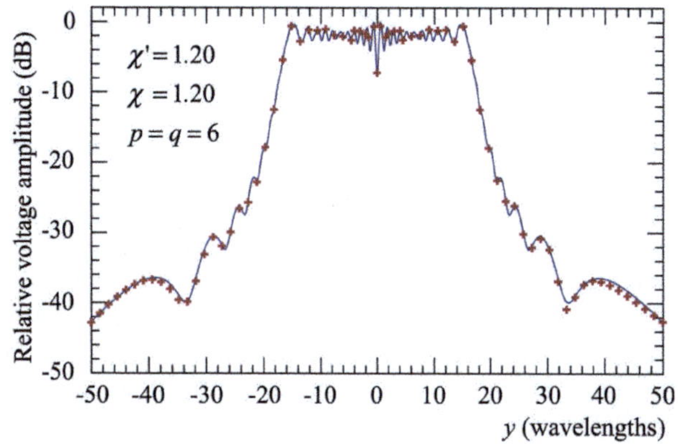

(a)

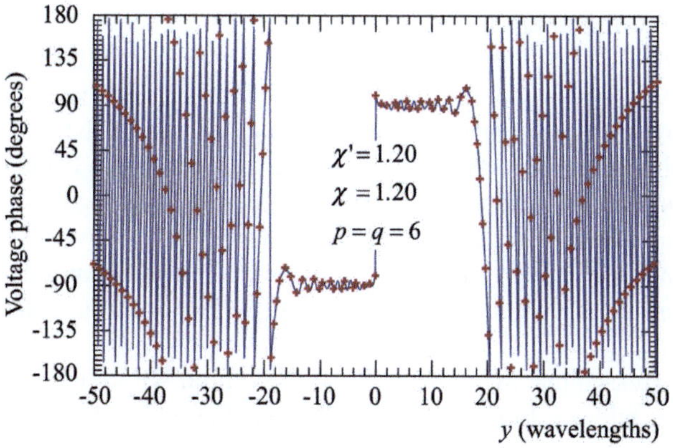

(b)

Figure 4.54 *Probe voltage V_ρ on the radial line at $\varphi = 90°$. Solid line: exact. Crosses: reconstructed from the NF samples acquired along the spiral. (a) Amplitude, (b) Phase.*

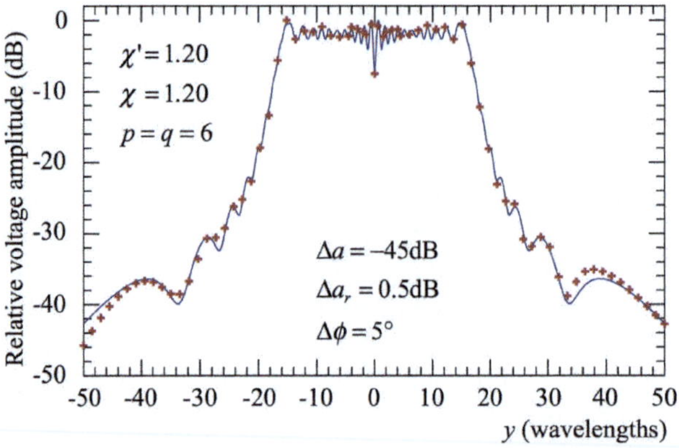

Figure 4.55 *Amplitude of the probe voltage V_ρ on the radial line at $\varphi = 90°$. Solid line: exact. Crosses: interpolated from error affected samples.*

(normalized to its maximum value on the plane) shown in Figure 4.56a and b. The evaluation has been accomplished by comparing the interpolated values of V_ρ with those directly computed on a close grid in the central zone of the scanning area in order to ensure the availability of the required guard samples. As can be observed, these errors decrease up to very low values on increasing the number of the considered closest samples and/or the oversampling factor. The OSI algorithm (4.54) and the relations (4.34) have been finally applied to recover the massive input NF data for the classical NF–FF transformation with PR scan [26] from the NR samples gathered along the spiral. The reconstruction of the antenna E-plane FF pattern is shown in Figure 4.57. As can be seen, the exact and recovered patterns are practically indistinguishable, thus further confirming the effectiveness of such a NF–FF transformation technique.

As previously stated, by using the double bowl modelling instead of the spherical one, it is possible to reduce the distance of the scanning plane, thus significantly reducing the truncation error for a given size of the scanning area. In order to highlight such a capability, the reconstruction of the E-plane FF pattern attained by employing the spherical modelling is shown in Figure 4.58. In such a case, the NF samples have been gathered on a spiral which covers a circular zone with the same radius, but lying now on a plane 20 λ away from the AUT. As can be clearly seen, the FF reconstruction results to be less accurate in the far out side lobe region.

It can be stressed that the number of the used spiral NF samples in the reported reconstructions relevant to the double bowl modelling is 8654 (8398 regular +256 extra samples) and, accordingly, is remarkably lower than those 126665 and 40401 required by the PP NF–FF transformation [32–34] and by the classical PR one [26], respectively.

The interested reader can find in [130] experimental results which assess the effectiveness of the described NF–FF transformation with planar spiral scanning also from the practical viewpoint.

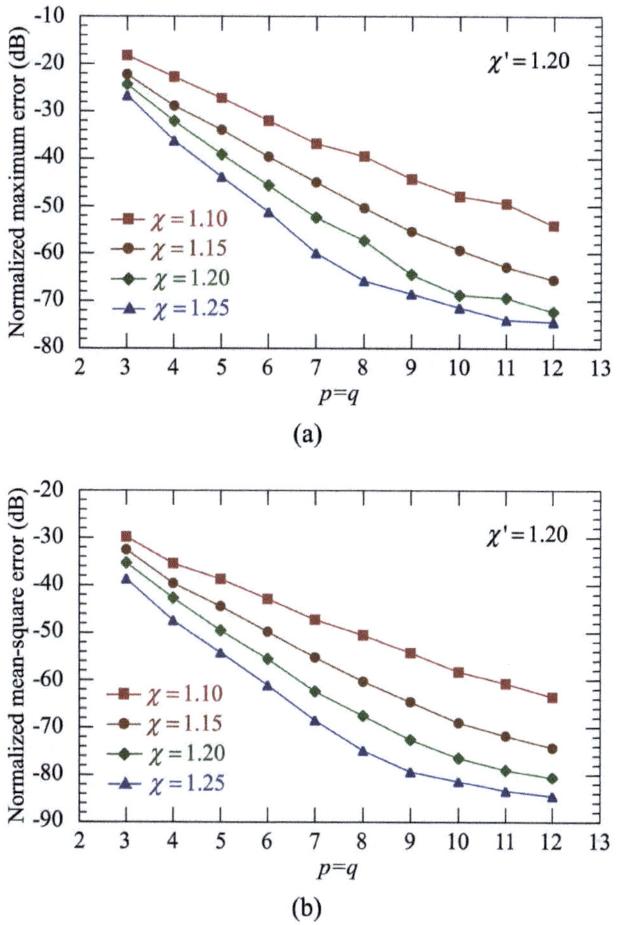

Figure 4.56 *Normalized errors in the reconstruction of the probe voltage V_ρ. (a) Maximum error, (b) Mean-square error.*

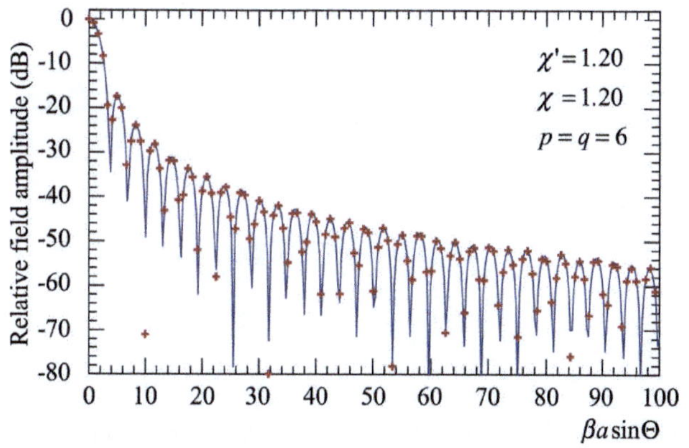

Figure 4.57 *E-plane pattern. Solid line: exact field. Crosses: reconstructed from the NF samples acquired along the spiral.*

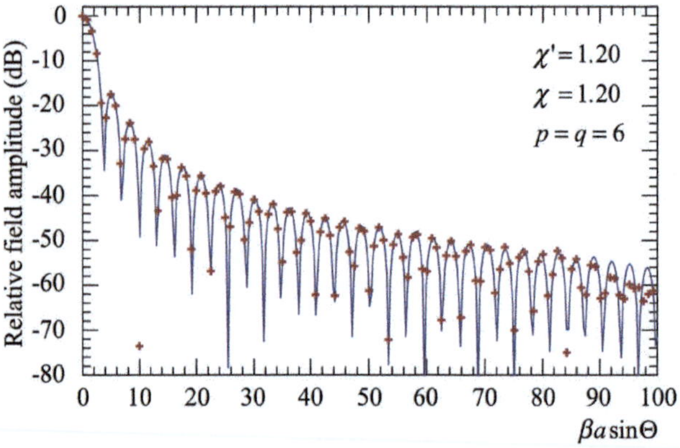

Figure 4.58 E-plane pattern. Solid line: exact field. Crosses: reconstructed from the NF samples acquired along the spiral by using the spherical modelling.

4.8.3 Non-redundant NF–FF transformation with planar spiral scanning: the oblate spheroidal modelling case

Let a quasi-planar AUT be modelled by an oblate spheroid Σ and let the voltage be measured by a non-directive, electrically small, probe which is scanning a planar spiral on a plane at distance d from the AUT centre (Figure 4.59). As in the previous subsection, the NR sampling representation, which allows the accurate reconstruction of the NF data required by the classical PR NF–FF transformation [26] from the NR voltage samples collected along the spiral, can be developed [129, 130] by suitably applying the unified theory of spiral scans for non-spherical antennas [124] (see Sect. 3.2). It is clear that also the use an oblate spheroid instead of a sphere as AUT modelling allows a significant mitigation of the truncation error for a given size of the scanning zone and a remarkable reduction in the number of required NF data and related measurement time.

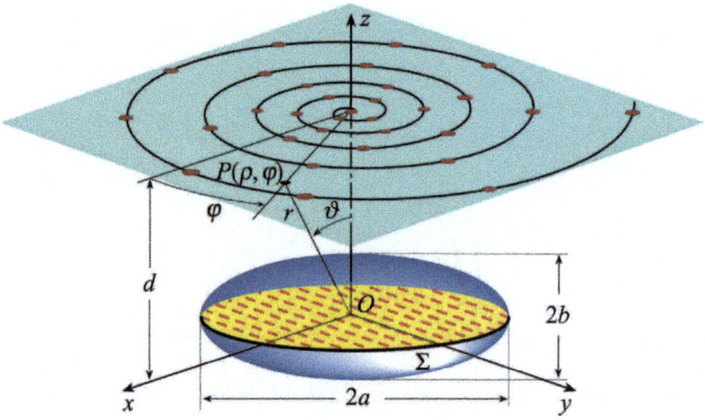

Figure 4.59 Planar spiral scanning: oblate spheroidal modelling

In the oblate spheroidal modelling case, the development of the probe voltage representation from its NR samples collected along the spiral is quite similar to that previously described for the double bowl modelling, so that only the differences will be underlined in the following. The spiral can be viewed as attained by projecting on the scanning plane, through the hyperbolas at ξ = constant (see Figure 2.7), the spiral wrapping, with the same step, the modelling oblate spheroid Σ. The bandwidth W_ξ, the phase function γ and the optimal parameter ξ for describing a radial line are given by the expressions (2.25), (2.27) and (2.28b), respectively.

The OSI formula (4.54) can be again applied for the accurate reconstruction of the probe and rotated probe voltages at any point lying in the scanning zone and, in particular, at those needed by the classical PR NF–FF transformation [26]. It is worth noting that, also in this case, the enlargement bandwidth factor χ', relevant to the representation along the spiral, must be properly increased nearby the pole to prevent a significant growth of the bandlimitation error in this zone [124, 130].

The experimental validation of this NF–FF transformation with planar spiral scan has been performed by means of the flexible NF measurement system available in the UNISA Antenna Characterization Lab, arranged in the PP scanning configuration (see Sect. 1.2). The AUT is the already considered E-plane monopulse antenna working at 10 GHz in the sum mode and the NF samples have been gathered by the employed probe, an open-ended WR90 rectangular waveguide, along a spiral spanning a circle of radius 110 cm on a plane at distance d = 16.5 cm from the horns apertures. An oblate spheroid with semi-axes a = 18.6 cm and b = 8.3 cm has been adopted as surface Σ modelling such an AUT.

In order to assess the accuracy of the OSI algorithm, the amplitude and phase of the recovered probe voltage V_ρ on the radial line at $\varphi = 90°$ are compared in Figure 4.60a and b with those directly measured at greater resolution. The reconstructions of the amplitude of the voltage V_φ on the radial lines at $\varphi = 30°$ and $\varphi = 0°$ are also shown in Figures 4.61 and 4.62. As can be clearly observed, there is a very good agreement between the reconstructed voltages and the measured ones, save for the zones characterized by very low voltage values, where the recovered voltages behaviour is smoother, due to the already mentioned filtering properties of the OSI functions. All these reconstructions have been got by using $\chi = 1.25$ and $p = q = 7$, which ensure a truncation error smaller than the measurement one [129]. As regards the choice of enlargement bandwidth factor χ', to guarantee a negligible aliasing error, a χ' value equal to 1.35 has been used save for the zone of the spiral corresponding to the 24 samples nearby the pole, wherein this factor has been further increased in such a way that the sample spacing is reduced by a factor 9.

At last, the OSI formula (4.54) and the relations (4.34) have been used to recover the 11025 PR NF data spaced by 0.45λ on a 140 cm × 140 cm sized square from the NF samples acquired along the spiral. The FF patterns in the E- and H-planes reconstructed from these data by applying the classical PR NF–FF transformation [26] are compared in Figure 4.63a and b with those obtained from the PR NF data

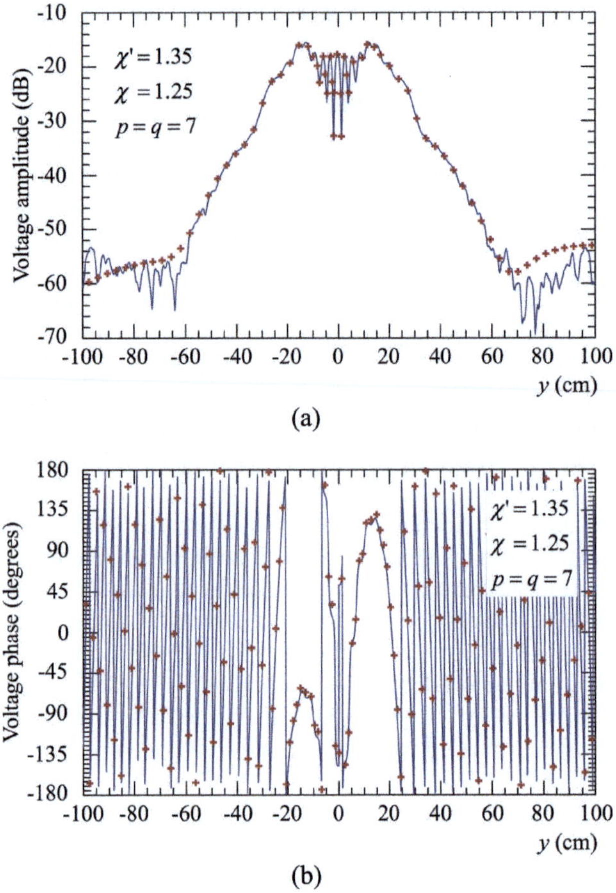

Figure 4.60 *Probe voltage V_ρ on the radial line at φ = 90°. Solid line: measured. Crosses: obtained by interpolating the spiral NF samples. (a) Amplitude, (b) Phase.*

directly measured on the same grid. As already observed in the PP and BP scannings, the agreement in the E-plane is very good, whereas a less accurate reconstruction occurs in the H-plane. This, as already stressed, is due to the fact that the far field radiated by an open-ended rectangular waveguide excited by a TE_{10} mode has only approximately a first-order azimuthal dependence [303]. On the contrary, when the spiral NF samples are acquired by using the hardware co-rotation, the reconstruction results to be very accurate (see Figure 4.64) also in the H-plane.

It is worthy to point out that the number of the NF samples acquired along the spiral is 2053 (1861 regular +192 extra samples) and, accordingly, remarkably less than that (33581) required by the PP NF–FF transformations [32–34] to cover the same scanning area.

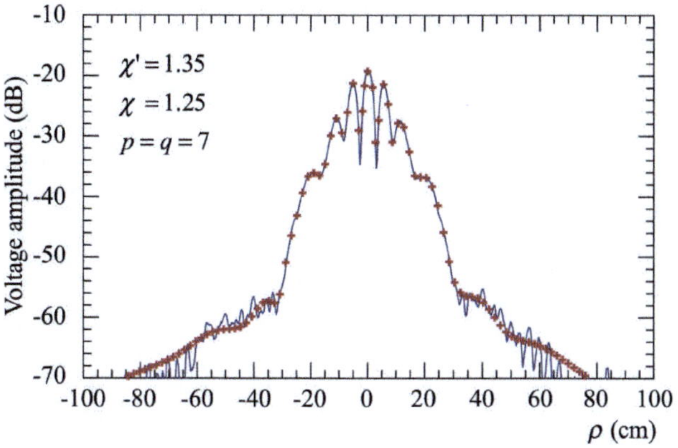

Figure 4.61 *Amplitude of the probe voltage V_φ on the radial line at $\varphi = 30°$. Solid line: measured. Crosses: obtained by interpolating the spiral NF samples.*

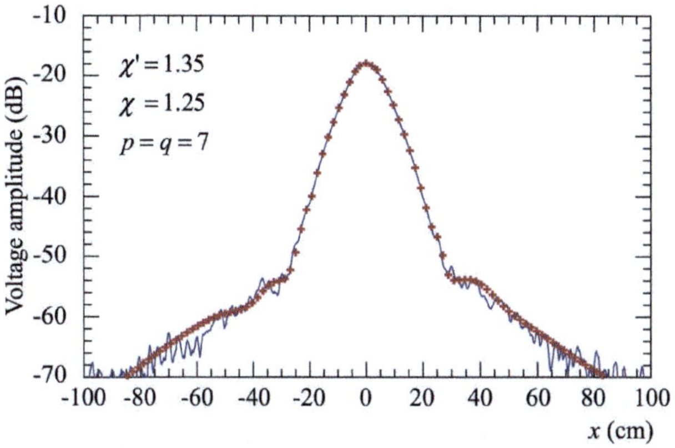

Figure 4.62 *Amplitude of the probe voltage V_φ on the radial line at $\varphi = 0°$. Solid line: measured. Crosses: obtained by interpolating the spiral NF samples.*

Appendix

4A.1 Plane wave expansion

The plane wave expansion (PWE) [16, 305] of the EM field radiated by an antenna will be briefly summarized in this appendix. As will be shown, the EM field $\boldsymbol{E}$, $\boldsymbol{H}$ of a monochromatic wave radiated by an antenna can be represented as a superposition of plane waves with different amplitudes which propagate in different directions.

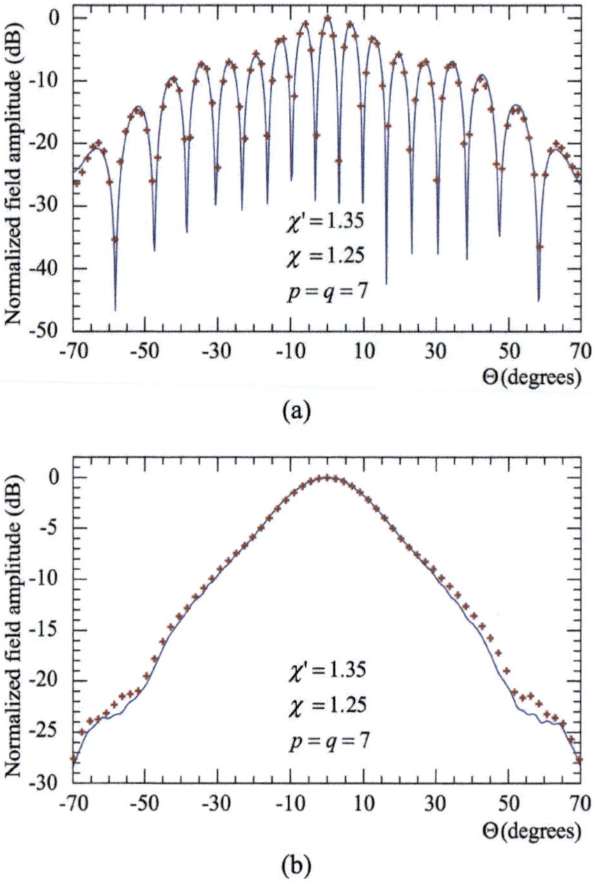

Figure 4.63 *FF patterns in the principal planes. Solid line: reconstructed from the plane-rectangular NF data. Crosses: reconstructed from the spiral NF samples with software co-rotation. (a) E-plane, (b) H-plane.*

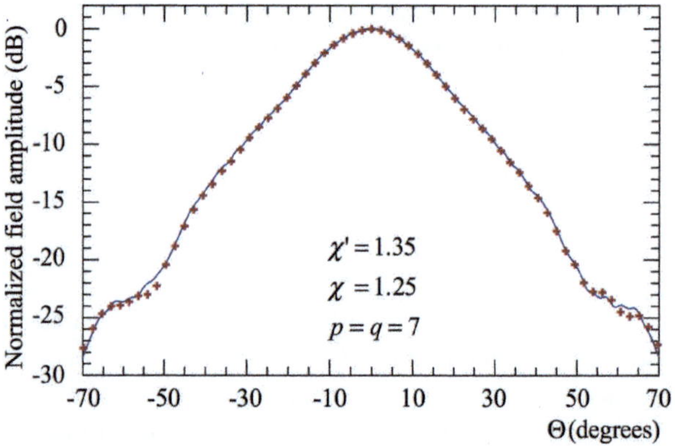

Figure 4.64 *H-plane pattern. Solid line: reconstructed from the plane-rectangular NF data. Crosses: reconstructed from the spiral NF samples with hardware co-rotation.*

Let the considered medium be linear, homogeneous, and isotropic. In such a medium, an EM field

$$\boldsymbol{E} = \hat{\boldsymbol{E}}\, \mathrm{e}^{-\mathrm{j}\boldsymbol{k}\cdot\boldsymbol{r}}; \qquad \boldsymbol{H} = \hat{\boldsymbol{H}}\, \mathrm{e}^{-\mathrm{j}\boldsymbol{k}\cdot\boldsymbol{r}} \tag{4A.1}$$

with $\hat{\boldsymbol{E}}$, $\hat{\boldsymbol{H}}$ constant complex vectors, represents a time-harmonic plane wave that propagates in the direction of the vector $\boldsymbol{k} = k_x\hat{x} + k_y\hat{y} + k_z\hat{z}$.

It is a simple task to verify that the operator ∇ transforms into $-\,\mathrm{j}\boldsymbol{k}$ for such a field. As a matter of fact, by taking into account the vector identity $\nabla \times (\boldsymbol{a}\,\phi) = \nabla\phi \times \boldsymbol{a} + \phi\nabla \times \boldsymbol{a}$ (where the symbol $(\times)$ indicates the vector product), it results:

$$\begin{aligned} \nabla \times \boldsymbol{E} = \nabla\,(\mathrm{e}^{-\mathrm{j}\boldsymbol{k}\cdot\boldsymbol{r}}) \times \hat{\boldsymbol{E}} &= \left(\hat{x}\frac{\partial}{\partial x}\mathrm{e}^{-\mathrm{j}\boldsymbol{k}\cdot\boldsymbol{r}} + \hat{y}\frac{\partial}{\partial y}\mathrm{e}^{-\mathrm{j}\boldsymbol{k}\cdot\boldsymbol{r}} + \hat{z}\frac{\partial}{\partial z}\mathrm{e}^{-\mathrm{j}\boldsymbol{k}\cdot\boldsymbol{r}}\right) \times \hat{\boldsymbol{E}} \\ &= -\mathrm{j}\,\boldsymbol{k} \times \hat{\boldsymbol{E}}\,\mathrm{e}^{-\mathrm{j}\boldsymbol{k}\cdot\boldsymbol{r}} \end{aligned} \tag{4A.2}$$

since $\nabla \times \hat{\boldsymbol{E}} = 0$. Accordingly, the Maxwell's equations (A.1) in a source-free region become:

$$\boldsymbol{k} \times \hat{\boldsymbol{E}} = \omega\,\mu\,\hat{\boldsymbol{H}} \tag{4A.3a}$$

$$\boldsymbol{k} \times \hat{\boldsymbol{H}} = -\omega\,\varepsilon\,\hat{\boldsymbol{E}} \tag{4A.3b}$$

$$\boldsymbol{k} \cdot \hat{\boldsymbol{E}} = 0 \tag{4A.3c}$$

$$\boldsymbol{k} \cdot \hat{\boldsymbol{H}} = 0 \tag{4A.3d}$$

From (4A.3a), it results that $\hat{\boldsymbol{H}}$ is related to $\hat{\boldsymbol{E}}$ by:

$$\hat{\boldsymbol{H}} = \frac{1}{\omega\,\mu}\,\boldsymbol{k} \times \hat{\boldsymbol{E}} \tag{4A.4}$$

so that $\hat{\boldsymbol{E}}$ and $\hat{\boldsymbol{H}}$ are not independent. Moreover, only two components of $\hat{\boldsymbol{E}}$ are independent. In fact, from (4A.3c), it results:

$$\hat{E}_z = -(\hat{E}_x k_x + \hat{E}_y k_y)/k_z \tag{4A.5}$$

From (4A.3a) and (4A.3b), it follows that

$$\boldsymbol{k} \times (\boldsymbol{k} \times \hat{\boldsymbol{E}}) = \omega\,\mu\,(\boldsymbol{k} \times \hat{\boldsymbol{H}}) = -\omega^2\,\varepsilon\,\mu\,\hat{\boldsymbol{E}} \tag{4A.6}$$

This last, by taking into account the vector identity $\boldsymbol{a} \times (\boldsymbol{b} \times \boldsymbol{c}) = \boldsymbol{b}\,(\boldsymbol{a} \cdot \boldsymbol{c}) - \boldsymbol{c}\,(\boldsymbol{a} \cdot \boldsymbol{b})$ and that $\boldsymbol{k} \cdot \hat{\boldsymbol{E}} = 0$, can be rewritten as:

$$\hat{\boldsymbol{E}}\,[(\boldsymbol{k} \cdot \boldsymbol{k}) - \omega^2\,\varepsilon\,\mu] = 0 \tag{4A.7}$$

It is evident that such an equation has a solution different from the trivial one $\hat{\boldsymbol{E}} = 0$ only if

$$\boldsymbol{k} \cdot \boldsymbol{k} = k_x^2 + k_y^2 + k_z^2 = \omega^2\,\varepsilon\,\mu = \beta^2 \tag{4A.8}$$

Accordingly, given the EM characteristics ε, μ of the medium, only two components of the vector $\boldsymbol{k}$, say k_x and k_y, are independent at a fixed frequency, the third component being related to them through the relation:

$$k_z = \begin{cases} \left(\beta^2 - k_x^2 - k_y^2\right)^{1/2}, & \text{if } k_x^2 + k_y^2 \le \beta^2 \\ -\mathrm{j}\left(k_x^2 + k_y^2 - \beta^2\right)^{1/2}, & \text{if } k_x^2 + k_y^2 > \beta^2 \end{cases} \tag{4A.9}$$

the choice of the minus sign in the second of (4A.9) is due to the need to guarantee that the plane wave (4A.1) is bounded at infinity.

Since the medium and the field equations are linear, the most general solution for the EM field in a homogeneous source-free region can be constructed as linear combination over all k_x and k_y of the plane waves (4A.1):

$$\boldsymbol{E}(x,y,z) = \int_{-\infty}^{+\infty}\int_{-\infty}^{+\infty} \hat{\boldsymbol{E}}(k_x,k_y)\,\mathrm{e}^{-\mathrm{j}(k_x x + k_y y + k_z z)}\,\mathrm{d}k_x\,\mathrm{d}k_y \tag{4A.10}$$

$$\boldsymbol{H}(x,y,z) = \int_{-\infty}^{+\infty}\int_{-\infty}^{+\infty} \frac{1}{\omega\,\mu}\,\boldsymbol{k}\times\hat{\boldsymbol{E}}(k_x,k_y)\,\mathrm{e}^{-\mathrm{j}(k_x x + k_y y + k_z z)}\,\mathrm{d}k_x\mathrm{d}k_y \tag{4A.11}$$

The complex amplitude function $\hat{\boldsymbol{E}}(\boldsymbol{k})$ is the so called plane wave spectrum of the field. It can be analytically determined if the tangential components E_x and E_y of the electric field $\boldsymbol{E}$ are known on a plane at $z = d$. As a matter of fact, from (4A.10), it results

$$E_{x,y}(x,y,d) = \int_{-\infty}^{+\infty}\int_{-\infty}^{+\infty} \left[\hat{E}_{x,y}(k_x,k_y)\,\mathrm{e}^{-\mathrm{j}k_z d}\right]\mathrm{e}^{-\mathrm{j}(k_x x + k_y y)}\,\mathrm{d}k_x\mathrm{d}k_y \tag{4A.12}$$

According to (4A.12), the tangential components E_x and E_y of the electric field are related to the corresponding ones of the spectrum by a double Fourier transform. Thus, by Fourier inverse transforming relation (4A.12), it results:

$$\hat{E}_{x,y}(k_x,k_y) = \frac{1}{4\pi^2}\mathrm{e}^{\mathrm{j}k_z d}\int_{-\infty}^{+\infty}\int_{-\infty}^{+\infty} E_{x,y}(x,y,d)\,\mathrm{e}^{\mathrm{j}(k_x x + k_y y)}\mathrm{d}x\mathrm{d}y \tag{4A.13}$$

The other component $\hat{E}_z$ of the spectrum can be found by means of (4A.5) and thus the plane wave spectrum of the field is fully determined.

According to relations (4A.10) and (4A.11), the EM field can therefore be expressed in a source-free region as a superposition of plane waves propagating in different directions.

Let the circumference Γ of equation $k_x^2 + k_y^2 = \beta^2$ be considered, it divides the plane $k_x k_y$ in two regions, an internal region G, wherein $k_x^2 + k_y^2 < \beta^2$, and an external one CG, where $k_x^2 + k_y^2 > \beta^2$. According to (4A.9), k_x, k_y, and k_z assume real values in G, whereas in CG, k_x and k_y are real and k_z is imaginary. Therefore, a point in

G is representative of a plane wave which propagates in a real direction (ϑ, φ) and contributes to the radiated field, whereas the plane waves relevant to points in CG are evanescent waves, which formally propagate in complex directions (of the invisible space) and decay exponentially on increasing z. These evanescent waves, obviously, do not contribute to the radiated far field, but only to the reactive near field.

4A.2 Relation between the antenna far field and the plane wave spectrum

This appendix is devoted to the determination of the relation between the field radiated in the FF region by an antenna and the plane wave spectrum. To this end, it is convenient to rewrite (4A.10) in the form:

$$\boldsymbol{E}(r, \vartheta, \varphi) = \int_{-\infty}^{+\infty}\int_{-\infty}^{+\infty} \hat{\boldsymbol{E}}(k_x, k_y)\, \mathrm{e}^{-\mathrm{j}(k_x \sin\vartheta\cos\varphi + k_y \sin\vartheta\sin\varphi + k_z\cos\vartheta)r}\mathrm{d}k_x \mathrm{d}k_y \qquad (4A.14)$$

By introducing the direction cosines $\xi = \sin\vartheta\cos\varphi$, $\eta = \sin\vartheta\sin\varphi$, and $\gamma = \cos\vartheta$ of the unit vector $\hat{r}$ and those $\xi' = k_x/\beta$, $\eta' = k_y/\beta$, and $\gamma' = k_z/\beta$ of the propagation vector $\boldsymbol{k}$, the expression (4A.14) becomes:

$$\boldsymbol{E}(r, \xi, \eta) = \beta^2 \int_{-\infty}^{+\infty}\int_{-\infty}^{+\infty} \hat{\boldsymbol{E}}(\xi', \eta')\, \mathrm{e}^{\mathrm{j}\beta r \psi(\xi, \eta, \xi', \eta')}\mathrm{d}\xi'\mathrm{d}\eta' \qquad (4A.15)$$

wherein

$$\psi(\xi, \eta, \xi', \eta') = -(\xi\xi' + \eta\eta' + \gamma\gamma') \qquad (4A.16)$$

At great distance ($\beta r \gg 1$) from the antenna, the integral in (4A.15) can be asymptotically evaluated by means of the stationary phase method [18, 306]. According to such a method (see Appendix A.4) an integral of the form

$$I(\Omega) = \int_{-\infty}^{+\infty}\int_{-\infty}^{+\infty} g(x, y)\, \mathrm{e}^{\mathrm{j}\Omega f(x,y)}\, \mathrm{d}x\mathrm{d}y, \qquad (4A.17)$$

where Ω is a large positive parameter, $g(x, y)$ is a regular real or complex function of the real variables x, y, and $f(x, y)$ is a regular real function, can be asymptotically evaluated in the hypothesis that $f(x, y)$ has an isolated first order stationary phase point at x_s, y_s, namely

$$\left.\frac{\partial f}{\partial x}\right|_{\substack{x=x_s\\y=y_s}} = f_x'(x_s, y_s) = 0; \qquad \left.\frac{\partial f}{\partial y}\right|_{\substack{x=x_s\\y=y_s}} = f_y'(x_s, y_s) = 0$$

but

$$\left.\frac{\partial^2 f}{\partial x^2}\right|_{\substack{x=x_s\\y=y_s}} = f_{xx}''(x_s, y_s) \neq 0; \; \left.\frac{\partial^2 f}{\partial y^2}\right|_{\substack{x=x_s\\y=y_s}} = f_{yy}''(x_s, y_s) \neq 0; \; \left.\frac{\partial^2 f}{\partial x \partial y}\right|_{\substack{x=x_s\\y=y_s}} = f_{xy}''(x_s, y_s) \neq 0$$

and the Hessian determinant

$$H[f(x_s,y_s)] = \begin{vmatrix} f''_{xx}(x_s,y_s) & f''_{xy}(x_s,y_s) \\ f''_{xy}(x_s,y_s) & f''_{yy}(x_s,y_s) \end{vmatrix} \neq 0$$

The approximated evaluation of $I(\Omega)$ for large values of Ω gives

$$I(\Omega) \underset{\Omega\to\infty}{\approx} g(x_s,y_s)\,\mathrm{e}^{\mathrm{j}\Omega f(x_s,y_s)} \frac{\mathrm{j}2\pi\delta}{\Omega\sqrt{|H[f(x_s,y_s)]|}} \tag{4A.18}$$

where

$$\delta = \begin{cases} 1 \text{ if } H[f(x_s,y_s)] > 0 \text{ and } f''_{xx}(x_s,y_s) > 0 \\ -1 \text{ if } H[f(x_s,y_s)] > 0 \text{ and } f''_{xx}(x_s,y_s) < 0 \\ -\mathrm{j} \text{ if } H[f(x_s,y_s)] < 0 \end{cases} \tag{4A.19}$$

Since

$$\frac{\partial\psi}{\partial\xi'} = -\xi + \frac{\gamma\xi'}{\sqrt{1-\xi'^2-\eta'^2}} = -\xi + \frac{\gamma\xi'}{\gamma'} \tag{4A.20a}$$

$$\frac{\partial\psi}{\partial\eta'} = -\eta + \frac{\gamma\eta'}{\sqrt{1-\xi'^2-\eta'^2}} = -\eta + \frac{\gamma\eta'}{\gamma'} \tag{4A.20b}$$

it can be easily recognized that $\psi(\xi,\eta,\xi',\eta')$ has a stationary phase point at $\xi'_s = \xi$, $\eta'_s = \eta$ and, accordingly, $\gamma'_s = \gamma$.

By differentiating relations (4A20), it results

$$\frac{\partial^2\psi}{\partial\xi'^2} = \gamma\frac{\gamma'^2+\xi'^2}{\gamma'^3};\qquad \frac{\partial^2\psi}{\partial\eta'^2} = \gamma\frac{\gamma'^2+\eta'^2}{\gamma'^3};\qquad \frac{\partial^2\psi}{\partial\xi'\partial\eta'} = \gamma\frac{\xi'\eta'}{\gamma'^3} \tag{4A.21}$$

and accordingly

$$\psi''_{\xi\xi}(\xi'_s,\eta'_s) = \frac{\gamma^2+\xi^2}{\gamma^2};\qquad \psi''_{\eta\eta}(\xi'_s,\eta'_s) = \frac{\gamma^2+\eta^2}{\gamma^2} \tag{4A.22a}$$

$$\psi''_{\xi\eta}(\xi'_s,\eta'_s) = \frac{\xi\eta}{\gamma^2};\qquad H[\psi(\xi'_s,\eta'_s)] = \frac{\gamma^2(\gamma^2+\xi^2+\eta^2)}{\gamma^4} = \frac{1}{\gamma^2} \tag{4A.22b}$$

By taking into account that

$$\psi(\xi,\eta,\xi'_s,\eta'_s) - (\xi^2+\eta^2+\gamma^2) = -1 \tag{4A.23}$$

and that $\delta = 1$, since $\psi''_{\xi\xi}(\xi'_s,\eta'_s) > 0$, the asymptotic evaluation of the integral in (4A.15) gives

$$I(\beta r) \underset{\beta r \to \infty}{\approx} \hat{\boldsymbol{E}}(\xi, \eta)\, \mathrm{e}^{-\mathrm{j}\beta r} \frac{\mathrm{j}2\pi\gamma}{\beta r} \tag{4A.24}$$

so that

$$\boldsymbol{E}(r, \xi, \eta) = 2\pi \mathrm{j}\beta \cos\vartheta\, \hat{\boldsymbol{E}}(\xi, \eta) \frac{\mathrm{e}^{-\mathrm{j}\beta r}}{r} \tag{4A.25}$$

Accordingly, the FF components of the electric field in the spherical coordinate system (r, ϑ, φ) are related to the corresponding components of the spectrum by:

$$E_r(r, \vartheta, \varphi) = 0 \tag{4A.26}$$

$$E_\vartheta(r, \vartheta, \varphi) = 2\pi\, \mathrm{j}\beta \cos\vartheta\, \hat{E}_\vartheta(\beta \sin\vartheta \cos\varphi, \beta \sin\vartheta \sin\varphi) \frac{\mathrm{e}^{-\mathrm{j}\beta r}}{r} \tag{4A.27}$$

$$E_\varphi(r, \vartheta, \varphi) = 2\pi \mathrm{j}\beta \cos\vartheta\, \hat{E}_\varphi(\beta \sin\vartheta \cos\varphi,\ \beta \sin\vartheta \sin\varphi) \frac{\mathrm{e}^{-\mathrm{j}\beta r}}{r} \tag{4A.28}$$

since, as will be shown, $\hat{E}_r = 0$.

On the other hand, the spherical components of the spectrum are related to the rectangular ones $\hat{E}_x, \hat{E}_y, \hat{E}_z$ by:

$$\hat{E}_r = \hat{E}_x \sin\vartheta \cos\varphi + \hat{E}_y \sin\vartheta \sin\varphi + \hat{E}_z \cos\vartheta \tag{4A.29}$$

$$\hat{E}_\varphi = -\hat{E}_x \sin\varphi + \hat{E}_y \cos\varphi \tag{4A.30}$$

$$\hat{E}_\vartheta = \hat{E}_x \cos\vartheta \cos\varphi + \hat{E}_y \cos\vartheta \sin\varphi - \hat{E}_z \sin\vartheta \tag{4A.31}$$

By taking into account that, according to (4A.5),

$$\hat{E}_z = -\frac{(\hat{E}_x \cos\varphi + \hat{E}_y \sin\varphi) \sin\vartheta}{\cos\vartheta} \tag{4A.32}$$

and substituting this last relation into (4A.29) and (4A.31), it results

$$\hat{E}_r = 0 \tag{4A.33}$$

$$\hat{E}_\vartheta = \frac{(\hat{E}_x \cos\varphi + \hat{E}_y \sin\varphi)}{\cos\vartheta} \tag{4A.34}$$

4A.3 Relevant to the probe-compensated PR NF–FF transformation

In this appendix, a very simple derivation of the formulas for the PR NF–FF transformation with probe compensation, which is valid when the mutual coupling effects between the probe and the AUT are negligible and the probe is reciprocal, is provided.

By expressing the electric field radiated by the AUT and incident on the probe as superposition of elementary plane waves (see (4A.10)), it can be easily verified that the open-circuit voltage received by the probe can expressed as:

$$V(x,y,d) = \int_{-\infty}^{+\infty}\int_{-\infty}^{+\infty} \hat{\boldsymbol{E}}\left(k_x, k_y\right) \cdot \boldsymbol{h}^r(\vartheta', \varphi')\, \mathrm{e}^{-\mathrm{j}k_z d}\, \mathrm{e}^{-\mathrm{j}(k_x x + k_y y)} \mathrm{d}k_x \mathrm{d}k_y \tag{4A.35}$$

where $\boldsymbol{h}^r(\vartheta', \varphi')$ is the receiving effective length [16, 298] of the probe relevant to the direction under which it sees the elementary plane wave $\hat{\boldsymbol{E}}(k_x, k_y)$.

By taking into account the relations between the AUT and the probe reference systems (see Figure 4A.1), it can be easily verified that $\vartheta' = \vartheta$, $\varphi' = -\varphi$ and, hence,

$$k_x' = \beta \sin\vartheta' \cos\varphi' = k_x; \qquad k_y' = \beta \sin\vartheta' \sin\varphi' = -k_y \tag{4A.36}$$

Therefore, relation (4A.35) can be rewritten in the form:

$$V(x,y,d) = \int_{-\infty}^{+\infty}\int_{-\infty}^{+\infty} \hat{\boldsymbol{E}}\left(k_x, k_y\right) \cdot \boldsymbol{h}\left(k_x, -k_y\right) \mathrm{e}^{-\mathrm{j}k_z d}\, \mathrm{e}^{-\mathrm{j}(k_x x + k_y y)}\, \mathrm{d}k_x \mathrm{d}k_y \tag{4A.37}$$

wherein it has been taken into account that, since the probe is reciprocal, its receiving and transmitting effective lengths coincide [16, 298], that is to say, $\boldsymbol{h}^r = \boldsymbol{h}^t = \boldsymbol{h}$.

By Fourier inverse transforming relation (4A.37), it follows that:

$$\hat{\boldsymbol{E}}\left(k_x, k_y\right) \cdot \boldsymbol{h}(k_x, -k_y) = \frac{1}{4\pi^2} \mathrm{e}^{\mathrm{j}k_z d} \int_{-\infty}^{+\infty}\int_{-\infty}^{+\infty} V\left(x,y,d\right) \mathrm{e}^{\mathrm{j}(k_x x + k_y y)}\, \mathrm{d}x\mathrm{d}y \tag{4A.38}$$

From this last, by expressing the electric field $\boldsymbol{E}'$ radiated by the probe in the FF region as function of its effective length [16, 298], namely

$$\boldsymbol{E}' = \mathrm{j}\frac{\zeta}{2\lambda r} I_0\, \mathrm{e}^{-\mathrm{j}\beta r} \boldsymbol{h} \tag{4A.39}$$

wherein ζ is the free-space intrinsic impedance and I_0 is the input current to the probe, it results:

$$\hat{\boldsymbol{E}}\left(k_x, k_y\right) \cdot \boldsymbol{E}'(k_x, -k_y) = \frac{\mathrm{j}\,\zeta I_0 \mathrm{e}^{-\mathrm{j}\beta r}}{2\lambda\, r} \frac{\mathrm{e}^{\mathrm{j}k_z d}}{4\pi^2} \int_{-\infty}^{+\infty}\int_{-\infty}^{+\infty} V\left(x,y,d\right) \mathrm{e}^{\mathrm{j}(k_x x + k_y y)}\, \mathrm{d}x\mathrm{d}y \tag{4A.40}$$

which, by taking into account (4A.25), can be rewritten in the form:

$$\boldsymbol{E}(\vartheta, \varphi) \cdot \boldsymbol{E}'(\vartheta, -\varphi) = -\frac{2\pi\beta\, \zeta I_0\, \mathrm{e}^{-\mathrm{j}2\beta r}}{2\lambda\, r^2} \frac{\mathrm{e}^{\mathrm{j}\beta d \cos\vartheta}}{4\pi^2} \cos\vartheta \int_{-\infty}^{+\infty}\int_{-\infty}^{+\infty} V\left(x,y,d\right) \mathrm{e}^{\mathrm{j}(k_x x + k_y y)}\, \mathrm{d}x\mathrm{d}y \tag{4A.41}$$

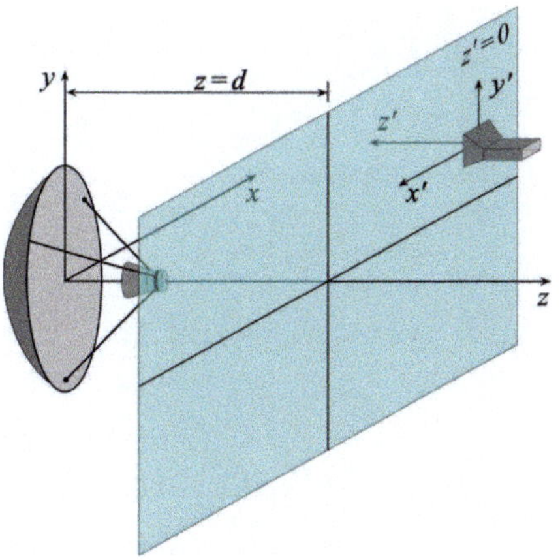

Figure 4A.1 AUT and probe reference systems

Since, as can be easily verified from Figure 4A.1, $\hat{\vartheta}' = -\hat{\vartheta}$ and $\hat{\varphi}' = \hat{\varphi}$, by putting in explicit form the scalar product in (4A.41), it follows that

$$E_\vartheta(\vartheta,\varphi)E'_\vartheta(\vartheta,-\varphi) - E_\varphi(\vartheta,\varphi)E'_\varphi(\vartheta,-\varphi)$$
$$= A\cos\vartheta\, e^{j\beta d\cos\vartheta}\int_{-\infty}^{+\infty}\int_{-\infty}^{+\infty} V(x,y,d)\, e^{j(k_x x + k_y y)}\, dx dy \quad (4A.42)$$

where

$$A = \frac{2\pi\,\beta\,\zeta\, I_0 e^{-j2\beta r}}{2\,\lambda\, r^2}\frac{1}{4\pi^2} \quad (4A.43)$$

Relation (4A.42) allows one to determine the FF spherical components E_ϑ and E_φ of the electric field radiated by the AUT from the probe-compensated NF measurements. As a matter of fact, by performing two independent sets of measurements, usually obtained rotating the probe by 90° around its longitudinal axis in the second set, the following linear system is obtained:

$$E_\vartheta(\vartheta,\varphi)E'_{\vartheta_p}(\vartheta,-\varphi) - E_\varphi(\vartheta,\varphi)E'_{\varphi_p}(\vartheta,-\varphi) = I_p \quad (4A.44a)$$

$$E_\vartheta(\vartheta,\varphi)E'_{\vartheta_r}(\vartheta,-\varphi) - E_\varphi(\vartheta,\varphi)E'_{\varphi_r}(\vartheta,-\varphi) = I_r \quad (4A.44b)$$

where I_p and I_r, related to the two-dimensional Fourier transforms of the probe and rotated probe voltages V_p and V_r, are given by:

$$I_p = A\cos\vartheta\, e^{j\beta d\cos\vartheta}\int_{-\infty}^{+\infty}\int_{-\infty}^{+\infty} V_p(x,y)\, e^{j\beta x\sin\vartheta\cos\varphi}\, e^{j\beta y\sin\vartheta\sin\varphi}\, dx dy \quad (4A.45a)$$

$$I_r = A\cos\vartheta\, \mathrm{e}^{\mathrm{j}\beta d\cos\vartheta} \int_{-\infty}^{+\infty}\int_{-\infty}^{+\infty} V_r(x,y)\, e^{\mathrm{j}\beta x\sin\vartheta\cos\varphi}\, \mathrm{e}^{\mathrm{j}\beta y\sin\vartheta\sin\varphi}\mathrm{d}x\mathrm{d}y \tag{4A.45b}$$

By solving the system (4A.44), it results:

$$E_\vartheta(\vartheta,\varphi) = \left(I_r E'_{\varphi_p}(\vartheta,-\varphi) - I_p E'_{\varphi_r}(\vartheta,-\varphi)\right)/\Delta \tag{4A.46a}$$

$$E_\varphi(\vartheta,\varphi) = \left(I_r E'_{\vartheta_p}(\vartheta,-\varphi) - I_p E'_{\vartheta_r}(\vartheta,-\varphi)\right)/\Delta \tag{4A.46b}$$

wherein

$$\Delta = E'_{\vartheta_r}(\vartheta,-\varphi)\; E'_{\varphi_p}(\vartheta,-\varphi) - E'_{\vartheta_p}(\vartheta,-\varphi)\, E'_{\varphi_r}(\vartheta,-\varphi) \tag{4A.47}$$

As already stressed in the body of this chapter, a commonly employed scanning probe is an open-ended rectangular waveguide. As shown by Yaghjian in [303], the spherical FF components of the electric field radiated by it, when a TE_{10} mode is propagating, are:

$$\begin{aligned} E_{\vartheta_p} &= f_\vartheta(\vartheta)\sin\varphi\frac{\mathrm{e}^{-\mathrm{j}\beta r}}{r} \\ &= A_E\frac{1+(k_z/\beta)\cos\vartheta}{1+(k_z/\beta)}\frac{\sin\left[\beta(b/2)\sin\vartheta\right]}{\beta(b/2)\sin\vartheta}\sin\varphi\,\frac{\mathrm{e}^{-\mathrm{j}\beta r}}{r} \end{aligned} \tag{4A.48}$$

$$\begin{aligned} E_{\varphi_p} &= f_\varphi(\vartheta)\cos\varphi\frac{\mathrm{e}^{-\mathrm{j}\beta r}}{r} = A_H\cos\left[\beta(a/2)\sin\vartheta\right] \\ &\times\left\{\frac{\cos\vartheta+(k_z/\beta)+\left[\cos\vartheta-(k_z/\beta)\right]\Gamma}{(\pi/2)^2-\left[\beta(a/2)\sin\vartheta\right]^2}+C_0\right\}\cos\varphi\frac{\mathrm{e}^{-\mathrm{j}\beta r}}{r} \end{aligned} \tag{4A.49}$$

wherein a and b are the waveguide dimensions along x and y (see Figure 4A.2),

$$A_E = A_H\left\{\frac{4}{\pi^2}\left[1+(k_z/\beta)+(1-(k_z/\beta))\,\Gamma\right]+C_0\right\} \tag{4A.50}$$

$$A_H = -\mathrm{j}b\beta^2 abE_0/8 \tag{4A.51}$$

E_0 and $k_z = [\beta^2-(\pi/a)^2]^{1/2}$ are the amplitude and the propagation constant of the TE_{10} mode, C_0 is a real constant, which can be numerically computed as shown in [303], and Γ is the reflection coefficient in correspondence of the waveguide aperture, whose measured values are provided in [303].

The FF components of the electric field radiated by the rotated probe can be easily determined by taking into account the relations between the reference system (x,y,z) of the rotated probe and that (x',y',z') of the probe, shown in Figure 4A.3. Accordingly, they can be obtained by substituting $\varphi+\pi/2$ in place of φ in the relations (4A.48) and (4A.49) relevant to the probe, thus getting:

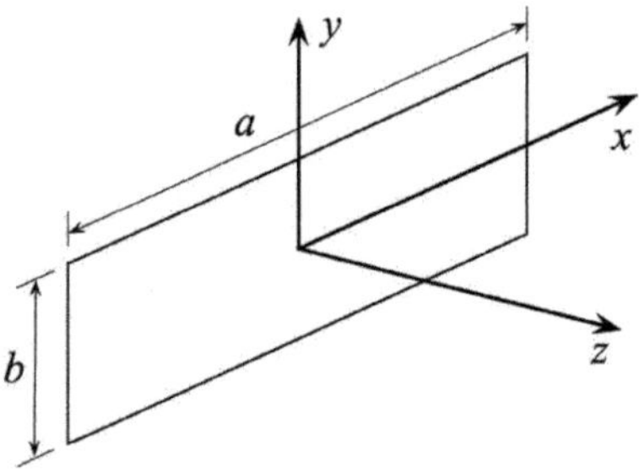

Figure 4A.2 Probe

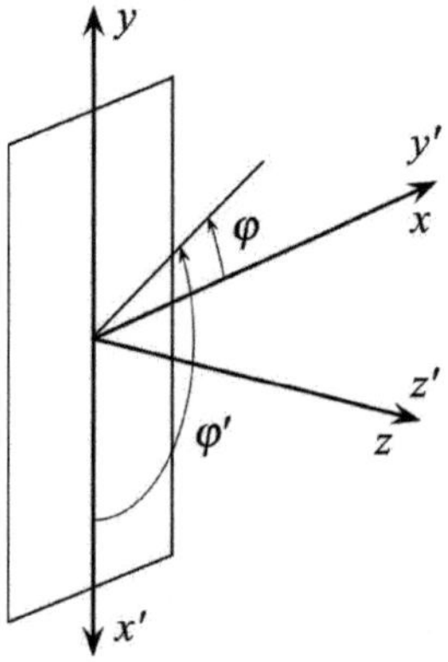

Figure 4A.3 Rotated probe

$$E_{\vartheta_r} = f_\vartheta(\vartheta)\cos\varphi\frac{e^{-j\beta r}}{r} \tag{4A.52}$$

$$E_{\varphi_r} = -f_\varphi(\vartheta)\sin\varphi\frac{e^{-j\beta r}}{r} \tag{4A.53}$$

4A.4 On the software co-rotation

In this appendix, it will be shown that a probe exhibiting only a first-order azimuthal dependence in its radiated far field, namely such that

$$E_{\vartheta_p} = f_\vartheta(\vartheta)\sin\varphi\frac{e^{-j\beta r}}{r}; \qquad E_{\varphi_p} = f_\varphi(\vartheta)\cos\varphi\frac{e^{-j\beta r}}{r} \tag{4A.54}$$

can be used without co-rotation, since the voltages V_p and V_r (measured by the probe and the rotated probe with co-rotation – see Figure 4.27) can be evaluated from the knowledge of the voltages V_φ and V_ρ (measured without co-rotation – see Figure 4.28) through the relations:

$$V_p = V_\varphi\cos\varphi - V_\rho\sin\varphi; \qquad V_r = V_\varphi\sin\varphi + V_\rho\cos\varphi \tag{4A.55}$$

For this kind of probe, as shown in the previous appendix, the FF components E_{ϑ_r}, E_{φ_r} of the electric field radiated by the rotated probe are given by relations (4A.52) and (4A.53). According to the above results and taking into account relation (4A.39), the components along ϑ and φ of the effective lengths of the probe and rotated probe can be expressed as:

$$h_{p_\vartheta}(\vartheta,\varphi) = Cf_\vartheta(\vartheta)\sin\varphi; \qquad h_{p_\varphi}(\vartheta,\varphi) = Cf_\varphi(\vartheta)\cos\varphi \tag{4A.56a}$$

$$h_{r_\vartheta}(\vartheta,\varphi) = Cf_\vartheta(\vartheta)\cos\varphi; \qquad h_{r_\varphi}(\vartheta,\varphi) = -Cf_\varphi(\vartheta)\sin\varphi \tag{4A.56b}$$

wherein $C = 2\lambda/(j\zeta I_0)$.

In order to demonstrate relations (4A.55), it is convenient to subdivide the AUT in N smaller portions such that the probe is situated in their FF region and to introduce, besides the Cartesian reference system $S(x, y, z)$ of the AUT, two additional reference systems: the reference system $S'(x', y', z')$ relevant to the probe with co-rotation and that $S''(x'', y'', z'')$ relevant to the probe without co-rotation (see Figures 4A.4 and 4A.5). By denoting with (ϑ, φ) the direction under which the probe is seen in S, with (ϑ_n, φ_n) the direction under which it is seen from the nth portion of the AUT, and with $(\vartheta'_n, \varphi'_n)$ and $(\vartheta''_n, \varphi''_n)$ the directions under which such a portion is seen from the probe in S' and S'', the following relations hold:

$$\varphi'_n = -\varphi_n; \quad \varphi''_n = \varphi - \varphi_n; \quad \vartheta''_n = \vartheta'_n = \vartheta_n \tag{4A.57}$$

According to (4.20), the open-circuit voltages received by the probe and the rotated probe with co-rotation are given by:

$$V_p = \sum_{n=1}^{N} \boldsymbol{E}^i_n \cdot \boldsymbol{h}_p(\vartheta'_n, \varphi'_n); \qquad V_r = \sum_{n=1}^{N} \boldsymbol{E}^i_n \cdot \boldsymbol{h}_r(\vartheta'_n, \varphi'_n) \tag{4A.58}$$

whereas those without co-rotation are given by:

$$V_\varphi = \sum_{n=1}^{N} \boldsymbol{E}^i_n \cdot \boldsymbol{h}_p(\vartheta''_n, \varphi''_n); \qquad V_\rho = \sum_{n=1}^{N} \boldsymbol{E}^i_n \cdot \boldsymbol{h}_r(\vartheta''_n, \varphi''_n) \tag{4A.59}$$

In relations (4A.58) and (4A.59),

$$\boldsymbol{h}_{p,r}(\vartheta'_n, \varphi'_n) = h_{p,r\vartheta}(\vartheta'_n, \varphi'_n)\,\hat{\vartheta}' + h_{p,r\varphi}(\vartheta'_n, \varphi'_n)\,\hat{\varphi}' \tag{4A.60a}$$

$$\boldsymbol{h}_{p,r}(\vartheta''_n, \varphi''_n) = h_{p,r\vartheta}(\vartheta''_n, \varphi''_n)\,\hat{\vartheta}' + h_{p,r\varphi}(\vartheta''_n, \varphi''_n)\,\hat{\varphi}' \tag{4A.60b}$$

since

$$\hat{\vartheta}'' = \hat{\vartheta}'; \qquad \hat{\varphi}'' = \hat{\varphi}' \tag{4A.61}$$

By taking into account relations (4A.56), and (4A.57), it is easy to verify that

$$h_{p\vartheta}(\vartheta'_n, \varphi'_n) = h_{p\vartheta}(\vartheta_n, -\varphi_n) = -C f_\vartheta(\vartheta_n) \sin\varphi_n \tag{4A.62a}$$

$$h_{p\varphi}(\vartheta'_n, \varphi'_n) = h_{p\varphi}(\vartheta_n, -\varphi_n) = C f_\varphi(\vartheta_n) \cos\varphi_n \tag{4A.62b}$$

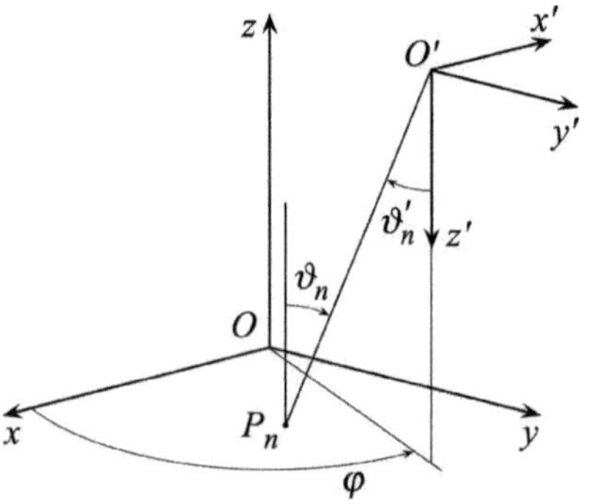

Figure 4A.4 Reference systems of the AUT and probe with co-rotation

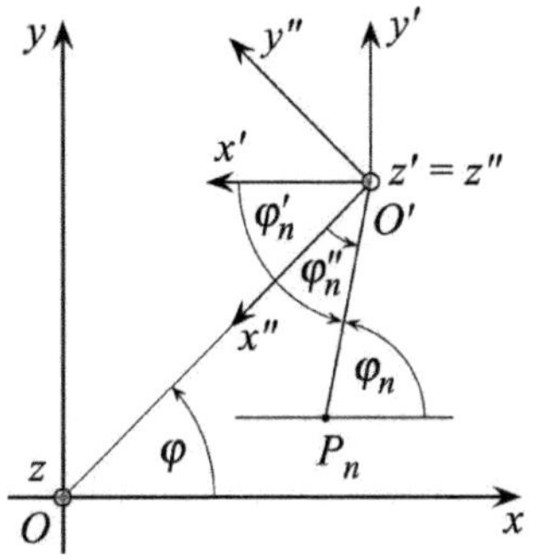

Figure 4A.5 Reference systems of the probe without and with co-rotation

$$h_{r_\vartheta}\left(\vartheta_n', \varphi_n'\right) = h_{r_\vartheta}\left(\vartheta_n, -\varphi_n\right) = C f_\vartheta\left(\vartheta_n\right)\cos\varphi_n \tag{4A.62c}$$

$$h_{r_\varphi}\left(\vartheta_n', \varphi_n'\right) = h_{r_\varphi}\left(\vartheta_n, -\varphi_n\right) = C f_\varphi\left(\vartheta_n\right)\sin\varphi_n \tag{4A.62d}$$

By making explicit the scalar products in relations (4A.58), it results:

$$V_p = C\sum_{n=1}^{N}\left(-E^i_{n_{\vartheta'}} f_\vartheta\left(\vartheta_n\right)\sin\varphi_n + E^i_{n_{\varphi'}} f_\varphi\left(\vartheta_n\right)\cos\varphi_n\right) \tag{4A.63}$$

$$V_r = C\sum_{n=1}^{N}\left(E^i_{n_{\vartheta'}} f_\vartheta\left(\vartheta_n\right)\cos\varphi_n + E^i_{n_{\varphi'}} f_\varphi\left(\vartheta_n\right)\sin\varphi_n\right) \tag{4A.64}$$

By taking into account relations (4A.56), and (4A.57), it is easy to verify that

$$h_{p_\vartheta}\left(\vartheta_n'', \varphi_n''\right) = h_{p_\vartheta}\left(\vartheta_n, \varphi - \varphi_n\right) = C f_\vartheta\left(\vartheta_n\right)\sin\left(\varphi - \varphi_n\right) \tag{4A.65a}$$

$$h_{p_\varphi}\left(\vartheta_n'', \varphi_n''\right) = h_{p_\varphi}\left(\vartheta_n, \varphi - \varphi_n\right) = C f_\varphi\left(\vartheta_n\right)\cos\left(\varphi - \varphi_n\right) \tag{4A.65b}$$

$$h_{r_\vartheta}\left(\vartheta_n'', \varphi_n''\right) = h_{r_\vartheta}\left(\vartheta_n, \varphi - \varphi_n\right) = C f_\vartheta\left(\vartheta_n\right)\cos\left(\varphi - \varphi_n\right) \tag{4A.65c}$$

$$h_{r_\varphi}\left(\vartheta_n'', \varphi_n''\right) = h_{r_\varphi}\left(\vartheta_n, \varphi - \varphi_n\right) = -C f_\varphi\left(\vartheta_n\right)\sin\left(\varphi - \varphi_n\right) \tag{4A.65d}$$

By making explicit the dot products in relations (4A.59), it results:

$$V_\varphi = C\sum_{n=1}^{N}\left(E^i_{n_{\vartheta'}} f_\vartheta\left(\vartheta_n\right)\sin\left(\varphi - \varphi_n\right) + E^i_{n_{\varphi'}} f_\varphi\left(\vartheta_n\right)\cos\left(\varphi - \varphi_n\right)\right) \tag{4A.66}$$

$$V_\rho = C\sum_{n=1}^{N}\left(E^i_{n_{\vartheta'}} f_\vartheta\left(\vartheta_n\right)\cos\left(\varphi - \varphi_n\right) - E^i_{n_{\varphi'}} f_\varphi\left(\vartheta_n\right)\sin\left(\varphi - \varphi_n\right)\right) \tag{4A.67}$$

Therefore,

$$\begin{cases} V_\varphi = V_p\cos\varphi + V_r\sin\varphi \\ V_\rho = -V_p\sin\varphi + V_r\cos\varphi \end{cases} \tag{4A.68}$$

By solving such an equation system, relations (4A.55) are finally obtained.

Chapter 5

Near-field to far-field transformations in cylindrical scanning geometry

5.1 Introduction

The near-field to far-field (NF–FF) transformation techniques in cylindrical scanning geometry [52–65, 113–126, 142], due to their own features, are commonly employed to characterize elongated antennas, which concentrate their radiated field mainly in the plane orthogonal to their predominant dimension and usually exhibit a main beam, which is narrow in elevation and broad in azimuth. As a matter of fact, these transformations make possible to accurately reconstruct the antenna far field except for the regions specified by the spherical polar caps corresponding to the height of the measurement cylinder (see Sect. 5.4).

The cylindrical scanning is achieved by matching a rotational movement of the antenna under test (AUT) and a linear motion of the probe. Commonly, it is carried out by mounting the AUT on a turntable with vertical rotation axis and using a vertical linear positioner for the probe motion. It is so possible to collect the NF amplitude and phase data on a cylindrical grid obtained as intersections of azimuthal rings and cylinder generatrices (see Figure 5.1). Then, the antenna FF pattern can

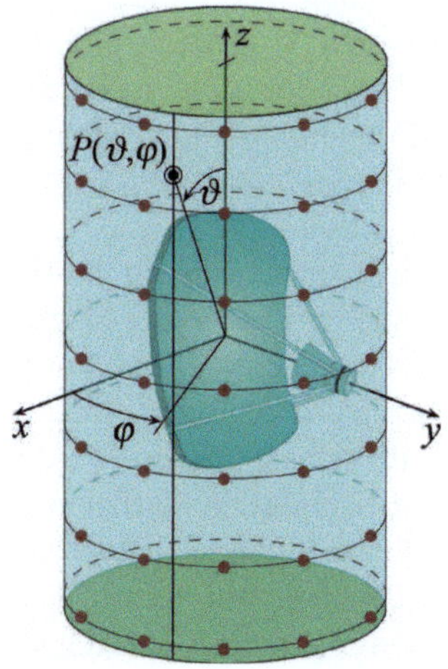

Figure 5.1 Cylindrical scanning

be evaluated from the knowledge of these data, acquired into two different probe orientations to account for the directional effects of the measuring probe. Two main substantially equivalent approaches have been proposed in literature to compensate these directional effects. The former has been developed by Leach and Paris in [52] by applying the Lorentz reciprocity theorem, the latter, proposed by Yaghjian in [53], makes use of the source scattering matrix formulation.

For reader's convenience, the probe-uncompensated NF–FF transformation formulas are first derived in Sect. 5.2. Then, the ideal probe assumption is removed and the probe-compensated formulas as derived in [52] are given in Sect. 5.3.

The non-redundant (NR) sampling representation of radiated electromagnetic (EM) fields [104, 105] have been exploited to develop effective probe-compensated NF–FF transformations [59–63], which require a non-redundant, i.e., minimum, number of samples, which would stay finite also in the case of an unbounded scanning cylinder. Accordingly, these techniques allow a significant reduction of the amount of needed NF data and, as a consequence, of the measurement time, with respect to that [52] required by the classical scanning. In fact, in such a case, the number of samples on each ring decreases moving from the central to the peripheral rings, whereas the spacing between two consecutive rings increases. The so acquired NR samples are then interpolated via optimal sampling interpolation (OSI) formulas to reconstruct the massive NF data needed by the classical transformation [52]. Apart from the approaches [59, 62], where the AUT is considered as enclosed in a sphere (spherical AUT modelling), effective AUT modellings for elongated antennas (see Sect. 5.5) are employed, thus further reducing the number of samples with respect to the approach in [59] and allowing one to consider cylinders with radii smaller than one half the AUT maximum size.

As already stressed, a measurement time saving can be also obtained by making faster the NF data acquisition and this can be achieved, as suggested by Rahmat-Samii *et al.* in [112], by using continuous and synchronized movements of the AUT and probe positioners and by collecting on fly the data. In particular, in a cylindrical NF facility, the helicoidal scanning (see Figure 5.2) is realized by synchronizing the rotational movement of the AUT turntable and the linear motion of the probe positioner. The approach in [142] adopts the same azimuthal and linear spacings as in the classical NF–FF transformation with cylindrical scanning [52] and, hence, it is referred in the following (see Sect. 5.6) as half-wavelength helicoidal scanning. In such a case, by properly modifying the FF reconstruction process [52], no intermediate interpolation is required for recovering the classical NF data. As shown in Sect. 5.7, the application of the unified theory of spiral scannings for volumetric [122] and non-volumetric [124] AUTs has allowed one to develop efficient NR NF–FF transformations with helicoidal scanning [114–126], where the massive NF data needed by the classical NF–FF transformation [52] are recovered via OSI expansions from the helicoidal NR samples. A further measurement time saving with respect to the approach [142] is so attained since, being based on the NR sampling representations, they require a remarkably reduced number of NF data. As a matter of fact, the step of the helix and the samples spacing along it increase when moving away from the centre plane ($z = 0$).

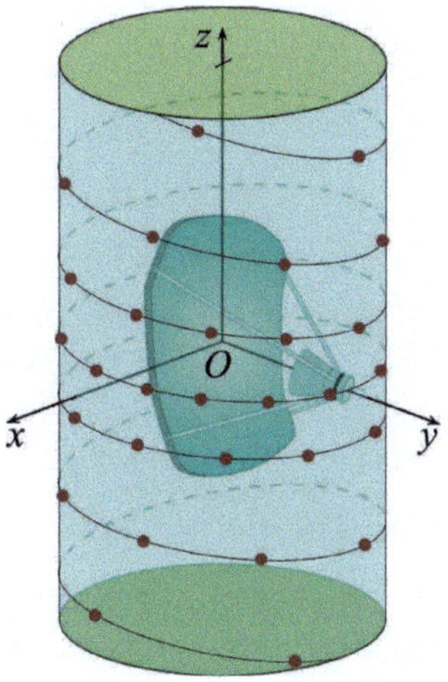

Figure 5.2 *Helicoidal scanning*

5.2 Classical cylindrical NF–FF transformation without probe compensation

As shown in the Appendix 5A.1, the cylindrical components of the electric field at any point P, located in the region external to the smallest cylinder enclosing the radiating antenna, can be represented as superposition of elementary cylindrical waves (see relations (5A.25), (5A.26), and (5A.27)). Accordingly, the tangential components of the electric field radiated by the AUT can be expressed on the scanning cylinder of radius d (see Figure 5.1) as [52]:

$$E_z(\varphi, z) = \sum_{\nu=-\infty}^{\infty} \int_{-\infty}^{\infty} b_\nu(\eta) \frac{\Lambda^2}{\beta} H_\nu^{(2)}(\Lambda d) \, e^{j\nu\varphi} \, e^{-j\eta z} \, d\eta \tag{5.1}$$

$$E_\varphi(\varphi, z) = \sum_{\nu=-\infty}^{\infty} \int_{-\infty}^{\infty} \left[b_\nu(\eta) \frac{\eta\nu}{\beta d} H_\nu^{(2)}(\Lambda d) - a_\nu(\eta) \left. \frac{d}{d\rho} H_\nu^{(2)}(\Lambda\rho) \right|_{\rho=d} \right] \times e^{j\nu\varphi} \, e^{-j\eta z} d\eta \tag{5.2}$$

wherein a_ν and b_ν are the cylindrical wave expansion (CWE) coefficients, $H_\nu^{(2)}(\Lambda\rho)$ is the Hankel function of second kind and order ν, and $\Lambda = \left(\beta^2 - \eta^2\right)^{1/2}$, β being the wavenumber.

The CWE coefficients can be determined [52] by Fourier inverse transforming (5.1) and (5.2), thus getting:

$$b_\nu(\eta) \frac{\Lambda^2}{\beta} H_\nu^{(2)}(\Lambda d) = \frac{1}{4\pi^2} \int_{-\infty}^{\infty} \int_{-\pi}^{\pi} E_z(\varphi, z) \, e^{-j\nu\varphi} \, e^{j\eta z} \, d\varphi \, dz \tag{5.3}$$

$$b_\nu(\eta)\frac{\eta\nu}{\beta d}H_\nu^{(2)}(\Lambda d) - a_\nu(\eta)\frac{\mathrm{d}}{\mathrm{d}\rho}H_\nu^{(2)}(\Lambda\rho)\Big|_{\rho=d}$$
$$= \frac{1}{4\pi^2}\int_{-\infty}^{\infty}\int_{-\pi}^{\pi} E_\varphi(\varphi, z)\,\mathrm{e}^{-\mathrm{j}\nu\varphi}\,\mathrm{e}^{\mathrm{j}\eta z}\mathrm{d}\varphi\,\mathrm{d}z \tag{5.4}$$

As will be shown in the following, the Fourier transform integrals in (5.3) and (5.4) are efficiently evaluated via the fast Fourier transform (FFT) algorithm [300]. As regards the sampling spacings of the NF data in the classical cylindrical NF–FF transformation [52], they must satisfy the following conditions:

$$\Delta z \le \pi/\beta = \lambda/2; \qquad \Delta\varphi \le \pi/\beta a' = \lambda/2a' \tag{5.5}$$

λ being the wavelength and a' the radius of the smallest cylinder containing the AUT. As a matter of fact, the CWE coefficients are zero for $|\eta| > \beta$ and $|\nu| > \beta a'$, when the measurement cylinder is situated in a region of space where the EM field does not contain evanescent waves. Therefore, according to the two-dimensional Nyquist sampling theorem [299], the EM field of the AUT can be accurately reconstructed on the scanning cylinder from the knowledge of its samples, whose spacings satisfy relations (5.5). The condition on the maximum value of $|\eta|$ can be explained by taking into account that Λ becomes an imaginary number for $|\eta| > \beta$, whereas that relevant to the maximum value of $|\nu|$ is a straightforward consequence of the sharp bound (2.16) for the azimuthal bandwidth W_φ.

Once the CWE coefficients have been so determined, the FF components of the electric field radiated by the AUT in the spherical coordinates system (R, Θ, Φ) can be evaluated by substituting them in the expression of the CWE valid in the antenna FF region (see Appendix 5A.2), namely:

$$E_\Theta(R, \Theta, \Phi) = -\mathrm{j}2\beta\frac{\mathrm{e}^{-\mathrm{j}\beta R}}{R}\sin\Theta\sum_{\nu=-\infty}^{\infty}\mathrm{j}^\nu b_\nu(\beta\cos\Theta)\,\mathrm{e}^{\mathrm{j}\nu\Phi} \tag{5.6}$$

$$E_\Phi(R, \Theta, \Phi) = -2\beta\frac{\mathrm{e}^{-\mathrm{j}\beta R}}{R}\sin\Theta\sum_{\nu=-\infty}^{\infty}\mathrm{j}^\nu a_\nu(\beta\cos\Theta)\,\mathrm{e}^{\mathrm{j}\nu\Phi} \tag{5.7}$$

It is convenient to describe the employment of the FFT algorithm for computing the two-dimensional Fourier transforms in (5.3) and (5.4) from the acquired NF data. By taking explicitly into account the truncation due to the finite height $2h$ of the scanning cylinder, relations (5.3) and (5.4) can be rewritten in the form:

$$b_\nu(\eta)\frac{\Lambda^2}{\beta}H_\nu^{(2)}(\Lambda d) = \frac{1}{4\pi^2}\int_{-h}^{h}\int_{0}^{2\pi} E_z(\varphi, z)\,\mathrm{e}^{-\mathrm{j}\nu\varphi}\,\mathrm{e}^{\mathrm{j}\eta z}\,\mathrm{d}\varphi\,\mathrm{d}z \tag{5.8}$$

$$b_\nu(\eta)\frac{\eta\nu}{\beta d}H_\nu^{(2)}(\Lambda d) - a_\nu(\eta)\frac{\mathrm{d}}{\mathrm{d}\rho}H_\nu^{(2)}(\Lambda\rho)\Big|_{\rho=d}$$
$$= \frac{1}{4\pi^2}\int_{-h}^{h}\int_{0}^{2\pi} E_\varphi(\varphi, z)\,\mathrm{e}^{-\mathrm{j}\nu\varphi}\,\mathrm{e}^{\mathrm{j}\eta z}\,\mathrm{d}\varphi\,\mathrm{d}z \tag{5.9}$$

The integration with respect to φ is first considered. By taking into account that, according to conditions (5.5), $\Delta\varphi = \pi/M_1 = 2\pi/N$, where $M_1 \geq \beta a'$ is the highest order ν of the considered angular harmonics, it results:

$$F_\nu = \int_0^{2\pi} f(\varphi)\,\mathrm{e}^{-\mathrm{j}\nu\varphi}\mathrm{d}\varphi \approx \frac{2\pi}{N}\sum_{m=0}^{N-1} f\left(\frac{2\pi m}{N}\right)\mathrm{e}^{-\mathrm{j}2\pi\nu m/N} \tag{5.10}$$

By comparing this relation with (A.90), it can recognized that, save for the factor $2\pi/N$, the integration can be efficiently performed via a direct FFT algorithm.

The integration over z is now considered. According to conditions (5.5), $\Delta z = 2h/N_1 \leq \lambda/2$, where N_1 is the number of measurement rings on the scanning cylinder. By putting $z_n = -h + n\Delta z = \left(n - N_1/2\right)\Delta z$, it follows that:

$$G(\eta) = \int_{-h}^{h} g(z)\,\mathrm{e}^{\mathrm{j}\eta z}\mathrm{d}z \approx \Delta z \sum_{n=0}^{N_1-1} g(z_n)\,\mathrm{e}^{\mathrm{j}\eta(n-N_1/2)\Delta z} \tag{5.11}$$

When evaluating $G(\eta)$ at $\eta_i = i\Delta\eta = 2\pi i/(N_1\Delta z)$, it results:

$$G(\eta_i) \approx \frac{2h}{N_1}\,\mathrm{e}^{-\mathrm{j}\pi i}\sum_{n=0}^{N_1-1} g(z_n)\,\mathrm{e}^{\mathrm{j}2\pi ni/N_1} \tag{5.12}$$

By comparing this relation with (A.91), it can easily recognized that the integration with respect to z in (5.3) and (5.4) can be efficiently accomplished by means of an inverse FFT algorithm, provided that the attained results are multiplied by the factor $2h\,\mathrm{e}^{-\mathrm{j}\pi i}$.

In the light of the above considerations, the two-dimensional Fourier transforms in (5.3) and (5.4) can be effectively evaluated as follows: i) the NF data acquired on each ring are transformed via a direct FFT and then multiplied by $2\pi/N$; ii) the obtained results are transformed, for each index ν via an inverse FFT, and then multiplied by $2h\,\mathrm{e}^{-\mathrm{j}\pi i}$.

It is worth noting that, by applying the described procedure, the CWE coefficients a_ν and b_ν are evaluated in correspondence of the η values given by

$$\eta_i = i\Delta\eta = 2\pi i/(N_1\Delta z) \tag{5.13}$$

This allows one to obtain the FF components at the values of Θ specified by

$$\Theta_i = \cos^{-1}(\eta_i/\beta) = \cos^{-1}(i\lambda/N_1\Delta z) \tag{5.14}$$

In order to get the FF components with a greater resolution in Θ, a proper "zero-filling" of the NF data must be employed. In particular, to obtain the FF components in correspondence of the Θ values given by $\Theta_i = \cos^{-1}\left(i\lambda/N_2\Delta z\right)$, fictitious $(N_2 - N_1)/2$ rings with NF data all equal to zero must be considered at both the ends of the scanning cylinder.

It must be stressed that the summations in the relations (5.6) and (5.7) can be efficiently performed via an inverse FFT, thus obtaining the FF components at $\Theta = \Theta_i$ in correspondence of $2M_1$ values of the azimuthal angle Φ. In order to obtain a greater resolution in Φ, additional zero angular harmonics must be considered. In particular, to get the FF components at $2M_2$ values of Φ, the positive and negative harmonics must be both augmented with $M_2 - M_1$ zeros before performing the summations (5.6) and (5.7) through the FFT algorithm.

A two-dimensional OSI algorithm, which allows one to obtain the FF components E_Θ and E_Φ at a given direction Θ, Φ, is described in the following. It can be properly used to get the antenna radiation pattern at any cut plane Φ as function of the zenithal angle Θ. Its mathematical justification relies on the fact that the CWE coefficients $a_\nu(\eta)$ and $b_\nu(\eta)$, and then the AUT far field, are related to the Fourier transform (with respect to z) of the tangential components of the electric field on the scanning cylinder. Accordingly, if these last are practically negligible for $|z| > h$, as a consequence of the Nyquist sampling theorem [299], the FF components on a given FF meridian can be accurately reconstructed from the knowledge of their values at the sampling points spaced by

$$\Delta\eta \leq \pi/h \tag{5.15}$$

Therefore, the following OSI expansion (see (2A.32)) can be employed to accurately reconstruct them [295]:

$$E_{\Theta,\Phi}(\eta(\Theta), \Phi) = \sum_{n=n_0-q+1}^{n_0+q} E_{\Theta,\Phi}(\eta_n, \Phi)\, \Psi(\eta - \eta_n)\, \mathrm{sinc}\,[\pi(\eta - \eta_n)/\Delta\eta]^{*} \tag{5.16}$$

where

$$\eta_n = n\Delta\eta = 2\pi n/N_2\Delta z \tag{5.17}$$

$\mathrm{sinc}(\eta)$ is the $\sin(\eta)/\eta$ function, $n_0 = \mathrm{Int}(\eta/\Delta\eta)$ is the index of the sample nearest (on the left) to the output point, $2q$ is the number of the retained nearest samples, and $\Psi(\eta)$ is the Knab's Sampling Window function (2A.33), wherein $\bar{\eta} = q\Delta\eta$ and $\chi = N_2/N_1$.

It must be stressed that the $2M_2$ samples of the FF components, obtained via FFT from (5.6) and (5.7), can be interpolated to get their values at any point specified by the angle Φ on the FF Θ parallel. To this end, the OSI expansion (2.58) developed in [39, 106] and properly modified to deal with an even number of samples can be applied:

$$E_{\Theta,\Phi}(\Theta, \Phi) = \sum_{m=m_0-p+1}^{m_0+p} E_{\Theta,\Phi}(\Theta, \Phi_m)\, \Omega_M(\Phi - \Phi_m, \bar{\Phi})\, D^e_{M_2}(\Phi - \Phi_m) \tag{5.18}$$

where

$$\Phi_m = m\,\Delta\,\Phi = m\pi/M_2; \qquad M = M_2 - M_1; \qquad \bar{\Phi} = p\,\Delta\,\Phi \tag{5.19}$$

$m_0 = \mathrm{Int}(\Phi/\Delta\Phi)$ is the index of the sample nearest (on the left) to the output point, $2p$ is the number of the retained nearest samples, and $\Omega_M(\Phi, \bar{\Phi})$ is the Tschebyscheff Sampling function (2.62). Moreover,

$$D^e_{M_2}(\Phi) = \frac{\sin((2M_2 - 1)\Phi/2)}{2M_2 \sin(\Phi/2)} \tag{5.20}$$

*In the representation of the EM field radiated on the FF sphere, the factor $e^{-j\beta R}/R$ is, as usual, omitted.

is the Dirichlet function for an even number of samples [307].

The two-dimensional OSI formula, which allows one to obtain the FF components E_Θ and E_Φ at a given direction Θ, Φ, can be obtained by properly merging the one-dimensional expansions (5.16) and (5.18):

$$E_{\Theta,\Phi}(\eta(\Theta),\Phi) = \sum_{n=n_0-q+1}^{n_0+q} \Big\{ \Psi(\eta-\eta_n)\,\mathrm{sinc}\,[\pi(\eta-\eta_n)/\Delta\eta] \times \sum_{m=m_0-p+1}^{m_0+p} E_{\Theta,\Phi}(\eta_n,\Phi_m)\,\Omega_M(\Phi-\Phi_m,\bar{\Phi})\,D^e_{M_2}(\Phi-\Phi_m) \Big\} \quad (5.21)$$

5.3 Classical cylindrical NF–FF transformation with probe compensation

It is not a difficult matter to realize that it is not possible to acquire, by means a real (not ideal) probe, the tangential components E_z and E_φ of the NF radiated by the AUT on the scanning cylinder. As a matter of fact, during the scanning, the probe sees the centre of the antenna in different directions. Furthermore, also at a given position, the direction under which each portion of the AUT is viewed is different. Therefore, according to the discussion made at the beginning of Section 4.3, the response (measured voltage) of an actual probe is not proportional to the tangential components of the AUT electric field. Accordingly, the far field radiated by the AUT cannot be accurately reconstructed from the acquired NF data by means of the previously described uncompensated cylindrical NF–FF transformation.

The probe-compensated NF–FF transformation technique with cylindrical scanning, wherein the directional effects introduced by a real probe are suitably taken into account, has been originally developed in [52] by Leach and Paris. They, by properly applying the Lorentz reciprocity theorem, proved in a rigorous way that the coefficients a_ν and b_ν of the cylindrical wave expansion of the AUT field can be determined from the knowledge of: i) the two-dimensional Fourier transforms of the voltages V_p and V_r measured by the probe and the probe rotated of 90° around its longitudinal axis; ii) the CWE coefficients of the field radiated by the probe and the rotated probe, when working as transmitting antennas. The key relations [52] to determine the CWE coefficients a_ν and b_ν are:

$$a_\nu(\eta) = \frac{\beta^2}{\Lambda^2\Delta_\nu(\eta)}\Big[I_\nu(\eta)\sum_{m=-\infty}^{\infty} d'_m(-\eta)\,H^{(2)}_{\nu+m}(\Lambda d) - I'_\nu(\eta)\sum_{m=-\infty}^{\infty} d_m(-\eta)H^{(2)}_{\nu+m}(\Lambda d)\Big] \quad (5.22)$$

$$b_\nu(\eta) = \frac{\beta^2}{\Lambda^2\Delta_\nu(\eta)}\Big[I'_\nu(\eta)\sum_{m=-\infty}^{\infty} c_m(-\eta)\,H^{(2)}_{\nu+m}(\Lambda d) - I_\nu(\eta)\sum_{m=-\infty}^{\infty} c'_m(-\eta)\,H^{(2)}_{\nu+m}(\Lambda d)\Big] \quad (5.23)$$

$$I_{\nu}(\eta) = \int_{-\infty}^{\infty}\int_{-\pi}^{\pi} V_p(\varphi, z)\, e^{-j\nu\varphi} e^{j\eta z} d\varphi\, dz \tag{5.24a}$$

$$I'_{\nu}(\eta) = \int_{-\infty}^{\infty}\int_{-\pi}^{\pi} V_r(\varphi, z)\, e^{-j\nu\varphi}\, e^{j\eta z} d\varphi\, dz \tag{5.24b}$$

$$\Delta_{\nu}(\eta) = \sum_{m=-\infty}^{\infty} c_m(-\eta) H_{\nu+m}^{(2)}(\Lambda d) \sum_{m=-\infty}^{\infty} d'_m(-\eta) H_{\nu+m}^{(2)}(\Lambda d) - \sum_{m=-\infty}^{\infty} c'_m(-\eta) H_{\nu+m}^{(2)}(\Lambda d) \sum_{m=-\infty}^{\infty} d_m(-\eta) H_{\nu+m}^{(2)}(\Lambda d) \tag{5.25}$$

It is noteworthy that analogous results have been achieved by Yaghjian in [53] by employing the source scattering matrix formulation. As a matter of fact, as explicitly shown in [55], the expressions of the CWE coefficients a_ν and b_ν in [52] and [53] are practically the same except for a normalization constant.

In the above relations, c_m and d_m are the CWE coefficients of the field radiated by the probe, whereas c'_m and d'_m are those relevant to the rotated probe. These coefficients can be determined [52] from the measured amplitude and phase of the FF components radiated by the probe and rotated probe.

The two-dimensional Fourier transforms I_ν and I'_ν of the voltages V_p and V_r can be again effectively computed using the FFT algorithm and the NF grid spacings must still satisfy conditions (5.5).

Once the CWE coefficients a_ν and b_ν have been evaluated, the FF spherical components of the electric field radiated by the AUT can be obtained by substituting them in the relations (5.6) and (5.7).

It is worth noting that the same techniques described in the previous section to obtain the FF components with a greater resolution in Θ and Φ and the two-dimensional OSI formula (5.21) can be applied also in this case.

A numerical example, which points out the need of applying a probe-compensated NF–FF transformation technique to accurately recover the AUT far field from NF data acquired by a real probe, is reported in the following. The simulations refer to the field radiated by a uniform planar array of 0.6 λ spaced elementary Huygens sources polarized along the z axis and covering an elliptical zone in the plane $y = 0$, with major and minor semi-axes equal to 24 λ and 6 λ. An open-ended WR-90 rectangular waveguide, operating at the frequency of 10 GHz, is chosen as probe. The radius d of the scanning cylinder is equal to 12 λ and the height $2h$ is 160 λ. Figure 5.3 shows the amplitude of the probe output voltage V_p (solid line) on the cylinder generatrix at $\varphi = 90°$. In the same figure is shown also the amplitude of the NF z-component E_z (crosses). As can be seen, the plots are significantly different, thus confirming that the voltage measured by a real probe is not proportional to the tangential electric field component. The antenna FF pattern in the E-plane reconstructed from voltage data by the uncompensated and the probe-compensated NF–FF transformation techniques are reported in Figures 5.4 and 5.5, respectively. As can be seen, some discrepancies are clearly evident in the reconstruction obtained by applying the NF–FF transformation without probe compensation. On the other

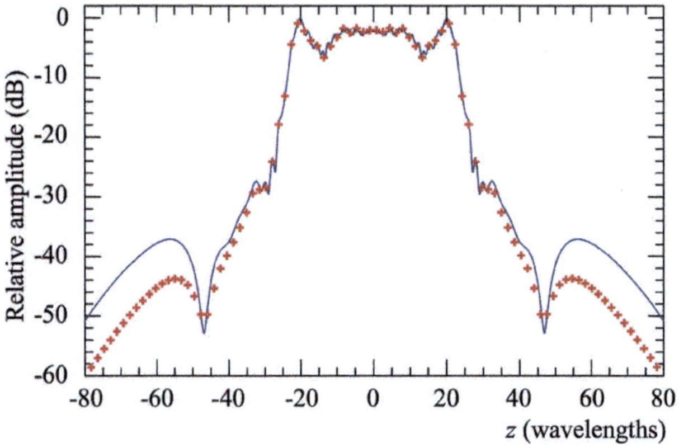

Figure 5.3 *Amplitude of the probe output voltage V_p (solid line) and of the NF component E_z (crosses) on the generatrix at $\varphi = 90°$*

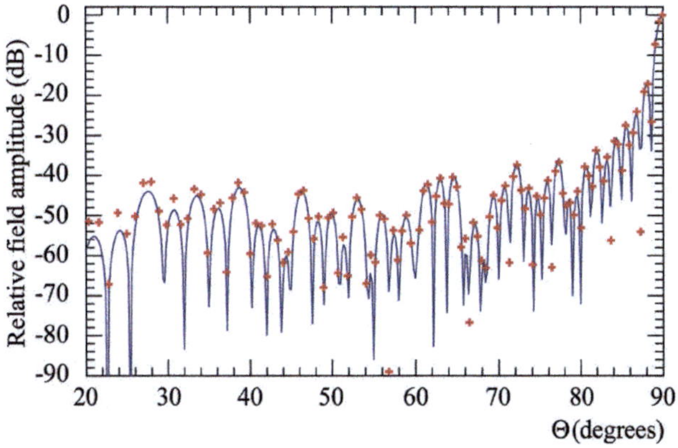

Figure 5.4 *E-plane pattern. Solid line: exact field. Crosses: reconstructed from voltage data by the uncompensated NF–FF transformation.*

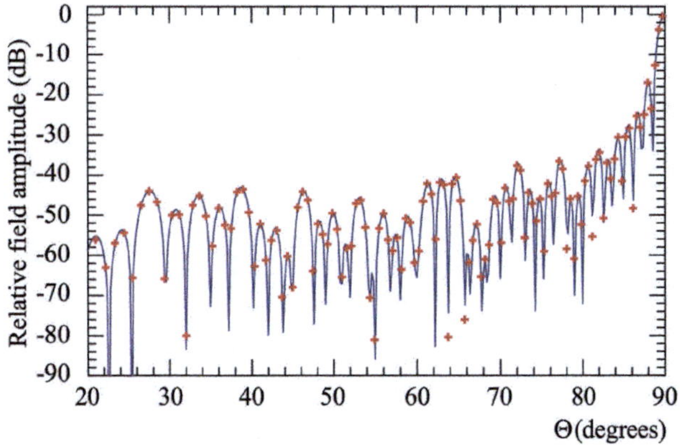

Figure 5.5 *E-plane pattern. Solid line: exact field. Crosses: reconstructed from voltage data by the probe-compensated NF–FF transformation.*

hand, the field reconstructed via the probe-compensated NF–FF transformation and the exact one are practically indistinguishable.

5.4 Error due to the truncation of the scanning area

When the scanning area is truncated, as in the cylindrical and planar scans, the reconstructed far field is unavoidably affected by a truncation error, whose magnitude obviously depends on the level of the neglected NF data outside the measurement area. Aim of this section is just to show the effect of the scanning cylinder height on the accuracy of the recovered far field.

Let a cylindrical NF measurement facility be equipped with a rotating table, on which is mounted an AUT having dimension $2a$, and with a vertical linear scanner of height $2h = h_l + h_u$ at distance d from the rotation axis (see Figure 5.6).

A useful and practical rule to predict the angular region of validity of the reconstructed FF pattern is given by

$$\Theta_u \le \Theta \le \Theta_l \tag{5.26}$$

wherein the validity angles Θ_u and Θ_l are so defined:

$$\Theta_u = \frac{\pi}{2} - \alpha_u = \frac{\pi}{2} - \tan^{-1}\left(\frac{h_u - a}{d}\right); \quad \Theta_l = \frac{\pi}{2} + \alpha_l = \frac{\pi}{2} + \tan^{-1}\left(\frac{h_l - a}{d}\right) \tag{5.27}$$

The criterion of the validity angular region can be easily got by extending to the cylindrical NF–FF transformation that relevant to the plane-rectangular NF–FF transformation reported in [252], which was empirically found from a large number of NF measurements, performed by using different antennas and probe combinations [301], and determined by means of a theoretical analysis [302]. It is noteworthy that the discontinuity of the AUT near field at the edges of the scanning area can produce a ripple in the reconstructed FF pattern, which can be present even in the validity region.

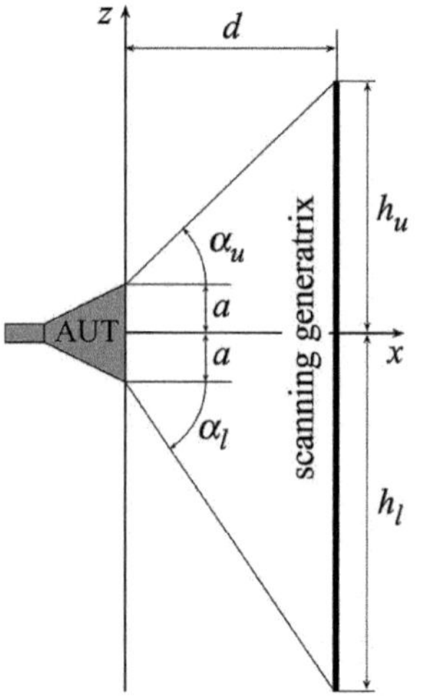

Figure 5.6 Geometry relevant to the evaluation of the validity angles

Of course, an enlargement of this angular region can be attained by decreasing the radius of the scanning cylinder. However, such a radius cannot be reduced beyond certain limits, otherwise the interactions between the AUT and the probe cannot be neglected any more. Moreover, as explicitly stressed when discussing the conditions (5.5) to be satisfied by the NF data spacings, the measurement cylinder must be situated in the radiating NF region, wherein the evanescent waves of the AUT field are negligible. In practice, during the scanning, a minimum distance of at least 4–5 wavelengths between the probe and AUT has to be ensured.

Some reconstruction examples are reported in the following in order to show the effect of the scanning cylinder height on the accuracy of the recovered far field.

The first set of figures refers to the same planar array and scanning cylinder radius d considered in the previous section. In particular, Figure 5.7a is relevant to the reconstruction obtained when the cylinder height is 100 λ, whereas Figure 5.7b

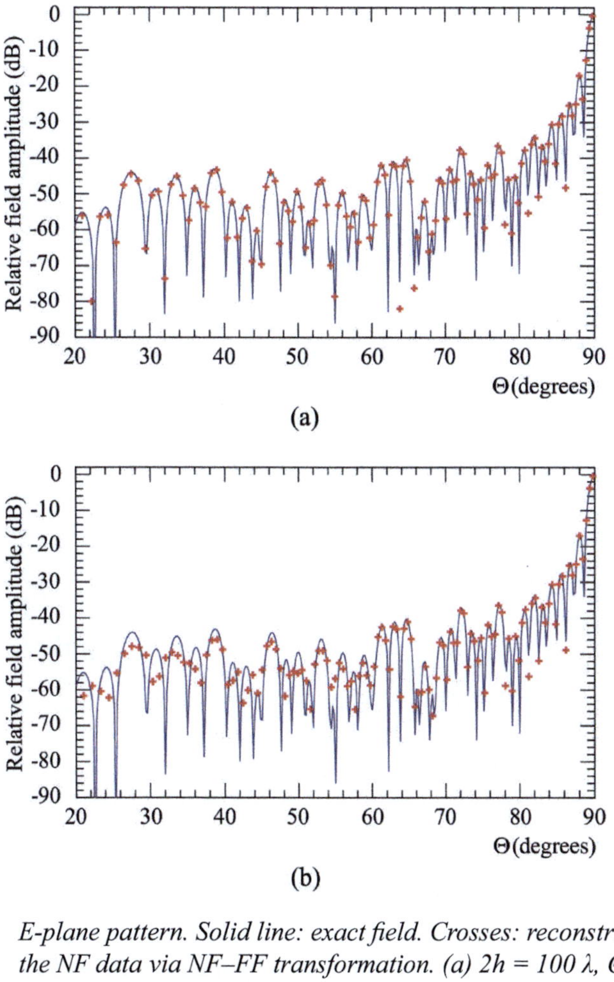

Figure 5.7 *E-plane pattern. Solid line: exact field. Crosses: reconstructed from the NF data via NF–FF transformation. (a)* $2h = 100\,\lambda$, $\Theta_u = 24.8°$, *(b)* $2h = 60\,\lambda$, $\Theta_u = 63.4°$.

shows that attained by reducing to 60 λ the height of the scanning zone. As can be seen, both the FF reconstructions result to be accurate in the validity region.

The same comment holds for Figure 5.8a and b, which are relevant to the same previously considered scanning zones, but to a uniform planar rectangular array having dimensions $8\lambda \times 48\lambda$. Its elements are 0.6λ spaced elementary Huygens sources polarized along the z axis. It is worthy to note that, in the case of the elliptical array, the FF reconstruction obtained when using the shortened cylinder is accurate even beyond the validity angle Θ_u. This is mainly due to the faster decrease of the NF level for such an antenna (see Figure 5.9).

The last set of figures refers to an antenna obtained from that considered in the second set by changing the phases of the array elements in order to get the main lobe maximum at $\Theta = 40°$. Also in such a case, an accurate pattern reconstruction is attained in the validity region (see Figure 5.10a). It could be convenient to consider

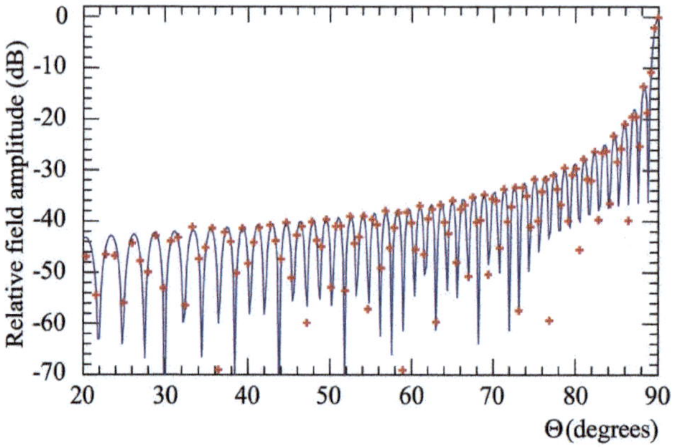

(a)

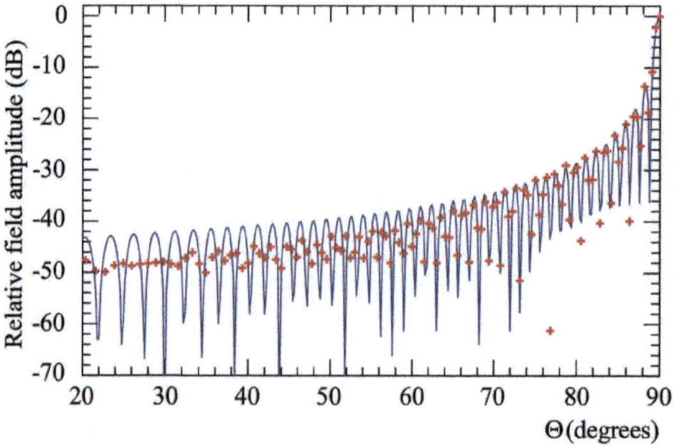

(b)

Figure 5.8 *E-plane pattern. Solid line: exact field. Crosses: reconstructed from the NF data via NF–FF transformation. (a) $2h = 100\ \lambda$, $\Theta_u = 24.8°$, (b) $2h = 60\ \lambda$, $\Theta_u = 63.4°$.*

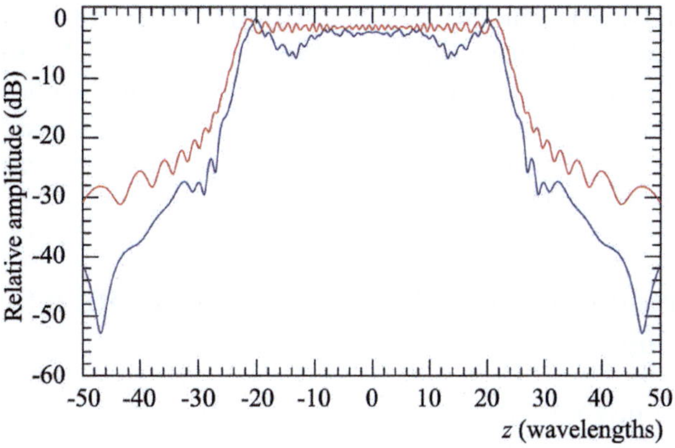

Figure 5.9 *Amplitudes of the probe output voltages on the generatrix at φ = 90°. Blue solid line: elliptic AUT. Red solid line: rectangular AUT.*

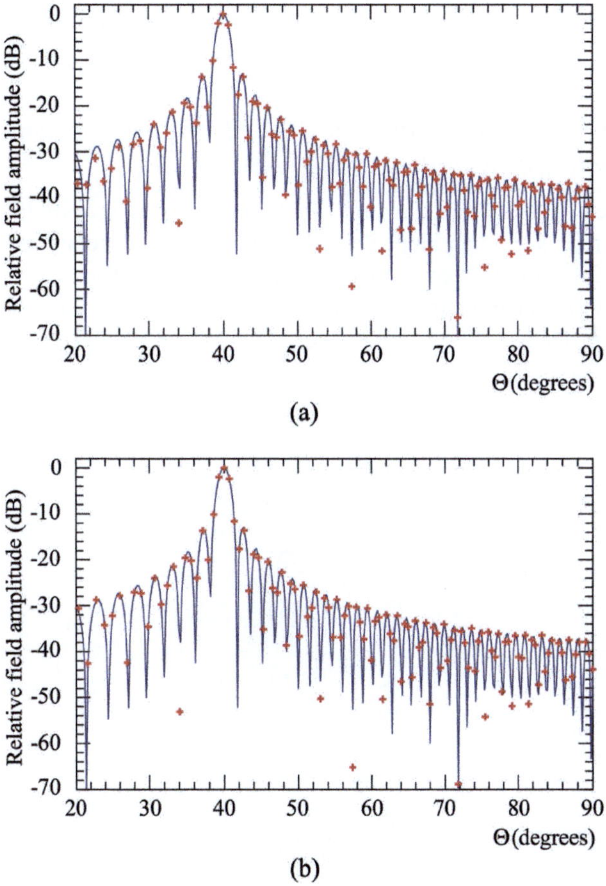

Figure 5.10 *E-plane pattern. Solid line: exact field. Crosses: reconstructed from the NF data via NF–FF transformation. (a) 2h =100 λ, Θ_u = 24.8°, (b) 2h = $h_l + h_u$ = 30 λ+70 λ, Θ_u = 14.6°.*

a scanning zone such that the angular range of good FF reconstruction covers all the region where the field is more significant. This can be obtained by using an asymmetric NF scanning, which, with the same extension, allows one to decrease the angle Θ_u. Figure 5.10b shows the pattern reconstruction obtained when the scanning zone extends from $-30\ \lambda$ to $70\ \lambda$. As can be seen, a very good reconstruction is so achieved in the zone of major interest.

5.5 Non-redundant NF–FF transformations with cylindrical scanning

In this section, two NR NF–FF transformations with cylindrical scanning will be described. Both these transformations are particularly tailored for elongated AUTs, as usually are those characterized in a cylindrical NF facility. In the former [61, 105], the AUT is considered as enclosed in a prolate spheroid, whereas in the latter [60], it is modelled by a rounded cylinder. At last, an efficient NF–FF transformation to evaluate the far field directly from the acquired NR NF data [60, 61, 63] is summarized.

5.5.1 Non-redundant cylindrical NF–FF transformation: the prolate spheroidal modelling case

Let an electrically small, non-directive, probe be acquiring the NF data radiated on a cylinder of radius d by an elongated AUT enclosed in a prolate spheroid Σ with major and minor semi-axes equal to a and b (see Figure 5.11).

As shown in [41] and already stressed in Chapter 2, the voltage measured by a non-directive probe has the same effective spatial bandwidth of the field radiated by

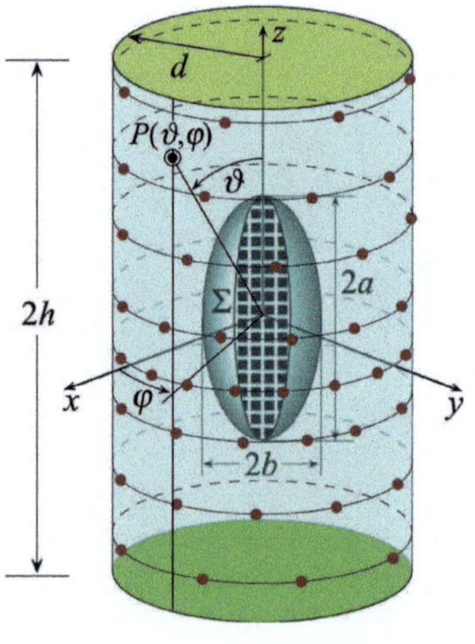

Figure 5.11 NR cylindrical scanning: prolate spheroidal modelling

the AUT, so that the NR sampling representations [104, 105] and related OSI expansions can be applied to the probe reduced voltage $\tilde{V}(\xi) = V(\xi)\,e^{j\gamma(\xi)}$.

According to the results reported in Sect. 2.2, the bandwidth W_ξ, the phase function γ and the optimal parameter ξ for describing a cylinder generatrix are given by the expressions (2.25), (2.27) and (2.28a), respectively, whereas the azimuthal bandwidth W_φ is fixed by the relation (2.29a). The expression of the two-dimensional OSI expansion (2.71), which allows the fast and accurate reconstruction of the probe and rotated probe voltages at any point on the scanning cylinder, is reported in the following for reader's convenience:

$$V(\xi(\vartheta), \varphi) = e^{-j\gamma(\xi)} \times \sum_{n=n_0-q+1}^{n_0+q} \left\{ SF(\xi, \xi_n, \bar{\xi}, N, N'') \sum_{m=m_0-p+1}^{m_0+p} \tilde{V}(\xi_n, \varphi_{m,n})\, SF(\varphi, \varphi_{m,n}, \bar{\varphi}_n, M, M''_n) \right\} \tag{5.28}$$

where all symbols have the same meanings as in (2.68). It is noteworthy that $\vartheta_\infty(\xi_n)$, given by (2.30a), takes the place of the angle $\vartheta(\xi_n)$ in the expression (2.67) of the azimuthal enlargement bandwidth factor χ^*.

The OSI expansion (5.28) can be employed to accurately recover the values of the probe output voltages V_p and V_r at the points needed to perform the classical probe-compensated NF–FF transformation with cylindrical scanning [52]. It is so possible to get a probe-compensated NF–FF transformation with cylindrical scanning which makes use of a NR number of data.

Some numerical examples assessing the effectiveness of the technique are reported in the following. The simulations refer to the previously considered elliptical AUT and the chosen probe is again an open-ended WR-90 rectangular waveguide working at 10 GHz. The radius d of the scanning cylinder is equal to $12\,\lambda$ and its height $2h$ is $160\,\lambda$. The AUT has been modelled by a prolate spheroid with major and minor semi-axes equal to $24\,\lambda$ and $6\,\lambda$. Figure 5.12a and b show a representative reconstruction example of the amplitude and phase of the probe voltage V_p on the cylinder generatrix at $\varphi = 90°$. As can be seen, there is a very good agreement between the reconstructed voltage and the exact one. The stability of the algorithm, namely its robustness with respect to errors affecting the samples, has been investigated by corrupting the exact samples with random errors. These errors simulate a background noise (bounded to Δa dB in amplitude and with arbitrary phase) and uncertainties on the data of $\pm\Delta a_r$ dB in amplitude and $\pm\Delta\phi$ degrees in phase. As shown in Figure 5.13, the algorithm is stable. The accuracy of the OSI algorithm is assessed in a more quantitative way by the maximum and mean-square reconstruction error values shown in Figure 5.14a and b. They are normalized to the voltage maximum value over the cylinder and have been attained by comparing the interpolated values of V_p with those directly evaluated on a close grid in the central zone of the scanning cylinder, so that the existence of the required guard samples is guaranteed. As can be seen, they decrease up to very low values on increasing the number of retained samples and/or the

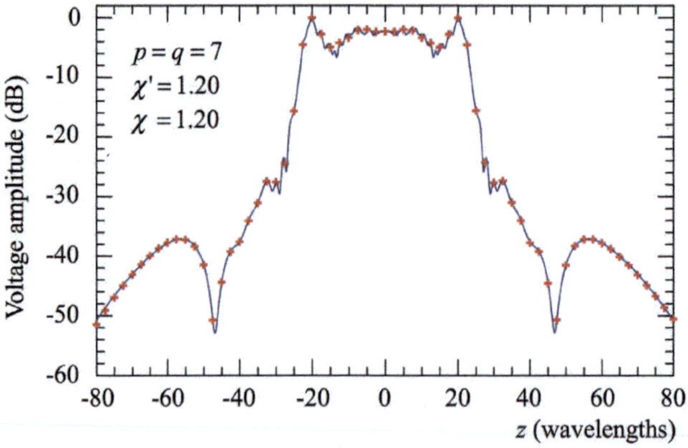

(a)

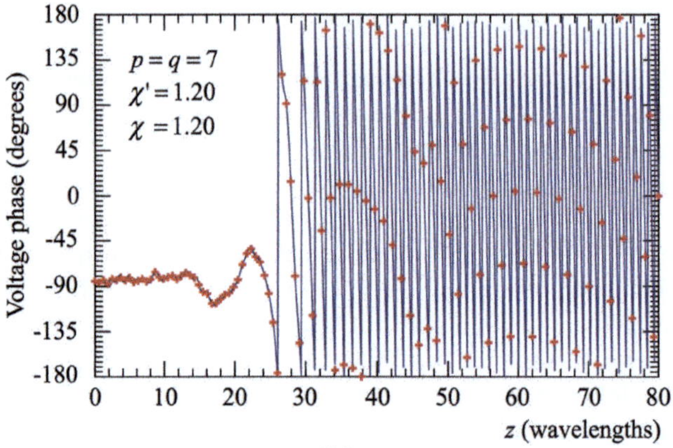

(b)

Figure 5.12 *Probe voltage V_p on the generatrix at $\varphi = 90°$. Solid line: exact. Crosses: obtained by interpolating the NR NF samples. (a) Amplitude, (b) Phase.*

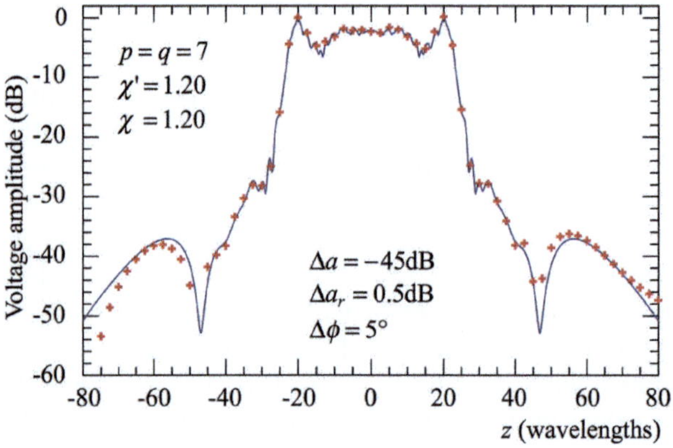

Figure 5.13 *Amplitude of the probe voltage V_p on the generatrix at $\varphi = 90°$. Solid line: exact. Crosses: interpolated from error affected samples.*

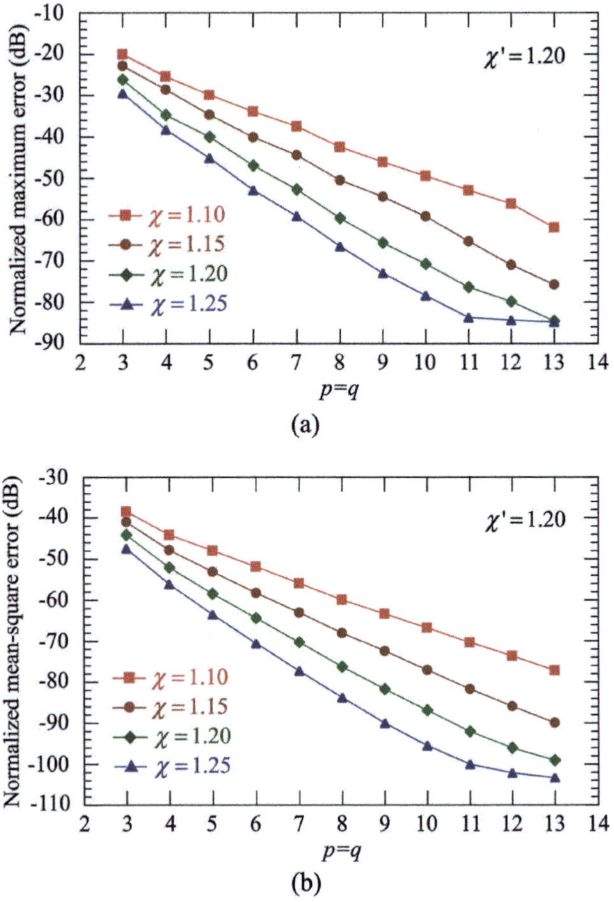

Figure 5.14 *Normalized errors in the reconstruction of the probe voltage V_p. (a) Maximum error, (b) Mean-square error.*

oversampling factor. At last, the OSI algorithm (5.28) has been applied to recover the NF data needed for the classical cylindrical NF–FF transformation [52]. The reconstructions of the antenna FF pattern in the principal planes are reported in Figure 5.15a and b. As can be seen, the exact and recovered patterns are practically indistinguishable.

It can be interesting to compare the number of used NF samples (13025) with that (40960) which would be required by applying the classical cylindrical NF–FF transformation [52]. In conclusion, the described NR cylindrical NF–FF transformation allows a remarkable reduction of the required NF data and related measurement time without any loss of accuracy in the recovery of the AUT far field.

The interested reader can find in [105] experimental results assessing the effectiveness of such a NF–FF transformation also from a practical viewpoint.

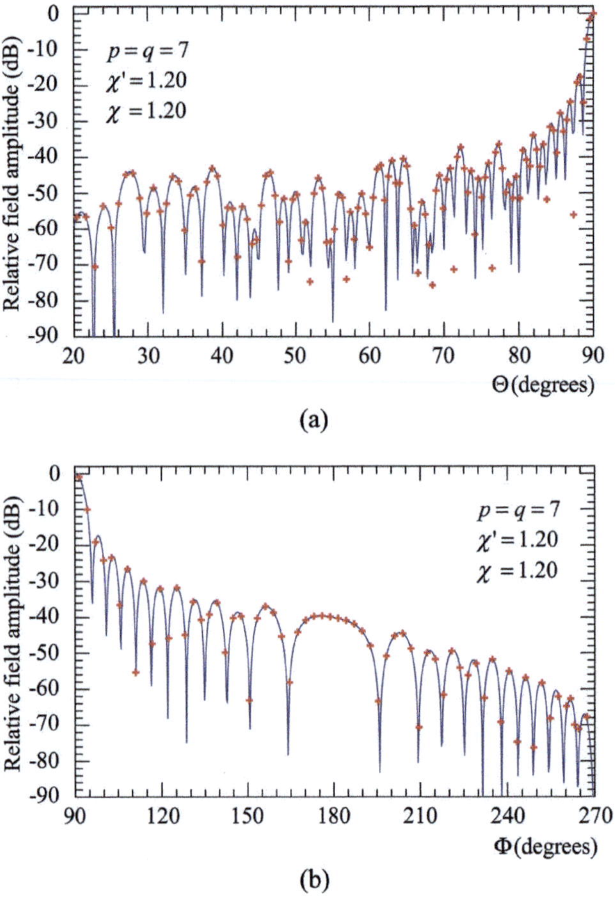

Figure 5.15 *FF patterns in the principal planes. Solid line: exact field. Crosses: reconstructed from NR NF samples. (a) E-plane, (b) H-plane.*

5.5.2 *Non-redundant cylindrical NF–FF transformation: the rounded cylinder modelling case*

Let an electrically small, non-directive, probe be acquiring the NF data radiated on a cylinder of radius d by a long AUT enclosed in a rounded cylinder Σ, namely, a cylinder of height h' terminated in two half-spheres of radius a' (see Figures 5.16 and 5.17). As in the previous subsection, the NR sampling representations [104] can be applied to the probe reduced voltage $\tilde{V}(\xi) = V(\xi)\,\mathrm{e}^{\mathrm{j}\gamma(\xi)}$.

According to the results presented in Sect. 2.3, the bandwidth W_ξ, the phase function γ and the optimal parameter ξ for describing a cylinder generatrix are given by the expressions (2.10), (2.8) and (2.11), here reported for reader's convenience.

$$W_\xi = \beta\ell'/2\pi; \quad \gamma = \frac{\beta}{2}[R_1 + R_2 + s_1' - s_2']; \quad \xi = \frac{\pi}{\ell'}[R_1 - R_2 + s_1' + s_2'] \quad (5.29)$$

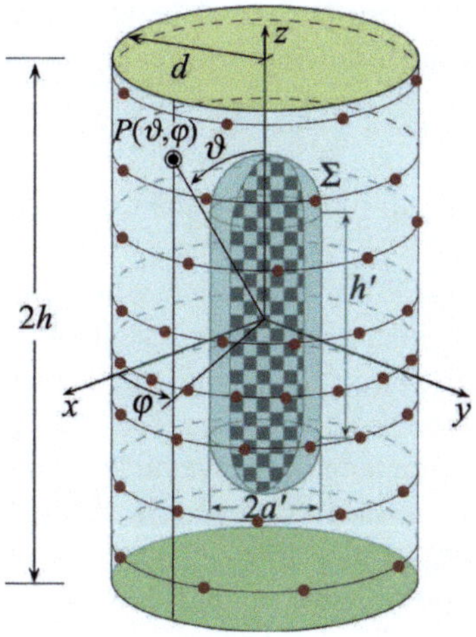

Figure 5.16 *NR cylindrical scanning: rounded cylinder modelling*

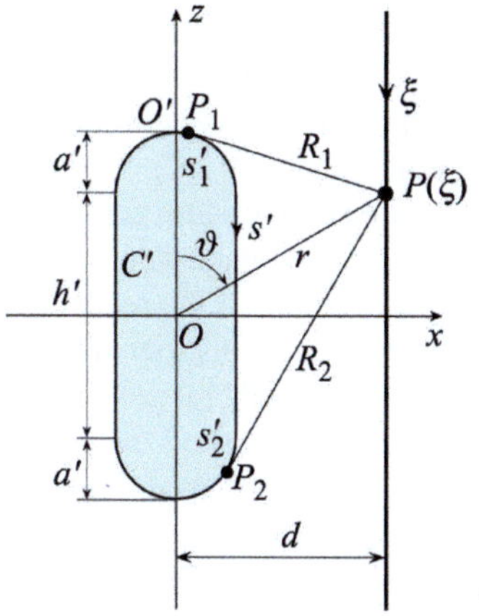

Figure 5.17 *Rounded cylinder modelling: relevant to the evaluation of ξ*

where $\ell' = 2\,(h' + \pi a')$ is the length of the curve C' intersection of Σ with a meridian plane. The explicit expressions of γ and ξ can be obtained by substituting the values of the distances $R_{1,2}$ and curvilinear abscissae $s'_{1,2}$ in (5.29). In such a case (see Figure 5.17), P_1 is always on the upper cap, whereas P_2 is on the lower one. Accordingly, R_1 and s'_1 are given by (2.31) and (2.32), whereas R_2 and s'_2 are given by (2.34) and (2.35).

As regards the evaluation of the maximum in (2.14), which allows the determination of the bandwidth W_φ, according to the results presented in Sect. 2.3, it is attained in correspondence of the z' value given by (2.38).

The OSI expansion (5.28) allows again the efficient reconstruction of the probe and rotated probe voltages at any point on the scanning cylinder and, therefore, can be employed to accurately recover the values of V_p and V_r at the points required by the classical NF–FF transformation with cylindrical scanning [52]. However, in order to have a symmetrical distribution of the scanning rings with respect to the plane $z = 0$, ξ_n is given by the following expression

$$\xi_n = \Delta\xi/4 + n\Delta\xi = \Delta\xi/4 + 2\pi n/(2N'' + 1) \tag{5.30}$$

instead of (2.59). Obviously, in such a case, the expression of n_0 is modified as follows: $n_0 = \text{Int}\,[(\xi - \Delta\xi/4)/\Delta\xi]$.

The described NR NF–FF transformation with cylindrical scanning has been experimentally validated by means of the flexible NF measurement system available in the laboratory of antenna characterization of the University of Salerno, which can be arranged to perform a cylindrical scanning (see Sect. 1.2).

An open-ended WR90 rectangular waveguide, whose end is tapered to minimize the diffraction effects, is used as probe. The AUT, located in the plane $x = 0$, is a very simple H-plane monopulse antenna, operating at 10 GHz in the difference mode. It has been realized (see Figure 5 of [120]) by using two pyramidal horns (8.9 × 6.8 cm) of Lectronic Research Labs at a distance of 26 cm (between centres) and a hybrid Tee. According to the described sampling representation, the AUT has been modelled as enclosed in the surface Σ formed by a cylinder of height $h' = 35$ cm ended in two half-spheres of radius $a' = 3.4$ cm. The probe voltages have been collected on a cylinder having $d = 16.6$ cm and $2h = 240$ cm. To assess the effectiveness of the two-dimensional OSI algorithm, the amplitude and phase of the reconstructed probe voltage V_p relevant to the generatrix at $\varphi = 0°$ are compared in Figure 5.18a and b with those directly measured on the same generatrix. Note that the comparison between the phase of the recovered voltage and the measured one is shown only in the range [–20 cm, 120 cm] to improve the readability. As can be seen, there is an excellent agreement between the reconstructed voltage (crosses) and the measured one (solid line), save for the peripheral zone (below about –60 dB), wherein the error is caused both by the truncation of the scanning zone and the environmental reflections. It is worthy to note that, due to the filtering properties of the interpolation functions, the spatial harmonics relevant to the noise sources outside the AUT spatial bandwidth are cut away. This results in a smoother behaviour of the reconstructed voltage with respect to the measured one. Note that the reported reconstructions have been obtained using $\chi' = 1.35$, $\chi = 1.20$, and $p = q = 8$. The overall effectiveness of the described NF–FF transformation technique is assessed by comparing (see Figure 5.19) the FF pattern in the principal plane H reconstructed from the acquired NR NF data with that obtained, by using the software package MI-3000, from the data directly measured on the classical cylindrical grid. The same software has been used to get the FF reconstruction from the NR NF data. To this end, the OSI algorithm has been employed for recovering the cylindrical data needed to carry out the NF–FF transformation. As can be seen, there is a very good agreement, thus confirming the effectiveness of the approach. For completeness, the comparison with direct FF measurements performed at the anechoic chamber of the University of Naples "Parthenope" is also shown in Figure 5.20. As can be seen, although the measurements have been carried out with quite different techniques and environmental conditions, a very good agreement results save for small discrepancies in the far-out side lobes region, due to the truncation of the scan zone.

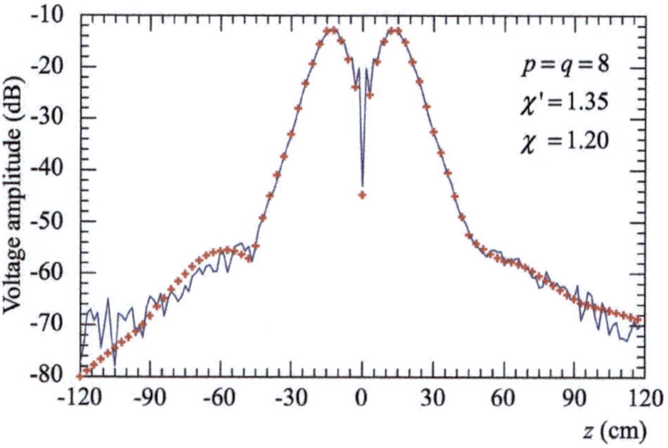

(a)

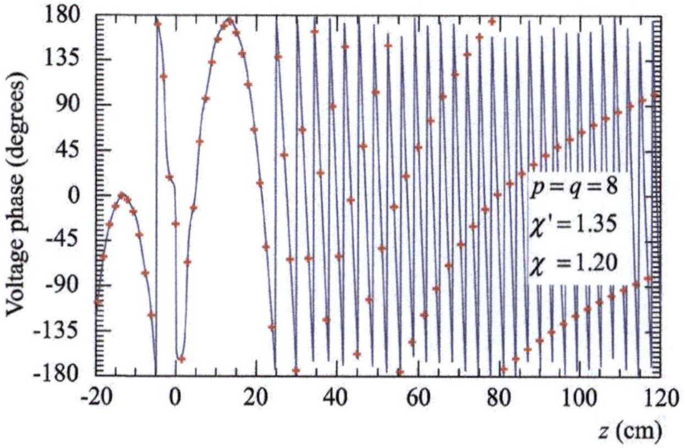

(b)

Figure 5.18 *Probe voltage V_p on the generatrix at $\varphi = 0°$. Solid line: measured. Crosses: reconstructed from the acquired NR NF samples. (a) Amplitude, (b) Phase.*

It is interesting to compare the number of data (1181) needed by the proposed NR NF–FF transformation with cylindrical scan with that (5760) required by MI software package to cover the same scanning zone. As shown, the described technique allows one to significantly reduce the number of needed NF data and related measurement time without losing the accuracy of the classical approach.

5.5.3 Direct non-redundant NF–FF transformation with cylindrical scanning

An efficient probe-compensated NF–FF transformation technique for evaluating the far field directly from the NR NF data acquired on the cylinder rings is described in this subsection. Also in such a case, it is assumed that a non-directive probe is

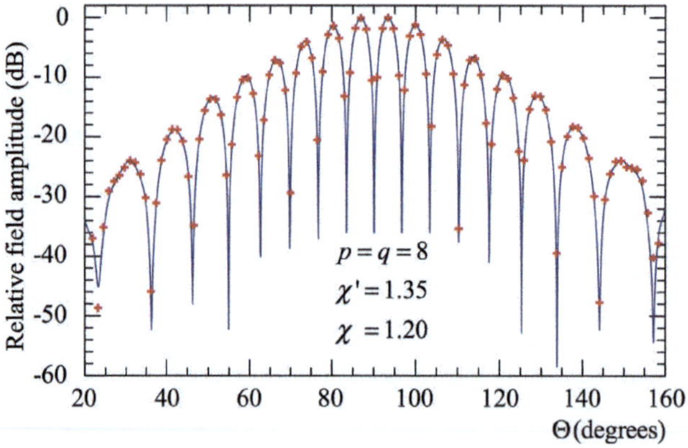

Figure 5.19 *H-plane pattern. Solid line: obtained from the NF data directly measured on the classical cylindrical grid. Crosses: reconstructed from the acquired NR NF samples.*

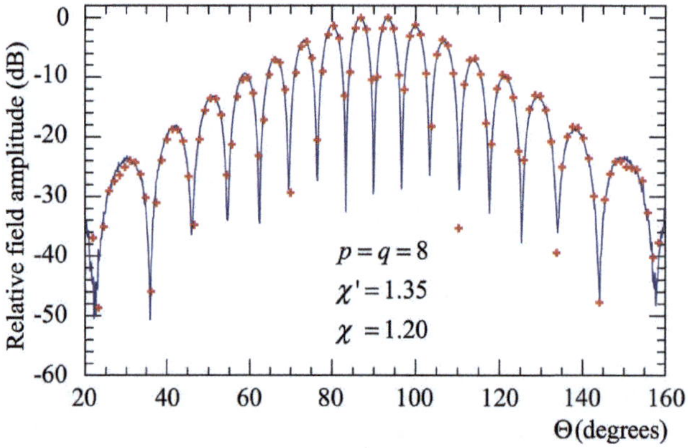

Figure 5.20 *H-plane pattern. Solid line: obtained from direct FF measurements. Crosses: reconstructed from the acquired NR NF samples.*

acquiring the NF data radiated on a cylinder of radius d by an elongated AUT, which is modelled by either a prolate spheroid or a rounded cylinder. By taking into account the OSI expansion (2.58) relevant to the probe voltage on the cylinder generatrix at φ, equation (5.24a) can be rearranged in the form:

$$I_\nu(\eta) = \sum_{n \in N_r} \int_{-\infty}^{\infty} \int_{-\pi}^{\pi} \tilde{V}_p(\xi_n, \varphi)\, Q(\xi(z), \xi_n, \bar{\xi}, N, N'')\, \mathrm{e}^{-\mathrm{j}\gamma(z)} \mathrm{e}^{-\mathrm{j}\nu\varphi} \mathrm{e}^{\mathrm{j}\eta z} \mathrm{d}\varphi\, \mathrm{d}z \quad (5.31)$$

wherein N_r is the set of indexes of the considered NF rings and $Q(\cdot, \cdot, \cdot, \cdot, \cdot) = SF(\cdot, \cdot, \cdot, \cdot, \cdot)$, if $|\xi(z) - \xi_n| \le q\Delta\xi$, or $Q(\cdot, \cdot, \cdot, \cdot, \cdot) = 0$, otherwise.

By expressing $\tilde{V}_p$ as a Fourier series in φ, namely,

$$\tilde{V}_p(\xi_n, \varphi) = \sum_{k=-M'_n}^{M'_n} \tilde{V}_p^k(\xi_n)\, \mathrm{e}^{\mathrm{j}k\varphi} \tag{5.32}$$

where M'_n is given by (2.55), relation (5.31) can be rewritten as:

$$I_\nu(\eta) = \sum_{n \in N_r} \tilde{V}_p^\nu(\xi_n)\, G_{n\eta} \tag{5.33}$$

wherein the coefficients $G_{n\eta}$ are given by:

$$G_{n\eta} = 2\pi \int_{z_i}^{z_f} SF(\xi(z)\,\xi_n, \bar{\xi}, N, N'')\, \mathrm{e}^{-\mathrm{j}\gamma(z)}\, \mathrm{e}^{\mathrm{j}\eta z}\, \mathrm{d}z \tag{5.34}$$

with $z_i = z(\xi_n + q\Delta\xi)$ and $z_f = z(\xi_n - q\Delta\xi)$. An expression quite analogous to (5.33) holds also for $I'_\nu(\eta)$.

It is noteworthy that the coefficients $G_{n\eta}$, depending only on the geometric parameters of the adopted AUT modelling and on the radius of the scanning cylinder, can be pre-calculated, once and for all, for given sets of antennas.

In the light of the above results, the far field can be effectively evaluated as follows: the acquired NR NF samples are multiplied by the phase factor $\mathrm{e}^{\mathrm{j}\gamma(\xi_n)}$ and the related Fourier series coefficients computed via the FFT algorithm; for each required value of Θ, the coefficients $G_{n\eta}$ relevant to the corresponding value of η are calculated or read, if available; the Fourier transforms $I_\nu(\eta)$ and $I'_\nu(\eta)$ are evaluated by performing the related summations; the CWE coefficients a_ν and b_ν are then determined by means of (5.22) and (5.23). The FF spherical components of the electric field radiated by the AUT can be finally obtained by substituting the CWE coefficients in relations (5.6) and (5.7), whose summations are again efficiently performed via FFT. From the efficiency point of view, it is convenient to employ such a technique to evaluate only the FF samples required to recover the antenna far field through an effective OSI expansion. To this end, the OSI expansion (2.76), properly modified to deal with even numbers of samples along the meridians and parallels, can be employed:

$$E_{\Theta,\Phi}(\Theta(\xi), \Phi) = \sum_{n=n_0-q+1}^{n_0+q} \Big\{ \Omega_{N_F}(\xi - \xi_n, \bar{\xi})\, D_{N''_F}^{e}(\xi - \xi_n) \times \sum_{m=m_0-p+1}^{m_0+p} E_{\Theta,\Phi}(\xi_n, \Phi_{m,n})\, \Omega_{M_n}(\Phi - \Phi_{m,n}, \bar{\Phi}_n)\, D_{M''_n}^{e}(\Phi - \Phi_{m,n}) \Big\} \tag{5.35}$$

where $n_0 = \mathrm{Int}[\xi/\Delta\xi]$ and $m_0 = \mathrm{Int}[\Phi/\Delta\Phi_n]$ are the indexes of the sample nearest (on the left) to the output point,

$$\xi_n = n\Delta\xi = n\pi/N''_F; \quad N''_F = 2\,[\mathrm{Int}(\chi N'/2) + 1]; \quad N' = \mathrm{Int}[\chi' W_\xi] + 1 \tag{5.36}$$

$$N_F = N''_F - N'; \quad \Phi_{m,n} = m\Delta\Phi_n = m\pi/M''_n; \quad M''_n = 2^i \geq \mathrm{Int}[M'_n] + 1 \tag{5.37}$$

$$M'_n = \mathrm{Int}[\chi^* W_\Phi(\xi_n)] + 1; \; M_n = M''_n - M'_n; \; \bar{\xi} = q\Delta\xi; \; \bar{\Phi}_n = p\Delta\Phi_n \tag{5.38}$$

$D^e_{L''}$, given by (5.20), is the Dirichlet function for an even number ($2L''$) of samples [307] and the other symbols have the same meanings as in (2.68). It is noteworthy that also in such a case, there is no need to extract the phase factor from the FF components, since it is constant on the FF sphere. The employment of an OSI expansion for an even number of samples on the FF parallels is a consequence of the use of a fast power of two FFT to perform the summations in (5.6) and (5.7), whereas the choice of N''_F according to (5.36) and, therefore, the employment of an analogous OSI expansions also on the FF meridians is required to have field samples on the FF parallel at $\Theta = 90°$.

It can be observed that, when the AUT is modelled by a rounded cylinder, the expressions of W_Φ and ξ relevant to the FF representation are simpler than those valid in the NF region. As a matter of fact [60], they become:

$$W_\Phi = W_\Phi(\Theta(\xi)) = \beta a' \sin(\Theta(\xi)); \quad \xi = \frac{\pi}{\ell'}\left[(1 - \cos\Theta)\, h' + 2a'\Theta\right] \tag{5.39}$$

ℓ' being the length of the curve $\mathcal{C}'$.

The former of (5.39) is a direct consequence of (2.15), whereas the latter can be easily derived by taking into account that, when the observation point is in the FF region, the two straight lines tangent to $\mathcal{C}'$ are parallel.

The interested reader can find the numerical validations of the direct NR NF–FF transformation techniques adopting the rounded cylinder and the prolate spheroidal modelling in [60] and [61], respectively. The experimental assessment of both these techniques is provided in [63].

As shown in [60, 61, 63], the described NF–FF transformation requires a remarkably reduced amount of NF samples and, accordingly, of measurement time with respect to the classical one [52]. However, the number of needed NF samples is slightly greater than those required by the NR transformations described in the previous two subsections. This is due to the need of acquiring, on each scanning ring, the number of NF samples required by the available FFT algorithm. For what concerns the comparison between the computational times required by this direct NR NF–FF transformation and the NR transformations described in the previous two subsections, it has been done in [59] with reference to the case of the spherical AUT modelling. The conclusion that, when using pre-calculated $G_{n\eta}$ values, the direct NR NF–FF transformation is more and more convenient stays valid for elongated antennas too. Accordingly, when the coefficients $G_{n\eta}$ are available, the moderate increase in the number of NF data required by the direct NR NF–FF transformation with respect to those needed by the other NR ones is a low price to pay. Moreover, the direct transformation allows a more flexible evaluation of the antenna FF pattern in any requested cut plane. In any case, it is always possible to evaluate the required NF data by interpolating, through the aforementioned one-dimensional OSI expansion along φ, the strictly NR ones.

5.6 Half-wavelength helicoidal scanning

The half-wavelength helicoidal scanning [142] results from the continuous and simultaneous linear movement of the probe and rotation of the AUT. The NF samples are collected on a circular helix with constant step Δz in z. The coordinates of a point P (see Figure 5.21) of the helix are given by the parametric equations:

$$\begin{cases} x = d\cos\phi \\ y = d\sin\phi \\ z = k\phi \end{cases} \tag{5.40}$$

where ϕ is the angular parameter describing the helix and $k = \Delta z / 2\pi$.

In such a scanning, the NF sample spacings are the same adopted in the classical cylindrical scanning [52], given by relations (5.5). Accordingly, the NF sampling points result to be aligned along the generatrices as in the conventional cylindrical scanning and, therefore, the so acquired data can be directly employed for reconstructing the far field without any intermediate interpolation.

The employment of the FFT algorithm for the efficient evaluation of the two-dimensional Fourier transforms I_v and I'_v of the voltages V_p and V_r from the NF data acquired by means of this helicoidal scanning is briefly described in the following by outlining the differences from the procedure reported in Sect. 5.2. In such a case, relations (5.24) can be so rewritten:

$$I_v(\eta) = \int_0^{2\pi} \int_{-h+k\varphi}^{h+k\varphi} V_p(\varphi, z)\, \mathrm{e}^{-\mathrm{j}v\varphi}\, \mathrm{e}^{\mathrm{j}\eta z}\, \mathrm{d}z\, \mathrm{d}\varphi \tag{5.41a}$$

$$I'_v(\eta) = \int_0^{2\pi} \int_{-h+k\varphi}^{h+k\varphi} V_r(\varphi, z)\, \mathrm{e}^{-\mathrm{j}v\varphi}\, \mathrm{e}^{\mathrm{j}\eta z}\, \mathrm{d}z\, \mathrm{d}\varphi \tag{5.41b}$$

Accordingly, in such a case the integration with respect to z must be carried out before that over φ. As regards this last integration, the considerations done in the

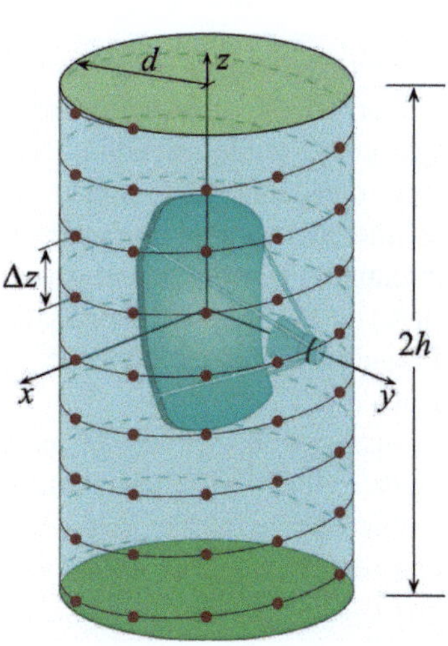

Figure 5.21 Half-wavelength helicoidal scanning

case of the cylindrical scanning still hold. On the contrary, particular attention must be paid when considering the integration over z. By taking explicitly into account the dependence of the integration limits on $\varphi_m = 2\pi m/N$, it can be easily verified that the analogous of (5.12) becomes:

$$G_m(\eta_i) \approx \frac{2h}{N_1} e^{-j\pi i} e^{j2\pi mi/NN_1} \sum_{n=0}^{N_1-1} g(z_n)\, e^{j2\pi ni/N_1} \tag{5.42}$$

where $z_n = (n - N_1/2 + m/N)\Delta z$, $N_1 = 2h/\Delta z$ being the number of the helix turns. As a consequence, save for the factor $2he^{-j\pi i} e^{j2\pi mi/NN_1}$, the integration with respect to z in relations (5.41) can be efficiently performed via an inverse FFT algorithm.

By taking into account the above discussion, it can be easily recognized that the FF reconstruction process must be so modified in order to allow the direct usage of the NF data collected through this helicoidal scanning: for each index m, these data are inverse transformed via FFT and then multiplied by the factor $2he^{-j\pi i} e^{j2\pi mi/NN_1}$; for each index i, the resulting output data are direct transformed via FFT and then multiplied by the factor $2\pi/N$.

In order to show the effectiveness of such a technique, a FF reconstruction example is reported in the following. The simulation refers to the same AUT, scanning cylinder, and probe considered in Sect. 5.3 with reference to the FF reconstruction shown in Figure 5.5, but the NF data are gathered now on a helix covering the scanning zone. The E-plane and H-plane FF patterns, obtained by applying this NF–FF transformation technique, are compared in Figure 5.22a and b with the exact ones. As can be seen, the agreement between the reconstructed and exact patterns is very good, thus assessing the effectiveness of the technique.

5.7 Non-redundant NF–FF transformations with helicoidal scanning

NR NF–FF transformations with helicoidal scanning will be presented in this section. They allow a further remarkable saving of measurement time with respect to the NR cylindrical NF–FF transformations described in Sect. 5.5, since the NR samples are gathered on fly and continuous and synchronized motions of the AUT rotating table and probe linear positioner are adopted. In particular, a NF–FF transformation using a uniform helicoidal scanning is summarized in subsection 5.7.1, whereas two NR NF–FF transformations with helicoidal scanning, tailored for elongated AUTs modelled by a prolate spheroid or a rounded cylinder, are presented in the subsections 5.7.2 and 5.7.3, respectively.

Although the NF–FF transformation with uniform helicoidal scanning requires a number of NF samples greater than those needed by the other two, its hardware implementation is simpler, since, being the step of the helix constant, the velocity of the linear positioner must not vary during the acquisition.

5.7.1 NF–FF transformation with uniform helicoidal scanning

Let a quasi-spherical AUT be considered as enclosed in a sphere Σ of radius a and let the voltage be acquired by an electrically small, non-directive, probe scanning a

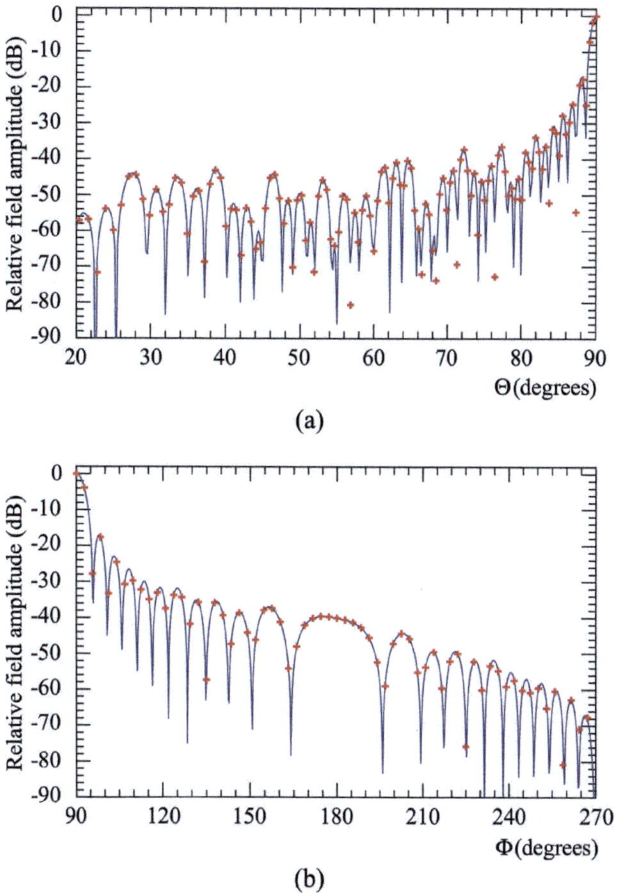

Figure 5.22 *FF patterns in the principal planes. Solid line: exact field. Crosses: reconstructed from the NF data acquired along the helix. (a) E-plane, (b) H-plane.*

helix, which wraps with constant step Δz in z a cylinder of radius $d > a$ (Figure 5.23). The Cartesian coordinates of a point Q on this helix are again given by (5.40), where ϕ is the angular parameter describing the helix and $k = \Delta z/2\pi$. However, in such a scanning, the NF sample spacings are no longer the same of the classical cylindrical scan [52]. In fact, the step Δz is now fixed according to the spatial bandwidth properties of EM fields [102] and, as a consequence, it grows when the cylinder radius increases and can become much greater than $\lambda/2$. Such a scanning [113] is attained, as in the case of the $\lambda/2$ helicoidal scan, by means of movements, at constant velocities, of the probe linear positioner and AUT rotating table.

A convenient way to determine the FF pattern of an AUT from a NR number of NF samples acquired along a proper helix with constant step in z is the development of an efficient interpolation algorithm for the voltage reconstruction over the cylindrical surface from these NR samples. In such a way, it is possible to accurately

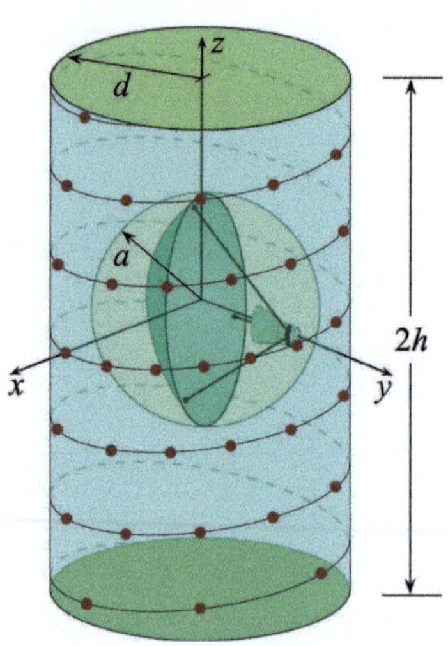

Figure 5.23 Uniform helicoidal scanning

reconstruct the NF data required by the classical NF–FF transformation with cylindrical scanning [52]. As outlined in Chapter 3, the two-dimensional OSI expansion to determine the probe voltage on the cylindrical surface from the collected helicoidal samples can be, in such a case, obtained:

i. by choosing the helix step coincident with the sampling spacing needed to interpolate the NF data along a cylinder generatrix according to the spatial bandwidth properties of EM fields [102], i.e.,

$$\Delta z = \frac{\pi d}{\chi \chi' \beta a} \tag{5.43}$$

ii. by developing a NR sampling representation of the voltage along the helix.

In such a way, the voltage at a point P can be reconstructed by evaluating the intermediate samples (those in correspondence of the intersection points between the helix and the generatrix through P) via an interpolation along the helix and then interpolating, along the generatrix, the so determined intermediate samples.

The development of the NR sampling representation along the helix is now considered. The phase function ψ and the optimal parameter η to describe the helix are [104] given by (2.4) and (2.6), here reported for convenience:

$$\psi(\eta) = \frac{\beta}{2} \int_0^s \left[\max_{r'} \hat{R} \cdot \hat{t} + \min_{r'} \hat{R} \cdot \hat{t} \right] ds \tag{5.44a}$$

$$\eta = \frac{\beta}{2W_\eta} \int_0^s \left[\max_{r'} \hat{R} \cdot \hat{t} - \min_{r'} \hat{R} \cdot \hat{t} \right] \mathrm{d}s \tag{5.44b}$$

It can be easily verified that the maximum and minimum values of $\hat{R} \cdot \hat{t}$ occur [113] at the tangency points $P_{1,2}$ (see Figure 5.24) of the sphere with the straight lines passing through the point Q on the helix and belonging to the plane specified by the unit vectors $\hat{t}$ (tangent to the helix at Q) and $\hat{r}$ (pointing from the origin to Q).

With the aid of Figure 5.24, it can be easily recognized that:

$$(\hat{R}_1 + \hat{R}_2)/2 = \hat{r} \sin\alpha = \hat{r}\sqrt{1 - a^2/r^2} \tag{5.45a}$$

$$(\hat{R}_1 - \hat{R}_2)/2 = \hat{n} \cos\alpha = \hat{n}(a/r) \tag{5.45b}$$

where $\hat{R}_{1,2}$ are the unit vectors pointing from $P_{1,2}$ to Q and $\hat{n}$ is a unit vector orthogonal to $\hat{r}$ and parallel to the plane identified by $\hat{r}$ and $\hat{t}$. Accordingly,

$$(\hat{R}_1 - \hat{R}_2) \cdot \hat{t}/2 = (\hat{n} \cdot \hat{t})(a/r) = (a/r)\cos(\pi/2 - \delta) = (a/r)\sin\delta \tag{5.46}$$

where δ is the angle between $\hat{r}$ and $\hat{t}$.

By substituting (5.45a) into (5.44a) and taking into account that $\mathrm{d}r = \hat{r} \cdot \hat{t}\,\mathrm{d}s$, it results:

$$\psi = \beta \int_0^r \sqrt{1 - a^2/r^2}\,\mathrm{d}r = \beta\sqrt{r^2 - a^2} - \beta a \cos^{-1}(a/r) \tag{5.47}$$

It is noteworthy that the expression (5.47) of the phase function ψ is different from that originally got in [113], but, since the difference is an additive constant value, these expressions are wholly equivalent from the practical viewpoint.

By taking into account the helix equations (5.40), it follows that

$$\mathrm{d}s = \sqrt{(\mathrm{d}x)^2 + (\mathrm{d}y)^2 + (\mathrm{d}z)^2} = \sqrt{d^2 + k^2}\,\mathrm{d}\phi \tag{5.48}$$

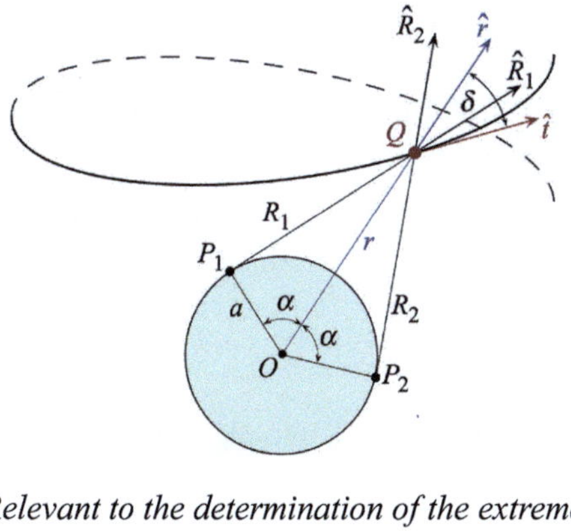

Figure 5.24 Relevant to the determination of the extreme values of $\hat{R} \cdot \hat{t}$

On the other hand,

$$\sin\delta = \sqrt{1-\cos^2\delta} = \sqrt{1-(\hat{r}\cdot\hat{t})^2} \tag{5.49}$$

where

$$\hat{r}\cdot\hat{t} = \frac{\mathrm{d}r}{\mathrm{d}s} = \frac{\mathrm{d}r}{\mathrm{d}\phi}\frac{\mathrm{d}\phi}{\mathrm{d}s} = \frac{1}{\sqrt{d^2+k^2}}\frac{\mathrm{d}r}{\mathrm{d}\phi} = \frac{k^2\phi}{\sqrt{d^2+k^2}\sqrt{d^2+k^2\phi^2}} \tag{5.50}$$

accordingly,

$$\sin\delta = \frac{d\sqrt{d^2+k^2+k^2\phi^2}}{\sqrt{d^2+k^2}\sqrt{d^2+k^2\phi^2}} \tag{5.51}$$

By taking into account the above relation, by substituting (5.46) and (5.48) into (5.44b), and choosing $W_\eta = \beta a$, it results:

$$\eta = \int_0^\phi \frac{d\sqrt{d^2+k^2+k^2\phi^2}}{d^2+k^2\phi^2}\,\mathrm{d}\phi \tag{5.52}$$

This last integral can be evaluated after lengthy but straightforward manipulations, thus obtaining:

$$\eta = \tau + \frac{\ln\left(\phi\sin\tau + \sqrt{(\phi\sin\tau)^2+1}\right)}{\tan\tau} + \tan^{-1}\left(\frac{-2\sin\tau\cos\tau}{2\cos^2\tau - 1 + (\phi\sin\tau + \sqrt{(\phi\sin\tau)^2+1})^2}\right) \tag{5.53}$$

wherein $\tau = \tan^{-1}(\Delta z/2\pi d) = \tan^{-1}(k/d)$.

In accordance with the above results, the voltage at a given point Q on the helix can be accurately reconstructed [113] from the NR samples gathered along the helix, by means of the OSI expansion (see (2A.32)):

$$V(\eta) = \mathrm{e}^{-\mathrm{j}\psi(\eta)} \sum_{m=m_0-p+1}^{m_0+p} \tilde{V}(\eta_m)\,\Psi(\eta-\eta_m)\,\mathrm{sinc}\left(\pi(\eta-\eta_m)/\Delta\eta\right) \tag{5.54}$$

wherein $m_0 = \mathrm{Int}(\eta/\Delta\eta)$ is the index of the sample closest (on the left) to the point Q, $2p$ is the number of the retained closest reduced voltages samples $\tilde{V}(\eta_m) = V(\eta_m)\,\mathrm{e}^{\mathrm{j}\psi(\eta_m)}$, and

$$\eta_m = m\Delta\eta = m\pi/(\chi\chi'\beta a) \tag{5.55}$$

Moreover, $\Psi(\eta)$ is the Knab's Sampling Window function (2A.33), where $\bar{\eta} = p\Delta\eta$ and $v = 1 - 1/\chi$.

The OSI expansion (5.54) can be employed to evaluate the intermediate voltages samples at the intersection points between the generatrix through P and the helix. The voltage at P can be, then, determined [113] by applying the OSI expansion:

$$V(s) = \mathrm{e}^{-\mathrm{j}\beta r} \sum_{n=n_0-q+1}^{n_0+q} \tilde{V}(s_n)\, \Psi(s - s_n)\, \mathrm{sinc}\left(\pi\left(s - s_n\right)/\Delta s\right) \tag{5.56}$$

where $n_0 = \mathrm{Int}\,[(s - s_0)/\Delta s]$, $s = z/d$, $\tilde{V}(s_n) = V(s_n)\,\mathrm{e}^{\mathrm{j}\beta r}$ [113] are the intermediate reduced voltage samples, $s_n = s_n(\varphi) = k\varphi/d + n\Delta s = s_0 + n\Delta s$ are their normalized abscissas, $\bar{s} = q\Delta s$, and the other symbols have the same meanings as in (5.54). It is so possible to accurately recover the NF data needed by the classical cylindrical NF–FF transformation [52] from the NR samples acquired on the helix.

The interested reader can find numerical tests assessing the effectiveness of this NF–FF transformation with uniform helicoidal scanning in [113].

5.7.2 Non-redundant helicoidal NF–FF transformation: the prolate spheroidal modelling case

Let an elongated AUT be considered as enclosed in a prolate spheroid Σ with major and minor semi-axes equal to a and b and let the voltage be acquired by an electrically small, non-directive, probe scanning a proper helix, which wraps a cylinder of radius $d > b$ (Figure 5.25). The unified theory of spiral scannings for non-spherical antennas [124] (see Sect. 3.2) can be applied to efficiently reconstruct the NF data needed by the classical probe-compensated NF–FF transformation with cylindrical scanning [52] from the NR voltage samples acquired on the helix [116]. In order to simplify the realization of the scanning from the practical viewpoint, it is convenient that the helix originates from a given point P_0 at the top of the cylinder. Accordingly, the spiral equations (3.23), when

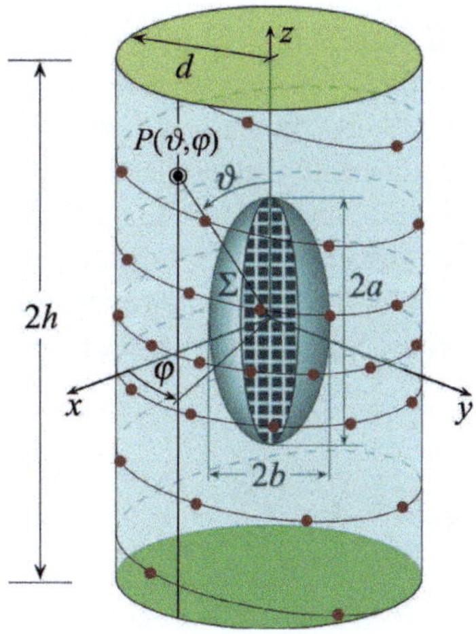

Figure 5.25 NR helicoidal scanning: prolate spheroidal modelling

imposing the passage through such a point on the generatrix at $\varphi = 0$ and taking into account that $r(\theta) = d/\sin\theta$ for a helix, become:

$$\begin{cases} x = d\cos(\phi - \phi_i) \\ y = d\sin(\phi - \phi_i) \\ z = d\cot\theta(\xi) \end{cases} \tag{5.57}$$

wherein ϕ is the angular parameter which describes the helix, ϕ_i is the value of ϕ at P_0, and $\xi = k\phi$. The constant k is determined by imposing that the helix step $2\pi k$ must be equal to the sampling spacing $\Delta\xi$ needed in the interpolation along a generatrix, which, according to (2.59), is given by $\Delta\xi = 2\pi/(2N'' + 1)$. Therefore, it results $k = 1/(2N'' + 1)$.

It is noteworthy that, by adopting the prolate spheroidal modelling instead of the spherical one [114, 115], it is possible to reduce the radius of the scanning cylinder, thus giving rise to a significant mitigation of the truncation error for a given size of the scanning zone. In addition, the reduced volumetric redundancy of this model allows a remarkable lowering of the number of required NF measurements.

It is worthy to note that the helix can be viewed as obtained by projecting on the measurement cylinder, through the hyperbolas at ξ = constant (see Figure 2.6), the spiral wrapping, with the same step, the modelling prolate spheroid Σ.

As outlined in Chapter 3, the NR sampling representation of the probe voltage from its samples acquired on the helix, whose step is equal to the sample spacing required for the interpolation along a generatrix, can be attained by developing a NR sampling voltage representation along such a helix.

The bandwidth W_ξ, the phase function γ, and the optimal parameter ξ to describe a cylinder generatrix are given by the expressions (2.25), (2.27), and (2.28a).

As regards the NR sampling representation along the helix, it can be determined by applying the heuristic procedure [124] described in Sect. 3.2, which parallels the rigorous procedure [122] valid when adopting the spherical AUT modelling. Accordingly, the parameter η for describing the helix is β/W_η times the curvilinear abscissa of the projecting point lying on the spiral which wraps the spheroid Σ, the related phase function ψ coincides with that γ relevant to a generatrix, and W_η is β/π times the length of the spiral wrapping Σ from the north to the south pole.

The expression of the two-dimensional OSI expansion (3.24), which allows the fast and accurate reconstruction of the probe and rotated probe voltages at any point on the cylinder, is reported in the following for reader's convenience:

$$V(\xi(\vartheta), \varphi) = e^{-j\gamma(\xi)} \times \sum_{n=n_0-q+1}^{n_0+q} \left[SF(\xi, \xi_n, \bar{\xi}, N, N'') \sum_{m=m_0-p+1}^{m_0+p} \tilde{V}(\eta_m)\, SF(\eta(\xi_n), \eta_m, \bar{\eta}, M, M'') \right] \tag{5.58}$$

where $m_0 = \mathrm{Int}[(\eta - \eta(\phi_i))/\Delta\eta]$,

$$\xi_n = \xi_n(\varphi) = \xi(\phi_i) + k\varphi + n\Delta\xi = \xi_0 + n\Delta\xi; \quad \eta_m = \eta(\phi_i) + m\Delta\eta \tag{5.59}$$

and all other symbols have the same meanings as in (3.24).

Some numerical tests assessing the effectiveness of this technique are reported in the following. The simulations refer to the previously considered elliptical AUT and the chosen probe is again an open-ended WR-90 rectangular waveguide working at 10 GHz. The helix wraps a cylinder having radius $d = 12\lambda$ and height $2h = 160\lambda$. The AUT has been modelled by a prolate spheroid with major and minor semi-axes equal to 24 λ and 6 λ, respectively. Figure 5.26a and b show the reconstruction of the amplitude and phase of the probe voltage V_p on the generatrix at $\varphi = 90°$. As can be observed, there is an excellent agreement between the reconstructed voltage and the exact one. The accuracy of the representation is assessed in a more quantitative way by the values of the maximum error in the reconstruction voltage V_p (normalized to the voltage maximum value on the cylinder) shown in Figure 5.27. The evaluation has been performed by comparing the interpolated values of V_p with those directly computed on a close grid in the central zone of the cylinder, so that the existence of

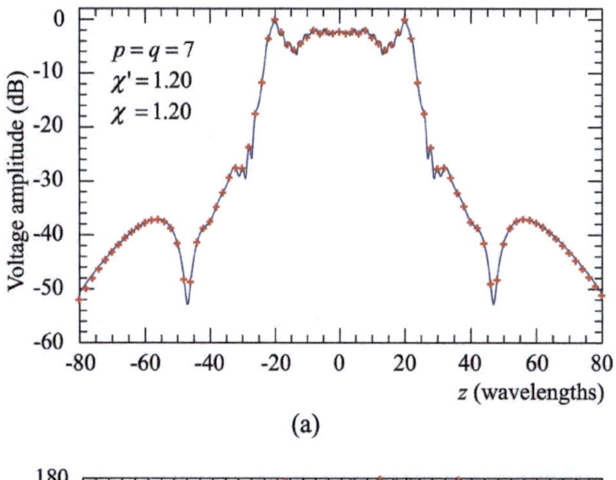

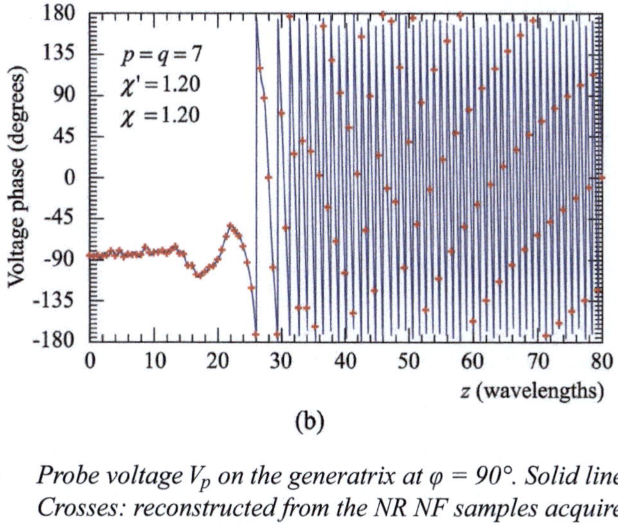

Figure 5.26 *Probe voltage V_p on the generatrix at $\varphi = 90°$. Solid line: exact. Crosses: reconstructed from the NR NF samples acquired along the helix. (a) Amplitude, (b) Phase.*

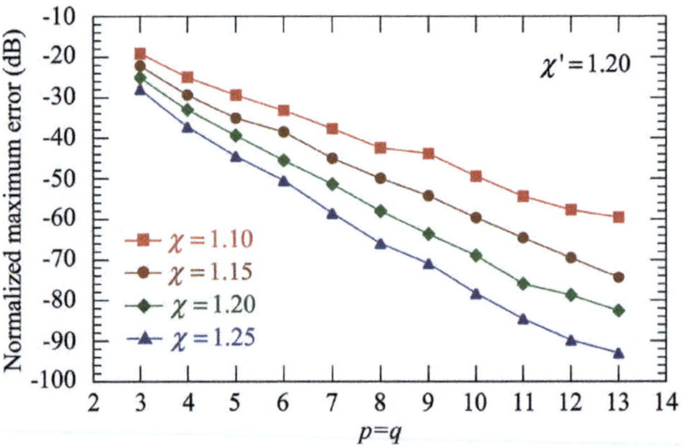

Figure 5.27 *Normalized maximum error in the reconstruction of the voltage V_p*

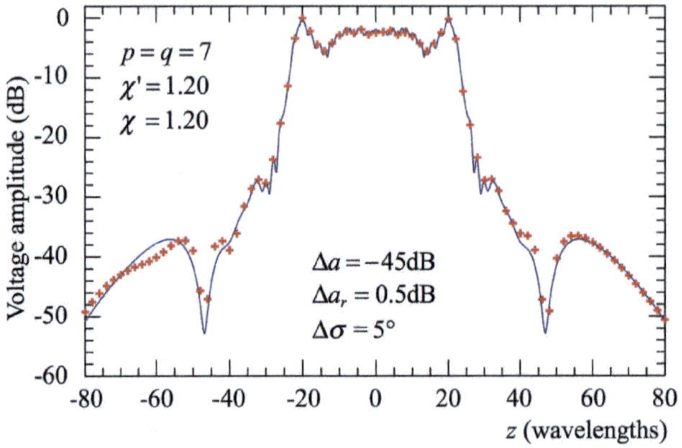

Figure 5.28 *Amplitude of the probe voltage V_p on the generatrix at $\varphi = 90°$. Solid line: exact. Crosses: reconstructed from error affected samples.*

the required guard samples is ensured. As can be seen, the maximum error decreases up to very low values on increasing the oversampling factor and/or the number of retained samples. The curves relevant to the mean-square error, not reported here for space saving, exhibit a quite analogous behaviour running about 20 dB down with respect to the maximum error ones. The algorithm stability with respect to errors affecting the samples has been investigated by corrupting the exact samples with random errors. These errors simulate a background noise (bounded to Δa dB in amplitude and with arbitrary phase) and uncertainties on the data of $\pm\Delta a_r$ dB in amplitude and $\pm\Delta\sigma$ degrees in phase. As shown in Figure 5.28, the algorithm is stable. At last, the OSI algorithm (5.58) has been applied to recover the NF data needed for the classical cylindrical NF–FF transformation [52]. The reconstructions of the antenna FF pattern in the principal planes are reported in Figure 5.29a and b. As can be seen, the exact and recovered patterns are practically indistinguishable.

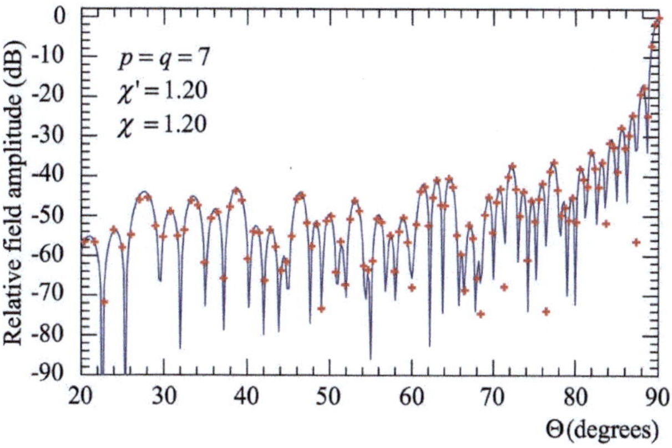

(a)

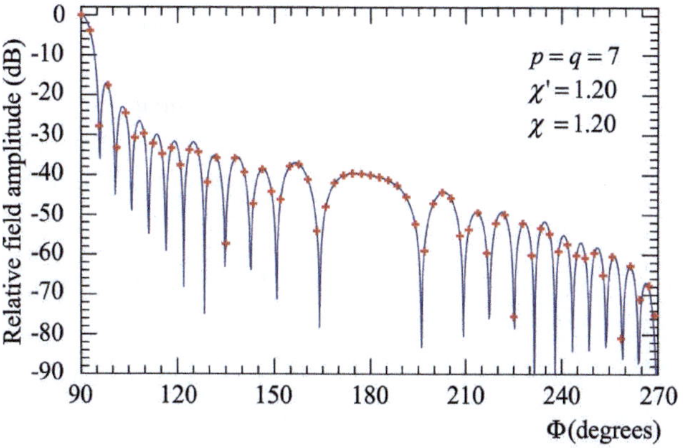

(b)

Figure 5.29 *FF patterns in the principal planes. Solid line: exact field. Crosses: reconstructed from NR NF samples acquired along the helix. (a) E-plane, (b) H-plane.*

It can be interesting to compare the number of used NF samples (12150) with that (40960), which would be required by applying the classical cylindrical NF–FF transformation [52] or that using the half-wavelength helicoidal scanning [142]. In conclusion, the described NR NF–FF transformation with helicoidal scanning allows a remarkable reduction of the required NF data and related measurement time without any loss of accuracy in the recovery of the AUT far field. The interested reader can find in [117] experimental results assessing the effectiveness of such a NF–FF transformation also from a practical viewpoint.

As stated, by adopting the prolate spheroidal modelling instead of the spherical one, it is possible to reduce the radius of the scanning cylinder, thus significantly reducing the truncation error for a given height of the cylinder. In order to highlight such a capability, the reconstruction of the E-plane FF pattern attained by employing the spherical modelling is shown in Figure 5.30. In such a case, the NF samples have

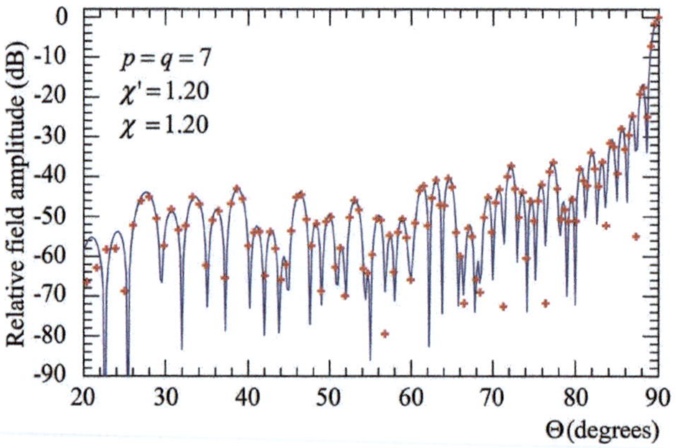

Figure 5.30 *E-plane pattern. Solid line: exact field. Crosses: reconstructed from NR NF samples acquired along the helix when using the spherical modelling.*

been gathered on a helix which wraps a cylinder having the same height but radius of 30 λ. As can be clearly observed, the FF reconstruction is less accurate in the far out side lobe region.

5.7.3 *Non-redundant helicoidal NF–FF transformation: the rounded cylinder modelling case*

Let a long AUT be considered as contained in a rounded cylinder Σ, i.e., a cylinder of height h' terminated in two half-spheres of radius a' and let the voltage be acquired by an electrically small, non-directive, probe scanning a proper helix wrapping a cylinder of radius $d > a'$ (see Figure 5.31). As in the previous subsection, the unified theory of spiral scannings for non-spherical antennas [124] can be applied to accurately determine the NF data needed by the classical probe-compensated NF–FF transformation with cylindrical scanning [52] from the NR voltage samples acquired on the helix [105, 118]. Also in such a case, it is convenient that the helix originates from a given point P_0 at the top of the cylinder, thus the helix equations are again given by the relations (5.57).

It is perhaps superfluous to stress that the use of the rounded cylinder modelling allows one to obtain a significant reduction both of the amount of needed NF measurements and of the truncation error when dealing with elongated antennas.

Since, when adopting the rounded cylinder modelling, the development of the NR sampling representation of the probe voltage from its samples acquired on the helix is quite similar to that previously described for the prolate spheroidal modelling, only the differences will be underlined in the following for space saving. The projecting curves at ξ = constant are no longer hyperbolas, but those shown in

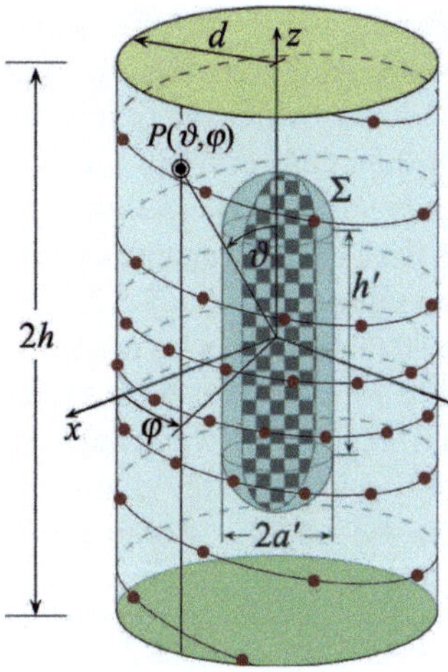

Figure 5.31 *NR helicoidal scanning: rounded cylinder modelling*

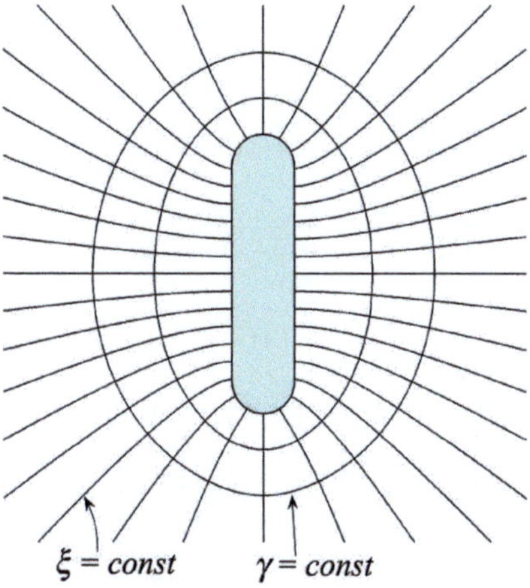

Figure 5.32 *Rounded cylinder modelling: curves at ξ = constant and γ = constant*

Figure 5.32. The bandwidth W_ξ, the phase function γ, and the optimal parameter ξ to describe a cylinder generatrix are now given by the expressions (5.29), where the involved parameters can be evaluated as described in Sect. 5.5.2.

The OSI expansion (5.58) can be again employed to efficiently reconstruct the probe and rotated probe voltages at any point on the cylinder and, in particular, at those needed by the classical NF–FF transformation with cylindrical scanning [52].

Some experimental results, assessing the effectiveness of such a NR NF–FF transformation with helicoidal scanning also from the practical of viewpoint and in part already published in [105], are shown in the following. The experimental proofs have been performed through the cylindrical NF facility available in the Antenna Characterization Lab of the University of Salerno (see Sect. 1.2).

The employed probe, the sizes of the scanning cylinder over which the helix is wrapping, and the AUT are the same as in Sect. 5.5.2. However, the H-plane monopulse AUT is working now at 10 GHz in the sum mode. As in Sect. 5.5.2, a rounded cylinder with $h' = 35$ cm and $a' = 3.4$ cm has been adopted as surface Σ modelling it. To assess the effectiveness of the two-dimensional OSI algorithm, the amplitude and phase of the probe voltage V_p relevant to the generatrix at $\varphi = 0°$, interpolated by using $\chi' = 1.35$, $\chi = 1.20$, and $p = q = 6$, are compared in Figure 5.33a and b with the

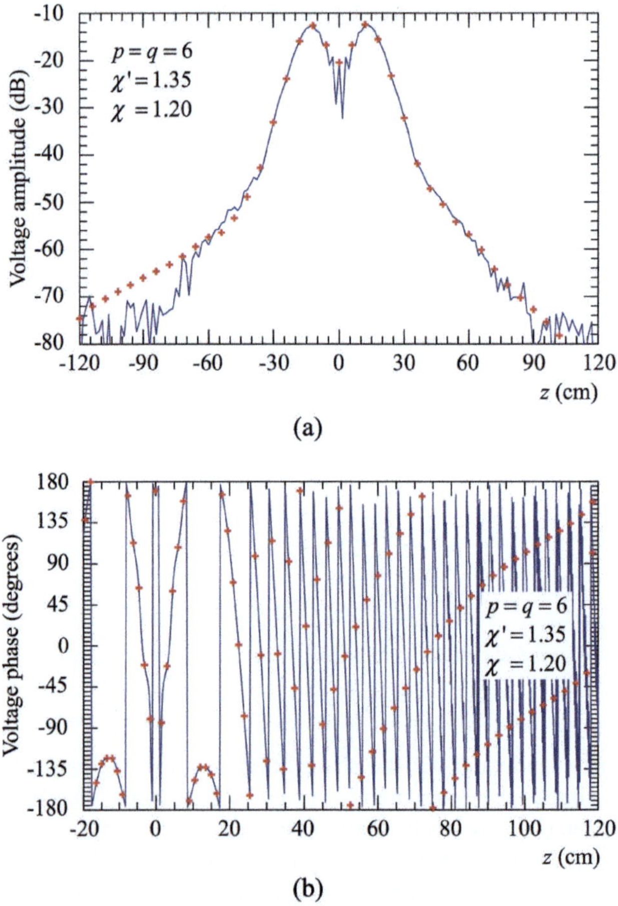

Figure 5.33 *Probe voltage V_p on the generatrix at $\varphi = 0°$. Solid line: measured. Crosses: reconstructed from the NR NF samples acquired along the helix. (a) Amplitude, (b) Phase.*

directly measured ones. Also in such a case, the comparison between the phase of the reconstructed voltage and the measured one is shown only in the range [–20 cm, 120 cm] in order to improve the readability. At last, the reconstruction of the amplitude of V_p relevant to the generatrix at $\varphi = 30°$ is shown in Figure 5.34. As can be seen, there is a very good agreement between the interpolated voltages and the measured ones, save for the zones characterised by very low voltage values, where the behaviour of reconstructed voltage is smoother, due to the already mentioned filtering properties of the OSI functions. The overall effectiveness of such a NF–FF transformation with helicoidal scanning is assessed by comparing (see Figure 5.35a and b) the FF patterns in the principal planes, reconstructed from the NR NF samples acquired on the helix, with those obtained from the NF data directly measured on the classical cylindrical grid. In both the cases, the software package MI-3000 has been used to get the FF reconstructions. The accuracy of this NF–FF transformation technique is further assessed by comparing (see Figure 5.36) the H-plane pattern reconstructed from the NF acquired on the helix with that directly obtained from FF measurements performed at the anechoic chamber of the University of Naples "Parthenope".

At last, the H-plane pattern relevant to the same monopulse antenna operating in the difference mode and reconstructed from the helicoidal NF data is compared with that (Figure 5.37) obtained from classical NF cylindrical measurements and with that (Figure 5.38) directly measured in the FF zone.

It is interesting to compare the number of data (948) needed by this NR NF–FF transformation with helicoidal scanning with that (5760) required by MI software package to cover the same scanning zone. As shown, the described technique allows one to significantly reduce the number of needed NF data and related measurement time, without losing the accuracy of the classical approach.

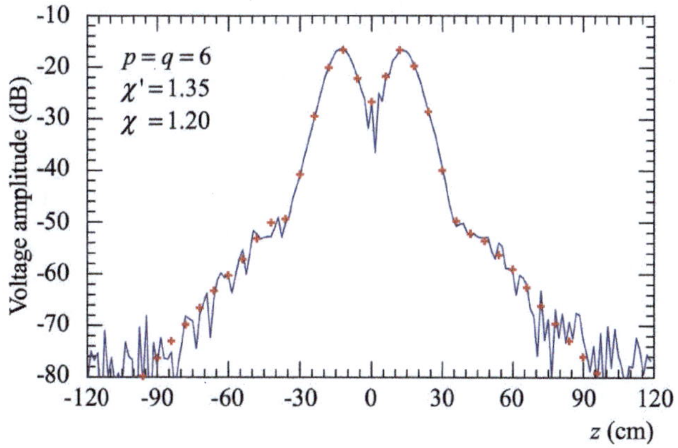

Figure 5.34 *Amplitude of the probe voltage V_p on the generatrix at $\varphi = 30°$. Solid line: measured. Crosses: reconstructed from the NR NF samples acquired along the helix.*

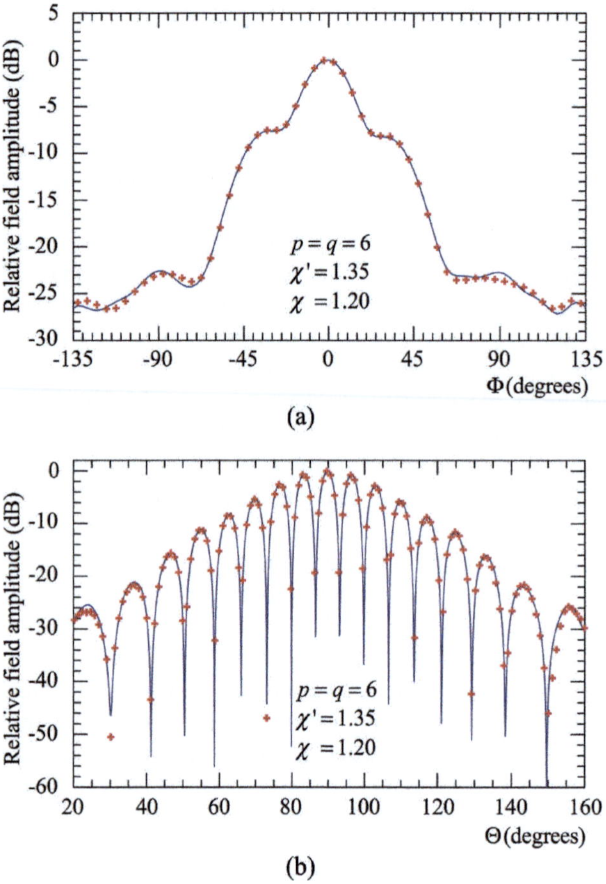

Figure 5.35 *FF patterns in the principal planes. Solid line: obtained from classical NF cylindrical measurements. Crosses: reconstructed from NR NF samples acquired along the helix. (a) E-plane, (b) H-plane.*

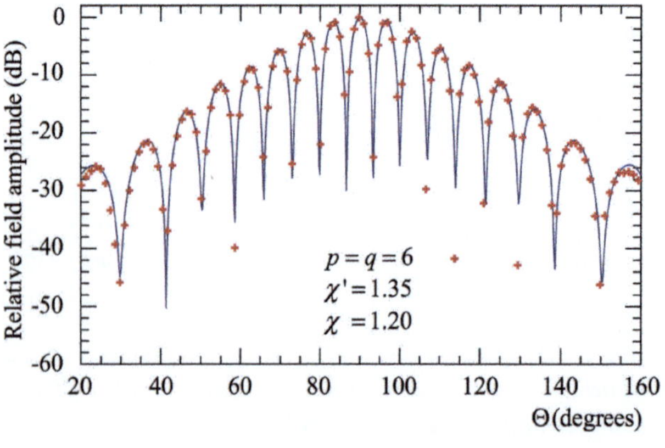

Figure 5.36 *H-plane pattern. Solid line: obtained from direct FF measurements. Crosses: reconstructed from NR NF samples acquired along the helix.*

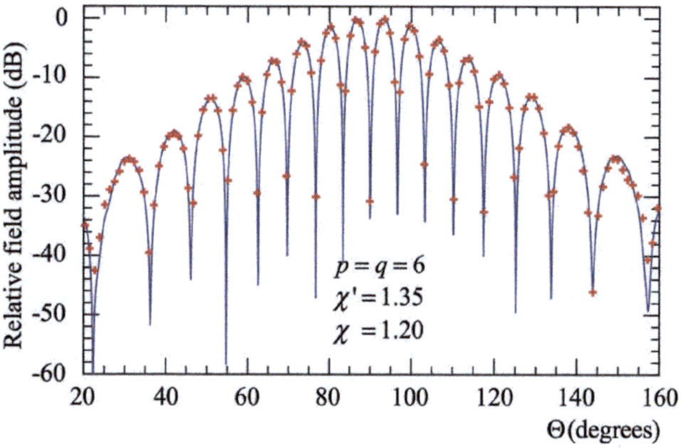

Figure 5.37 *Difference mode. H-plane pattern. Solid line: obtained from classical NF cylindrical measurements. Crosses: reconstructed from NR NF samples acquired along the helix.*

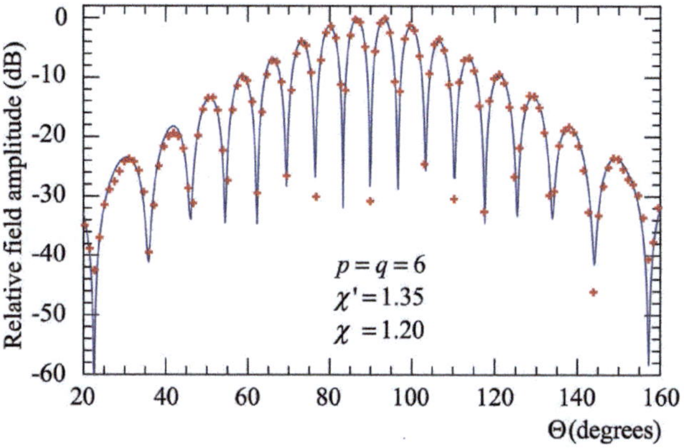

Figure 5.38 *Difference mode. H-plane pattern. Solid line: obtained from direct FF measurements. Crosses: reconstructed from NR NF samples acquired along the helix.*

Appendix

5A.1 Cylindrical wave expansion

The CWE will be briefly summarized in this appendix. In such an expansion the EM field is expressed as superposition of transverse electric (TE) and transverse magnetic (TM) cylindrical modes, elementary solutions of the homogeneous wave equation in the cylindrical coordinates (ρ, φ, z). As a matter of fact, in a homogeneous source-free region, an arbitrary EM field can be always represented as sum of a TE field and a TM field [308]. The expressions of these modes in a cylindrical coordinate system can be determined in a simple way by employing the vector potentials $\boldsymbol{A}$ and $\boldsymbol{F}$.

As shown in Appendix A.1, it results

$$\boldsymbol{E} = -\mathrm{j}\omega \boldsymbol{A} + \frac{\nabla\nabla \cdot \boldsymbol{A}}{\mathrm{j}\omega\varepsilon\mu} - \frac{1}{\varepsilon}\nabla \times F \tag{5A.1}$$

$$\boldsymbol{H} = \frac{1}{\mu}\nabla \times \boldsymbol{A} - \mathrm{j}\omega \boldsymbol{F} + \frac{\nabla\nabla \cdot \boldsymbol{F}}{\mathrm{j}\omega\varepsilon\mu} \tag{5A.2}$$

where the magnetic and electric vector potentials $\boldsymbol{A}$ and $\boldsymbol{F}$ can be determined by solving the vector Helmholtz equations (A.11) and (A.16), respectively. These last, being in such a case $\boldsymbol{J} = \boldsymbol{J}_m = 0$, reduce to

$$\nabla^2 \boldsymbol{A} + \beta^2 \boldsymbol{A} = 0 \tag{5A.3}$$

$$\nabla^2 \boldsymbol{F} + \beta^2 \boldsymbol{F} = 0 \tag{5A.4}$$

By taking into account (5A.1) and (5A.2), it can be easily verified that the field expressions of the modes TE with respect to z are determined [17] by putting

$$\boldsymbol{F} = \hat{z}\, F(\rho, \varphi, z) \tag{5A.5a}$$

$$\boldsymbol{A} = 0 \tag{5A.5b}$$

whereas those relevant to the TM to z modes are derived [17] by assuming

$$\boldsymbol{A} = \hat{z}\, A(\rho, \varphi, z) \tag{5A.6a}$$

$$\boldsymbol{F} = 0 \tag{5A.6b}$$

In the light of the above discussion, when considering the TE modes, relations (5A.1) and (5A.2) become

$$\boldsymbol{E} = -\frac{1}{\varepsilon}\nabla \times \boldsymbol{F} \tag{5A.7}$$

$$\boldsymbol{H} = -\mathrm{j}\omega \boldsymbol{F} + \frac{\nabla\nabla \cdot \boldsymbol{F}}{\mathrm{j}\omega\varepsilon\mu} \tag{5A.8}$$

Moreover, the vector Helmholtz equation (5A.4) reduces to the scalar one

$$\nabla^2 F(\rho, \varphi, z) + \beta^2 F(\rho, \varphi, z) = 0 \tag{5A.9}$$

whose elementary solution can be obtained by using the separation of variables method and, in the considered case of an observation region external to the smallest cylinder enclosing the AUT, is given by (Appendix A.2):

$$F(\rho, \varphi, z) = H_\nu^{(2)}(\Lambda \rho)\, \mathrm{e}^{\mathrm{j}\nu\varphi}\, \mathrm{e}^{-\mathrm{j}\eta z} \tag{5A.10}$$

where ν is an integer, η is a real number, $\Lambda = (\beta^2 - \eta^2)^{1/2}$, and $H_\nu^{(2)}(\Lambda\rho)$ is the Hankel function of second kind and order ν.

By expanding $\nabla \times \boldsymbol{F}$ in cylindrical coordinates, it results:

$$E_\rho = -\frac{1}{\varepsilon\rho}\frac{\partial F}{\partial \varphi} \tag{5A.11a}$$

$$E_\varphi = \frac{1}{\varepsilon}\frac{\partial F}{\partial \rho} \tag{5A.11b}$$

$$E_z = 0 \tag{5A.11c}$$

When considering TM modes, relations (5A.1) and (5A.2) become

$$\boldsymbol{E} = -\mathrm{j}\omega \boldsymbol{A} + \frac{\nabla\nabla\cdot\boldsymbol{A}}{\mathrm{j}\omega\varepsilon\mu} \tag{5A.12}$$

$$\boldsymbol{H} = \frac{1}{\mu}\nabla\times\boldsymbol{A} \tag{5A.13}$$

Moreover, the vector Helmholtz equation (5A.3) reduces to the scalar one

$$\nabla^2 A(\rho,\varphi,z) + \beta^2 A(\rho,\varphi,z) = 0 \tag{5A.14}$$

whose elementary solution is

$$A(\rho,\varphi,z) = H_\nu^{(2)}(\Lambda\rho)\,\mathrm{e}^{\mathrm{j}\nu\varphi}\,\mathrm{e}^{-\mathrm{j}\eta z} \tag{5A.15}$$

Since $\nabla\cdot\boldsymbol{A} = \partial A/\partial z$, it results

$$\nabla\nabla\cdot\boldsymbol{A} = \hat{\rho}\,\frac{\partial^2 A}{\partial\rho\partial z} + \hat{\varphi}\,\frac{1}{\rho}\frac{\partial^2 A}{\partial\varphi\partial z} + \hat{z}\,\frac{\partial^2 A}{\partial z^2} \tag{5A.16}$$

and, as a consequence

$$E_\rho = \frac{1}{\mathrm{j}\omega\varepsilon\mu}\frac{\partial^2 A}{\partial\rho\partial z} \tag{5A.17a}$$

$$E_\varphi = \frac{1}{\mathrm{j}\omega\varepsilon\mu}\frac{1}{\rho}\frac{\partial^2 A}{\partial\varphi\partial z} \tag{5A.17b}$$

$$E_z = \frac{1}{\mathrm{j}\omega\varepsilon\mu}\left(\frac{\partial^2 A}{\partial z^2} + \beta^2 A\right) \tag{5A.17c}$$

By taking into account (5A.10) and (5A.11), it can be verified that the electric field related to an elementary cylindrical TE wave is given by:

$$\boldsymbol{E}_{TE} = \frac{1}{\varepsilon}\left(-\frac{\mathrm{j}\nu}{\rho}H_\nu^{(2)}(\Lambda\rho)\;\hat{\rho} + \frac{\mathrm{d}}{\mathrm{d}\rho}H_\nu^{(2)}(\Lambda\rho)\;\hat{\varphi}\right)\mathrm{e}^{\mathrm{j}\nu\varphi}\,\mathrm{e}^{-\mathrm{j}\eta z} \tag{5A.18}$$

The electric field related to an elementary cylindrical TM wave can be determined in a similar way by taking into account (5A.15) and (5A.17), thus obtaining:

$$\boldsymbol{E}_{TM} = \frac{1}{\mathrm{j}\omega\varepsilon\mu}\Big(-\mathrm{j}\eta\frac{\mathrm{d}}{\mathrm{d}\rho}H_\nu^{(2)}(\Lambda\rho)\,\hat{\rho} + \frac{\nu\eta}{\rho}\,H_\nu^{(2)}(\Lambda\rho)\,\hat{\varphi} + (\beta^2-\eta^2)\,H_\nu^{(2)}(\Lambda\rho)\,\hat{z}\Big)\,\mathrm{e}^{\mathrm{j}\nu\varphi}\,\mathrm{e}^{-\mathrm{j}\eta z} \tag{5A.19}$$

Relations (5A.18) and (5A.19) can be rewritten as:

$$\boldsymbol{E}_{TE} = \left(-1/\varepsilon\right) \boldsymbol{M}_{\nu\eta} \tag{5A.20}$$

$$\boldsymbol{E}_{TM} = \left(\beta/\mathrm{j}\omega\varepsilon\mu\right)\boldsymbol{N}_{\nu\eta} \tag{5A.21}$$

wherein

$$\boldsymbol{M}_{\nu\eta}\left(\rho,\varphi,z\right) = \left(\frac{\mathrm{j}\nu}{\rho} H_{\nu}^{(2)}\left(\Lambda\rho\right)\hat{\rho} - \frac{\mathrm{d}}{\mathrm{d}\rho} H_{\nu}^{(2)}\left(\Lambda\rho\right)\hat{\varphi}\right) \mathrm{e}^{\mathrm{j}\nu\varphi}\,\mathrm{e}^{-\mathrm{j}\eta z} \tag{5A.22}$$

$$\boldsymbol{N}_{\nu\eta}\left(\rho,\varphi,z\right) = \left(-\frac{\mathrm{j}\eta}{\beta}\frac{\mathrm{d}}{\mathrm{d}\rho} H_{\nu}^{(2)}\left(\Lambda\rho\right)\hat{\rho} + \frac{\nu\eta}{\beta\rho} H_{\nu}^{(2)}\left(\Lambda\rho\right)\hat{\varphi} + \frac{\Lambda^2}{\beta} H_{\nu}^{(2)}\left(\Lambda\rho\right)\hat{z}\right) \mathrm{e}^{\mathrm{j}\nu\varphi}\,\mathrm{e}^{-\mathrm{j}\eta z} \tag{5A.23}$$

According to [308], the EM field in a homogeneous source-free region can be expressed as superposition of TE and TM elementary cylindrical waves. As a consequence, the electric field at a point $P(\rho,\varphi,z)$, located in the region external to smallest cylinder enclosing the radiating antenna, can be represented as a linear combination of the elementary fields (5A.20) and (5A.21) involving an integral over all η values and a sum over all ν:

$$\boldsymbol{E}\left(\rho,\varphi,z\right) = \sum_{\nu=-\infty}^{\infty}\int_{-\infty}^{\infty}\left[a_{\nu}(\eta)\,\boldsymbol{M}_{\nu\eta}(\rho,\varphi,z) + b_{\nu}(\eta)\,\boldsymbol{N}_{\nu\eta}(\rho,\varphi,z)\right]\mathrm{d}\eta^{\dagger} \tag{5A.24}$$

wherein a_ν and b_ν are the CWE coefficients,

As a consequence, the cylindrical components of the electric field at the point P are given by:

$$E_{\rho}(\rho,\varphi,z) = \sum_{\nu=-\infty}^{\infty}\int_{-\infty}^{\infty}\left[a_{\nu}(\eta)\frac{\mathrm{j}\nu}{\rho}H_{\nu}^{(2)}(\Lambda\rho) - \frac{\mathrm{j}\eta}{\beta}\,b_{\nu}(\eta)\frac{\mathrm{d}}{\mathrm{d}\rho}H_{\nu}^{(2)}(\Lambda\rho)\right] \times \mathrm{e}^{\mathrm{j}\nu\varphi}\,\mathrm{e}^{-\mathrm{j}\eta z}\mathrm{d}\eta \tag{5A.25}$$

$$E_{\varphi}(\rho,\varphi,z) = \sum_{\nu=-\infty}^{\infty}\int_{-\infty}^{\infty}\left[b_{\nu}(\eta)\frac{\eta\nu}{\beta\rho}H_{\nu}^{(2)}(\Lambda\rho) - a_{\nu}(\eta)\frac{\mathrm{d}}{\mathrm{d}\rho}H_{\nu}^{(2)}(\Lambda\rho)\right] \times \mathrm{e}^{\mathrm{j}\nu\varphi}\,\mathrm{e}^{-\mathrm{j}\eta z}\mathrm{d}\eta \tag{5A.26}$$

$$E_{z}(\rho,\varphi,z) = \sum_{\nu=-\infty}^{\infty}\int_{-\infty}^{\infty} b_{\nu}(\eta)\frac{\Lambda^2}{\beta}H_{\nu}^{(2)}(\Lambda\rho)\,\mathrm{e}^{\mathrm{j}\nu\varphi}\,\mathrm{e}^{-\mathrm{j}\eta z}\mathrm{d}\eta \tag{5A.27}$$

As described in the body of this chapter, the CWE can be properly used to determine the antenna far field from measurements made on a cylinder of radius d in the NF region. In particular, the orthogonality properties of the modes are exploited to get the modal expansion coefficients, from which the antenna far field is obtained by substituting them in the expression of the CWE valid in the FF region.

†The constants $-1/\varepsilon$ and $\beta/j\omega\varepsilon\mu$ have been incorporated in the coefficients $a_\nu(\eta)$ and $b_\nu(\eta)$.

5A.2 Cylindrical wave expansion valid in the antenna far-field region

This appendix is devoted to the determination of the CWE valid in the antenna FF region. The first step to derive its expression is to replace in the CWE the Hankel function and its first derivative with their asymptotic expansions

$$H_\nu^{(2)}(x) \underset{x\to\infty}{\approx} \sqrt{\frac{2\mathrm{j}}{\pi x}}\,\mathrm{j}^\nu \mathrm{e}^{-\mathrm{j}x} \tag{5A.28}$$

$$\frac{\mathrm{d}}{\mathrm{d}x} H_\nu^{(2)}(\alpha x) \underset{x\to\infty}{\approx} \sqrt{\frac{2\mathrm{j}\alpha}{\pi x}}\,\mathrm{j}^{\nu-1}\,\mathrm{e}^{-\mathrm{j}\alpha x} \tag{5A.29}$$

Note that (5A.29) can be easily derived by using (5A.28) and the recurrence relation:

$$\frac{\mathrm{d}}{\mathrm{d}x} H_\nu^{(2)}(\alpha x) = \alpha\, H_{\nu-1}^{(2)}(\alpha x) - \frac{\nu}{x} H_\nu^{(2)}(\alpha x) \tag{5A.30}$$

By taking into account (5A.28) and (5A.29), the following expressions for the cylindrical components of the antenna far field result:

$$E_\rho \underset{\rho\to\infty}{\approx} -\sum_{\nu=-\infty}^{\infty} \int_{-\infty}^{\infty} \frac{\mathrm{j}\eta}{\beta}\, b_\nu(\eta)\,\mathrm{j}^{\nu-1}\Lambda \sqrt{\frac{2\mathrm{j}}{\pi\Lambda\rho}}\, \mathrm{e}^{-\mathrm{j}\Lambda\rho}\, \mathrm{e}^{\mathrm{j}\nu\varphi}\, \mathrm{e}^{-\mathrm{j}\eta z}\, \mathrm{d}\eta \tag{5A.31a}$$

$$E_\varphi \underset{\rho\to\infty}{\approx} -\sum_{\nu=-\infty}^{\infty} \int_{-\infty}^{\infty} a_\nu(\eta)\,\mathrm{j}^{\nu-1}\Lambda \sqrt{\frac{2\mathrm{j}}{\pi\Lambda\rho}}\, \mathrm{e}^{-\mathrm{j}\Lambda\rho}\, \mathrm{e}^{\mathrm{j}\nu\varphi}\, \mathrm{e}^{-\mathrm{j}\eta z}\, \mathrm{d}\eta \tag{5A.31b}$$

$$E_z \underset{\rho\to\infty}{\approx} \sum_{\nu=-\infty}^{\infty} \int_{-\infty}^{\infty} \frac{\Lambda^2}{\beta} b_\nu(\eta)\,\mathrm{j}^{\nu} \sqrt{\frac{2\mathrm{j}}{\pi\Lambda\rho}}\, \mathrm{e}^{-\mathrm{j}\Lambda\rho}\, \mathrm{e}^{\mathrm{j}\nu\varphi}\, \mathrm{e}^{-\mathrm{j}\eta z}\, \mathrm{d}\eta \tag{5A.31c}$$

It is convenient to express relations (5A.31) in the spherical coordinates R, Θ, Φ by using the relations:

$$\rho = R\sin\Theta; \qquad z = R\cos\Theta; \qquad \varphi = \Phi \tag{5A.32}$$

thus getting

$$E_\rho \underset{R\to\infty}{\approx} -\sum_{\nu=-\infty}^{\infty} \int_{-\infty}^{\infty} \frac{\mathrm{j}\eta}{\beta} b_\nu(\eta)\,\mathrm{j}^{\nu-1}\Lambda \sqrt{\frac{2\mathrm{j}}{\pi\Lambda R\sin\Theta}}\, \mathrm{e}^{\mathrm{j}\nu\Phi}\, \mathrm{e}^{\mathrm{j}Rg(\eta)}\, \mathrm{d}\eta \tag{5A.33a}$$

$$E_\varphi \underset{R\to\infty}{\approx} -\sum_{\nu=-\infty}^{\infty} \int_{-\infty}^{\infty} a_\nu(\eta)\,\mathrm{j}^{\nu-1}\Lambda \sqrt{\frac{2\mathrm{j}}{\pi\Lambda R\sin\Theta}}\, \mathrm{e}^{\mathrm{j}\nu\Phi}\, \mathrm{e}^{\mathrm{j}Rg(\eta)}\, \mathrm{d}\eta \tag{5A.33b}$$

$$E_z \underset{R\to\infty}{\approx} \sum_{\nu=-\infty}^{\infty} \int_{-\infty}^{\infty} \frac{\Lambda^2}{\beta} b_\nu(\eta)\,\mathrm{j}^{\nu} \sqrt{\frac{2\mathrm{j}}{\pi\Lambda R\sin\Theta}}\, \mathrm{e}^{\mathrm{j}\nu\Phi}\mathrm{e}^{\mathrm{j}Rg(\eta)}\, \mathrm{d}\eta \tag{5A.33c}$$

wherein

$$g(\eta) = -\sqrt{\beta^2 - \eta^2}\sin\Theta - \eta\cos\Theta \tag{5A.34}$$

The second step is the asymptotic evaluation of the integrals in (5A.33) by means of the stationary phase method [306]. According to such a method (see Appendix A.4), an integral of the form

$$I(\Omega) = \int_{-\infty}^{\infty} F(\tau)\, e^{j\Omega f(\tau)}\, d\tau, \tag{5A.35}$$

where $F(\tau)$ is a regular complex function of the real variable τ, Ω is a large positive parameter, and $f(\tau)$ is a regular real function having an isolated first order stationary phase point at τ_s, i.e., $f'(\tau_s) = 0$, but $f''(\tau_s) \neq 0$, can be asymptotically evaluated. Its approximated evaluation for large values of Ω gives

$$I(\Omega) \underset{\Omega\to\infty}{\approx} \sqrt{\frac{2\pi}{\Omega\,|f''(\tau_s)|}} F(\tau_s)\, e^{j\Omega f(\tau_s)}\, e^{j\,\mathrm{sgn}[f''(\tau_s)]\pi/4} \tag{5A.36}$$

wherein the prime and double prime denote the first and second order derivative, respectively and sgn (x) is the sign of x function (see (A.67)).

It can be easily recognized that the integration over η in relations (5A.33) involves an integral of the form (5A.35). Since from (5A.34) it results

$$g'(\eta) = -\cos\Theta + \frac{\eta\sin\Theta}{\sqrt{\beta^2 - \eta^2}}, \tag{5A.37}$$

the stationary phase point occurs at $\eta_s = \beta\cos\Theta$. Accordingly,

$$g''(\eta_s) = \frac{\beta^2\sin\Theta}{(\beta^2 - \eta_s^2)^{3/2}} = \frac{1}{\beta\sin^2\Theta} \tag{5A.38}$$

$$e^{jRg(\eta_s)} = e^{-jR\left(\beta\cos^2\Theta + \sqrt{\beta^2 - \beta^2\cos^2\Theta}\sin\Theta\right)} = e^{-j\beta R} \tag{5A.39}$$

and, therefore,

$$\sqrt{\frac{2\pi}{R\,|g''(\eta_s)|}}\, e^{jRg(\eta_s)}\, e^{j\,\mathrm{sgn}[g''(\eta_s)]\pi/4} = \sqrt{\frac{2\pi j\beta\sin^2\Theta}{R}}\, e^{-j\beta R} \tag{5A.40}$$

On the other hand

$$\left.\sqrt{\frac{2j}{\pi\Lambda R\sin\Theta}}\right|_{\eta=\eta_s} = \sqrt{\frac{2j}{\pi\beta R\sin^2\Theta}} \tag{5A.41}$$

and, then,

$$\sqrt{\frac{2j}{\pi\beta R\sin^2\Theta}}\sqrt{\frac{2\pi j\,\beta\sin^2\Theta}{R}}\, e^{-j\beta R} = 2j\,\frac{e^{-j\beta R}}{R} \tag{5A.42}$$

In the light of the above discussion, by asymptotically evaluating the integrals in (5A.33), it results

$$E_\rho \underset{R\to\infty}{\approx} -\mathrm{j}2\beta \sin\Theta \cos\Theta \frac{\mathrm{e}^{-\mathrm{j}\beta R}}{R} \sum_{\nu=-\infty}^{\infty} \mathrm{j}^\nu b_\nu(\beta\cos\Theta)\,\mathrm{e}^{\mathrm{j}\nu\Phi} \tag{5A.43a}$$

$$E_\varphi \underset{R\to\infty}{\approx} -2\beta \sin\Theta \frac{\mathrm{e}^{-\mathrm{j}\beta R}}{R} \sum_{\nu=-\infty}^{\infty} \mathrm{j}^\nu a_\nu(\beta\cos\Theta)\,\mathrm{e}^{\mathrm{j}\nu\Phi} \tag{5A.43b}$$

$$E_z \underset{R\to\infty}{\approx} \mathrm{j}2\beta \sin^2\Theta \frac{\mathrm{e}^{-\mathrm{j}\beta R}}{R} \sum_{\nu=-\infty}^{\infty} \mathrm{j}^\nu b_\nu(\beta\cos\Theta)\,\mathrm{e}^{\mathrm{j}\nu\Phi} \tag{5A.43c}$$

From (5A.43) and by taking into account that the spherical components of a vector are related to the cylindrical ones by

$$E_R = E_\rho \sin\Theta + E_z \cos\Theta \tag{5A.44a}$$

$$E_\Theta = E_\rho \cos\Theta - E_z \sin\Theta \tag{5A.44b}$$

$$E_\Phi = E_\varphi \tag{5A.44c}$$

it follows that the spherical components of the antenna far field are given by:

$$E_R(R,\Theta,\Phi) = 0 \tag{5A.45a}$$

$$E_\Theta(R,\Theta,\Phi) = -\mathrm{j}2\beta \frac{\mathrm{e}^{-\mathrm{j}\beta R}}{R} \sin\Theta \sum_{\nu=-\infty}^{\infty} \mathrm{j}^\nu b_\nu(\beta\cos\Theta)\,\mathrm{e}^{\mathrm{j}\nu\Phi} \tag{5A.45b}$$

$$E_\Phi(R,\Theta,\Phi) = -2\beta \frac{\mathrm{e}^{-\mathrm{j}\beta R}}{R} \sin\Theta \sum_{\nu=-\infty}^{\infty} \mathrm{j}^\nu a_\nu(\beta\cos\Theta)\,\mathrm{e}^{\mathrm{j}\nu\Phi} \tag{5A.45c}$$

These last are the expression of the CWE valid in the antenna FF region.

Chapter 6
Near-field to far-field transformations in spherical scanning geometry

6.1 Introduction

The near-field to far-field (NF–FF) transformation techniques in spherical scanning geometry [64–101, 122–126, 131–141] are conveniently employed to obtain the full radiation pattern reconstruction of the antenna under test (AUT) from a single set of NF measurements, even though the involved analytical complexity and computational effort grow considerably as compared to those required by the NF–FF transformations in planar and cylindrical scanning geometries.

In the spherical scanning, the acquisition of the NF amplitude and phase data on the wanted lattice specified by the intersections of parallels and meridians (Figure 6.1) can be accomplished by using quite different spherical NF facilities [76]. In the roll-over-azimuth facility (see Figure 6.2), the probe is stationary and the scanning is performed by mounting the AUT on a roll positioner (providing the rotation around the φ axis), which is anchored via a L-shaped bracket to an azimuth turntable (allowing the rotation around the ϑ axis). In the elevation-over-azimuth facility (Figure 6.3), the probe is still stationary and the NF measurement points are reached by a rotation, around the φ axis, of the AUT mounted on an elevation positioner in turn placed on an azimuth turntable, which allows the AUT rotation also around the ϑ axis. On the other hand, in the azimuth-over-elevation facility (Figure 6.4), the AUT is attached to an azimuth turntable (now providing the rotation along φ), placed on an elevation positioner (allowing the rotation along ϑ). In the arch-over-azimuth facility, the scanning is accomplished by rotating along φ the AUT by means of an azimuth turntable and by moving the probe along an arc to reach the requested ϑ positions of the NF measurement points.

Regardless the adopted NF measurement facility, the NF amplitude and phase data, acquired on the prescribed grid into two different probe orientations to account for the directional effects of the measuring probe, are employed to evaluate the antenna FF pattern by means of a probe-compensated spherical NF–FF transformation technique.

For reader's convenience, the classical probe-uncompensated NF–FF transformation with spherical scanning [76] and its modified version [93], which takes into account the spatial bandlimitation properties [102] of radiated electromagnetic (EM)

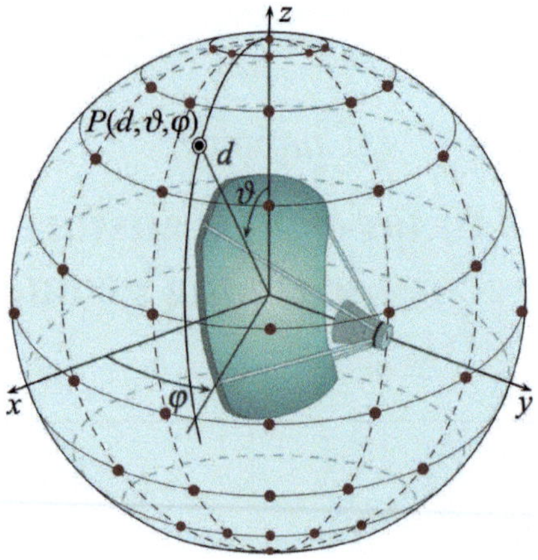

Figure 6.1 Spherical scanning

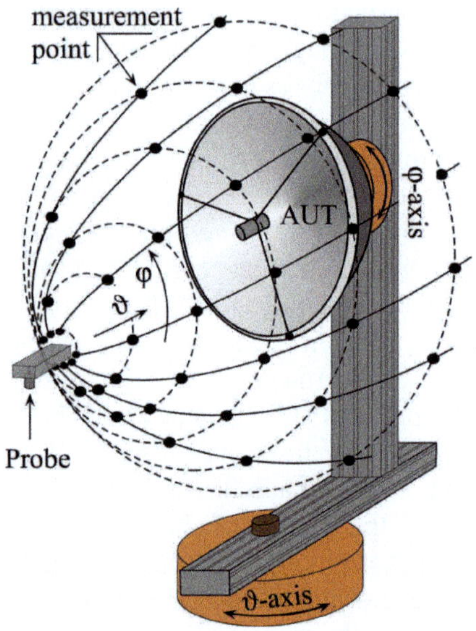

Figure 6.2 Roll-over-azimuth spherical NF facility

fields, are first derived (see Sect. 6.2). The key steps of the classical [76] and modified [95] probe-compensated NF–FF transformations with spherical scanning, valid when the probe exhibits only a first-order azimuthal dependence in its radiated far field, are then described. The compensation of the directional effects of higher-order probes has been also tackled in the open literature [78–89].

In the classical spherical NF–FF transformation [76], the number of required NF data is fixed by the order of the highest spherical wave to be considered in the spherical wave expansion (SWE), which in turn is related to the radius of the minimum

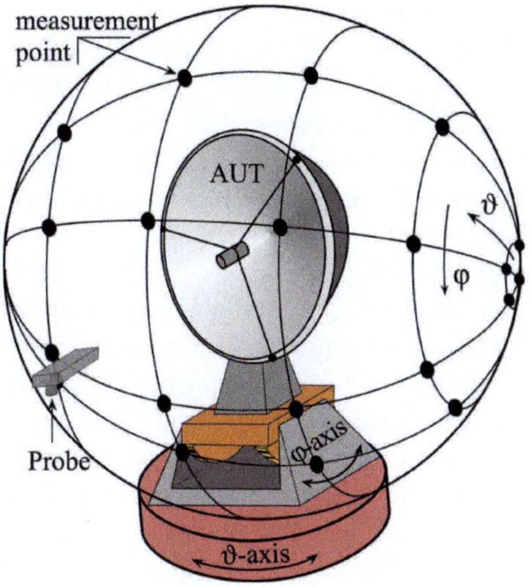

Figure 6.3 Elevation-over-azimuth spherical NF facility

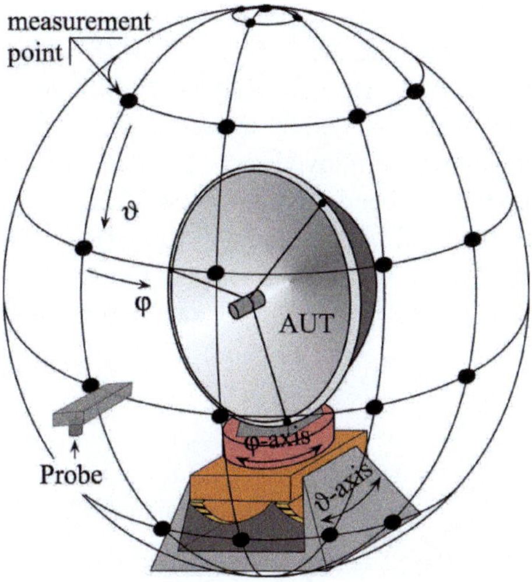

Figure 6.4 Azimuth-over-elevation spherical NF facility

sphere with its centre at the origin* of the adopted reference system and enclosing the AUT (see also Sects. 6.2 and 6.3). It can be easily recognized that, when practical constraints do not make possible to mount the AUT in centred configuration, i.e., with its geometric centre at the origin, the radius of the minimum sphere increases as well as the number of NF data to be acquired. To overcome such a shortcoming

*The origin of the adopted reference system coincides obviously with the centre of the scanning sphere.

several approaches, wherein the spherical wave functions are referred no longer to the origin but to the centre of the AUT, have been proposed in [90–92].

The application of the non-redundant (NR) sampling representations of radiated EM fields [104, 105] to the voltage acquired by a non-directive probe has allowed the development of effective probe-compensated NF–FF transformation techniques [94–97], which require a non-redundant, i.e., minimum, number of samples (see Sect. 6.4). They make use of optimal sampling interpolation (OSI) formulas to determine the input NF data for the classical spherical NF–FF transformation from the acquired NR samples. Accordingly, these techniques allow a significant reduction of the amount of required NF data and, as a consequence, of the measurement time with respect to that needed by the classical NF–FF transformation [76]. This reduction results to be particularly remarkable in the case of non-spherical AUTs, which, having one or two predominant dimensions, can be effectively modelled by prolate or oblate spheroids [94, 97] or by very flexible surfaces (rounded cylinder or double bowl) [95, 96] to further reduce the redundancy as compared to the use of the spherical modelling. These techniques have been then extended in [99–101] to the case of non-centred AUTs mounted with an offset along the rotational symmetry axis, by showing that the number of required samples is practically the same as in the case of the centred mounting.

The recently developed NR NF–FF transformation techniques with spherical spiral scanning [122–126, 131–138] (see Figure 6.5) can allow a further significant reduction of the measurement time. As a matter of fact, they require a minimum number of NF samples which, following the Rahmat-Samii's idea [112], are collected on fly along spherical spirals obtained by making continuous and synchronized the movements of the positioners employed to perform the spherical scanning. These NF–FF transformations have been developed (see Sect. 6.5) by properly applying the unified theory of spiral scannings for volumetric [122] and non-volumetric [124] antennas in dependence of the geometry of the AUT and make use of two-dimensional OSI expansions to recover the NF data needed by the classical spherical

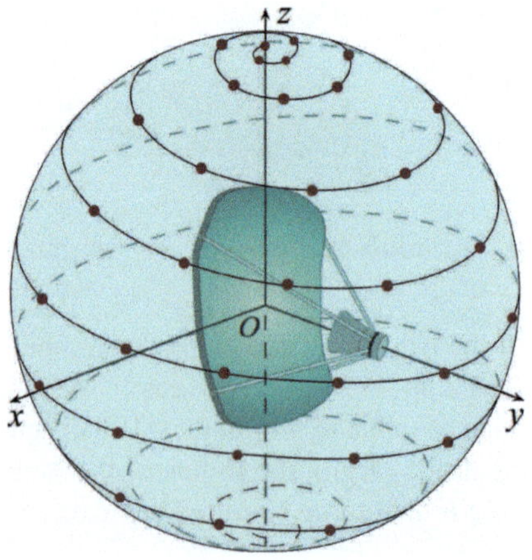

Figure 6.5 Spherical spiral scanning

NF–FF transformation from the acquired NR samples. NR NF–FF transformations with spherical spiral scanning accounting for a non-centred mounting of the AUT have been also developed in [139–141].

6.2 Classical spherical NF–FF transformation without probe compensation

In this section, the classical spherical NF–FF transformation without probe compensation [76] and its rearranged version in [93], modified by properly taking into account the spatial band-limitation properties [102] of the radiated EM fields, will be summarized.

As shown in the Appendix 6A.1, the electric field at any point $P(r, \vartheta, \varphi)$, located in the region external to the smallest sphere enclosing the radiating antenna, can be represented as superposition of elementary TE and TM spherical waves (see (6A.32) and (6A.34)). Accordingly, the tangential electric field $\boldsymbol{E}_t$ radiated by the AUT on the scanning sphere of radius d (Figure 6.1) can be expressed, see (6A.45), as:

$$\boldsymbol{E}_t = \beta \sum_{n=1}^{N_{max}} \sum_{m=-n}^{n} \left[a_{1nm}\, g_{1n}(\beta r)\, \boldsymbol{F}_{1nm}(\vartheta, \varphi) + a_{2nm}\, g_{2n}(\beta r)\, \boldsymbol{F}_{2nm}(\vartheta, \varphi)\right] \tag{6.1}$$

wherein β is the wavenumber, a_{1nm}, a_{2nm} are the SWE coefficients, and

$$g_{1n}(\beta r) = h_n^{(2)}(\beta r); \qquad g_{2n}(\beta r) = \frac{1}{\beta r}\frac{\mathrm{d}}{\mathrm{d}(\beta r)}\left(\beta r\, h_n^{(2)}(\beta r)\right) \tag{6.2}$$

$h_n^{(2)}(\beta r)$ being the spherical Hankel function of second kind and order n.

According to relations (6A.32), (6A.34), (6A.33b), (6A.35b), and (6A.44), the spherical vector wave functions $\boldsymbol{F}_{1nm}$ and $\boldsymbol{F}_{2nm}$ in (6.1) are given by:

$$\boldsymbol{F}_{1nm}(\vartheta, \varphi) = \boldsymbol{f}_{1nm}(\vartheta)\, \mathrm{e}^{\mathrm{j}m\varphi}; \qquad \boldsymbol{F}_{2nm}(\vartheta, \varphi) = \boldsymbol{f}_{2nm}(\vartheta)\, \mathrm{e}^{\mathrm{j}m\varphi} \tag{6.3}$$

wherein

$$\boldsymbol{f}_{1nm}(\vartheta) = \left(\frac{-m}{|m|}\right)^m \frac{1}{\sqrt{2\pi n(n+1)}}\left[\frac{\mathrm{j}m}{\sin\vartheta}\,\bar{P}_n^{|m|}(\cos\vartheta)\hat{\vartheta} - \frac{\mathrm{d}}{\mathrm{d}\vartheta}\bar{P}_n^{|m|}(\cos\vartheta)\hat{\varphi}\right] \tag{6.4}$$

$$\boldsymbol{f}_{2nm}(\vartheta) = \left(\frac{-m}{|m|}\right)^m \frac{1}{\sqrt{2\pi n(n+1)}}\left[\frac{\mathrm{d}}{\mathrm{d}\vartheta}\bar{P}_n^{|m|}(\cos\vartheta)\hat{\vartheta} + \frac{\mathrm{j}m}{\sin\vartheta}\bar{P}_n^{|m|}(\cos\vartheta)\hat{\varphi}\right] \tag{6.5}$$

$\bar{P}_n^m(\cos\vartheta)$ being the normalized associated Legendre function (see (A.56)) as defined by Belousov [310].

In the classical approach, the order of the highest spherical wave to be considered in the SWE (6.1) is usually fixed according to the empirical minimum sphere rule:

$$N_{\max} = \text{Int}(\beta a) + 10 \tag{6.6}$$

wherein a is the radius of the smallest sphere able to contain the AUT. Whereas, in the modified approach [93], it is rigorously determined by the spatial band-limitation properties of radiated EM fields [102] and, accordingly,

$$N_{\max} = \text{Int}(\chi'\beta a) + 1 \tag{6.7}$$

χ' being the enlargement bandwidth factor (slightly greater than unity for antennas whose sizes are large with respect to the wavelength λ).

The SWE coefficients a_{1nm}, a_{2nm} can be determined from the knowledge of the tangential electric field $\boldsymbol{E}_t$ on the scanning sphere by exploiting the orthonormality properties of the spherical vector wave functions (see Appendix 6A.1). In fact, by taking into account (6A.40), (6A.41), (6A.43), and (6A.44), from the relation (6.1), it follows that:

$$\langle \boldsymbol{E}_t, \boldsymbol{F}_{1nm} \rangle = \int_0^{\pi} \int_{-\pi}^{\pi} \boldsymbol{E}_t(d, \vartheta, \varphi) \cdot \boldsymbol{F}^*_{1nm}(\vartheta, \varphi) \sin\vartheta \, d\varphi \, d\vartheta = \beta a_{1nm} g_{1n}(\beta d) \tag{6.8a}$$

$$\langle \boldsymbol{E}_t, \boldsymbol{F}_{2nm} \rangle = \int_0^{\pi} \int_{-\pi}^{\pi} \boldsymbol{E}_t(d, \vartheta, \varphi) \cdot \boldsymbol{F}^*_{2nm}(\vartheta, \varphi) \sin\vartheta \, d\varphi \, d\vartheta = \beta a_{2nm} g_{2n}(\beta d) \tag{6.8b}$$

where the asterisk (*) denotes the complex conjugation and the symbol (·) indicates the scalar product. Accordingly, it results:

$$a_{1,2nm} = \frac{1}{\beta g_{1,2n}(\beta d)} \int_0^{\pi} \int_{-\pi}^{\pi} \boldsymbol{E}_t(d, \vartheta, \varphi) \cdot \boldsymbol{f}^*_{1,2nm}(\vartheta) \, e^{-jm\varphi} \sin\vartheta \, d\varphi \, d\vartheta \tag{6.9}$$

The efficient evaluation of the SWE coefficients a_{1nm}, a_{2nm} can be carried out, as will be shown in the following, by using the fast Fourier transform (FFT) algorithm [300] to compute the integrals in (6.9). As regards the NF data sampling spacings, in the classical spherical NF–FF transformation [76], they are:

$$\Delta\vartheta \leq 2\pi/(2N_{\max} + 1); \qquad \Delta\varphi = \Delta\vartheta \tag{6.10}$$

When the radius a' of the smallest cylinder containing the AUT and having its axis coincident with the z one is smaller than the radius a of the minimum sphere, the number of the NF data along φ can be reduced and it results [76]:

$$\Delta\varphi \leq 2\pi/(2M + 1) \tag{6.11}$$

where $M = \text{Int}(\beta a') + 10$. However, when dealing with directive AUTs pointing in the equatorial plane, that is at $\vartheta = 90°$, the sampling spacing along φ must still be chosen equal to that along ϑ [9].

In the modified approach [93], the spacings $\Delta\vartheta$ and $\Delta\varphi$ are rigorously determined according to the spatial band-limitation properties of radiated EM fields [102]. As a consequence, $N_{\max}$ in (6.10) is chosen according to the relation (6.7), whereas M in (6.11) is determined by the azimuthal bandwidth of the AUT and, therefore, is given by

$$M = \text{Int}\left(\chi^* \beta a \sin \vartheta\right) + 1 \tag{6.12}$$

wherein

$$\chi^* = \chi^*\left(\vartheta\right) = 1 + \left(\chi' - 1\right)\left[\sin\ \vartheta\right]^{-2/3} \tag{6.13}$$

is the azimuthal enlargement bandwidth factor (see (2A.30)). That is, the number of NF data to be acquired on the scanning parallels decreases as one moves from the equator towards the poles.

Once the SWE coefficients have been determined, the electric field $\boldsymbol{E}$ radiated by the AUT in the FF region can be determined by substituting them in the expression (6A.47) of the SWE valid in such a region:

$$\boldsymbol{E}(R, \Theta, \Phi) = \frac{e^{-j\beta R}}{R} \sum_{n=1}^{N_{\max}} \sum_{m=-n}^{n} \left[j^{n+1} a_{1nm} \boldsymbol{f}_{1nm}\left(\Theta\right) + j^{n} a_{2nm} \boldsymbol{f}_{2nm}\left(\Theta\right)\right] e^{jm\Phi} \tag{6.14}$$

It is noteworthy that, in the above expression, the spherical coordinate system (R, Θ, Φ) has been adopted to identify an observation point in the FF region.

The efficient evaluation of the integrals in (6.9) is now described. The integration over φ can be performed by expanding, on each scanning parallel, the tangential electric field $\boldsymbol{E}_t$ in Fourier series with respect to φ via the FFT algorithm, i.e.,

$$\boldsymbol{E}_t(d, \vartheta, \varphi) = \sum_{k=-M}^{M} \boldsymbol{G}_k(\vartheta)\, e^{jk\varphi} \tag{6.15}$$

Such an expansion is then substituted in (6.9), thus obtaining:

$$a_{1,2nm} = \frac{2\pi}{\beta g_{1,2n}\left(\beta d\right)} \int_0^{\pi} \boldsymbol{G}_m\left(\vartheta\right) \cdot \boldsymbol{f}^*_{1,2nm}\left(\vartheta\right) \sin\vartheta\, d\vartheta \tag{6.16}$$

In this last relation, the integration can be efficiently performed by expanding, $\boldsymbol{G}_m(\vartheta)$ and $\boldsymbol{f}^*_{1,2nm}(\vartheta)$ in Fourier series with respect to ϑ via the FFT. By substituting these expansions in (6.16), it results:

$$a_{1,2nm} = \frac{2\pi}{\beta g_{1,2n}\left(\beta d\right)} \sum_{i=-N_{\max}}^{N_{\max}} \sum_{l=-N_{\max}}^{N_{\max}} \boldsymbol{G}_{mi} \cdot \boldsymbol{f}^*_{1,2nml} \int_0^{\pi} e^{j(i-l)\vartheta} \sin\vartheta\, d\vartheta \tag{6.17}$$

It must be stressed that, in order to compute the Fourier series coefficients of $\boldsymbol{G}_m(\vartheta)$ and $\boldsymbol{f}^*_{1,2nm}(\vartheta)$, it is necessary to extend $\boldsymbol{G}_m(\vartheta)$ and $\boldsymbol{f}^*_{1,2nm}(\vartheta)$ from the range $[0, \pi]$ to the $[-\pi, \pi]$ one. This can be achieved by taking into account that $\boldsymbol{f}^*_{1,2nm}(\vartheta)$ is even when m is odd and odd when m is even and that $\boldsymbol{G}_m(\vartheta)$ exhibits the same parity as $\boldsymbol{f}^*_{1,2nm}(\vartheta)$. As a matter of fact, from relations (6.4) and (6.5), it is clear that $\boldsymbol{f}^*_{1,2nm}(\vartheta)$ has the same parity of the functions $m\bar{P}_n^{|m|}(\cos\vartheta)/\sin\vartheta$ and $d\bar{P}_n^{|m|}(\cos\vartheta)/d\vartheta$, and, hence, is even when m is odd and vice versa. Moreover, by substituting relations (6.3) into (6.1) and inverting the summation order, it results:

$$\boldsymbol{E}_t = \beta \sum_{m=-N_{\max}}^{N_{\max}} \sum_{\substack{n=|m| \\ (n \neq 0)}}^{N_{\max}} \left[a_{1nm}\, g_{1n}(\beta r)\, \boldsymbol{f}_{1nm}(\vartheta) + a_{2nm}\, g_{2n}(\beta r)\, \boldsymbol{f}_{2nm}(\vartheta)\right] \mathrm{e}^{\mathrm{j}m\varphi} \tag{6.18}$$

From the comparison of this last with (6.15), it follows that:

$$\boldsymbol{G}_m(\vartheta) = \beta \sum_{\substack{n=|m| \\ (n \neq 0)}}^{N_{\max}} \left[a_{1nm}\, g_{1n}(\beta r) \boldsymbol{f}_{1nm}(\vartheta) + a_{2nm}\, g_{2n}(\beta r)\, \boldsymbol{f}_{2nm}(\vartheta)\right] \tag{6.19}$$

which demonstrates that $\boldsymbol{G}_m(\vartheta)$ and $\boldsymbol{f}^*_{1,2nm}(\vartheta)$ have the same parity.

It is noteworthy that, to exploit the numerical efficiency of the FFT, the number of the scanning parallels to be considered and the number of NF samples on them must be the first power of 2, or power of 2, 3, and 5 (in dependence of the available FFT algorithm) greater or equal to $N_{\max}$ and $2M$, respectively.

By inverting the summation order, the SWE (6.14) valid in the antenna FF region can be rewritten in the following form:

$$\begin{aligned}\boldsymbol{E}(R, \Theta, \Phi) &= \frac{\mathrm{e}^{-\mathrm{j}\beta R}}{R} \sum_{m=-N_{\max}}^{N_{\max}} \sum_{\substack{n=|m| \\ (n \neq 0)}}^{N_{\max}} \left[\mathrm{j}^{n+1} a_{1nm} \boldsymbol{f}_{1nm}(\Theta) + \mathrm{j}^n a_{2nm} \boldsymbol{f}_{2nm}(\Theta)\right] \mathrm{e}^{\mathrm{j}m\Phi} \\ &= \frac{\mathrm{e}^{-\mathrm{j}\beta R}}{R} \sum_{m=-N_{\max}}^{N_{\max}} C_m(\Theta)\, \mathrm{e}^{\mathrm{j}m\Phi} \end{aligned} \tag{6.20}$$

which allows one to efficiently evaluate the antenna far field at the considered zenithal angle Θ by carrying out the summation by means of the FFT algorithm.

From the numerical point of view, it is convenient to apply the above procedure for evaluating only the FF samples needed by the OSI expansion (2.76), properly modified to deal with an even number of samples on the FF parallels:

$$\begin{aligned}\mathcal{F}_{\Theta,\Phi}(\Theta, \Phi) = \sum_{n=n_0-q+1}^{n_0+q} & \Big\{ \Omega_N(\Theta - \Theta_n, \bar{\Theta})\, D_{N''}(\Theta - \Theta_n) \\ & \times \sum_{m=m_0-p+1}^{m_0+p} \mathcal{F}_{\Theta,\Phi}(\Theta_n, \Phi_{m,n})\, \Omega_{M_n}(\Phi - \Phi_{m,n}, \bar{\Phi}_n) D^e_{M''_n}(\Phi - \Phi_{m,n}) \Big\}\end{aligned} \tag{6.21}$$

where $\mathcal{F}(\Theta, \Phi) = R\, \mathrm{e}^{\mathrm{j}\beta R} \boldsymbol{E}(R, \Theta, \Phi)$ is the far-field pattern function, $2q, 2p$ are the numbers of the considered samples $\mathcal{F}_{\Theta,\Phi}(\Theta_n, \Phi_{m,n})$ nearest to the output point, $n_0 = \mathrm{Int}(\Theta/\Delta\Theta)$, $m_0 = \mathrm{Int}(\Phi/\Delta\Phi)$, and

$$N'' = \mathrm{Int}(\chi N') + 1; \qquad N'' = \mathrm{Int}(\chi' \beta a) + 1; \qquad N = N'' - N' \tag{6.22a}$$

$$M''_n \geq \mathrm{Int}(\chi M'_n) + 1; \qquad M'_n = \mathrm{Int}(\chi^* \beta a' \sin\Theta_n); \qquad M_n = M''_n - M'_n \tag{6.22b}$$

$$\Theta_n = n\Delta\Theta = 2\pi n/(2N' + 1); \qquad \Phi_{m,n} = m\Delta\Phi_n = m\pi/M''_n \tag{6.22c}$$

$$\chi^* = \chi^*(\Theta) = 1 + (\chi' - 1)[\sin\Theta]^{-2/3}; \quad \bar{\Theta} = q\Delta\Theta; \quad \bar{\Phi}_n = p\Delta\Phi_n \tag{6.22d}$$

$\mathrm{Int}(x)$ denoting the greatest integer less than or equal to x, and $\chi > 1$ being the over-sampling factor, which allows the control of the OSI truncation error. Moreover, $D_{L''}(\alpha)$ and $\Omega_L(\alpha, \bar{\alpha})$ are the Dirichlet (2.53) and the Tschebyscheff Sampling function (2.62), respectively, and $D^e_{M''_n}(\Phi)$, given by (5.20), is the Dirichlet function for an even number ($2M''_n$) of samples [307].

The two-dimensional OSI expansion (6.21) has been obtained by properly matching that relevant to a FF meridian, based on the spherical modelling of the antenna, and that relevant to a FF parallel, derived by taking into account that in such a case $W_\Phi = \beta\, a' \sin\Theta$, where $2a'$ is the antenna maximum transverse dimension (see (2.15)). Obviously, OSI expansions, which require a lower number of FF samples, can be derived (see Chapter 2) when using more effective modellings tailored to the actual geometry of the considered AUT.

It is noteworthy that the samples $\mathcal{F}_{\Theta,\Phi}(\Theta_n, \Phi_{m,n})$ relevant to the FF parallel at Θ_n are efficiently obtained from (6.20) by an inverse FFT, where the number $2M''_n$ of considered harmonics is the first power of 2, or power of 2, 3, and 5 (in dependence of the used FFT algorithm) greater or equal to $2\,[\mathrm{Int}(\chi M'_n) + 1]$.

As well known, the total power P_{rad} radiated by the AUT is given by:

$$P_{rad} = \frac{1}{2\zeta} \int_0^{\pi} \int_{-\pi}^{\pi} |\mathcal{F}(\Theta, \Phi)|^2 \sin\Theta \, \mathrm{d}\Phi \, \mathrm{d}\Theta \tag{6.23}$$

where ζ is the free-space impedance and, according to (6.14) and (6.3),

$$\mathcal{F}(\Theta, \Phi) = \sum_{n=1}^{N_{\max}} \sum_{m=-n}^{n} \left[\mathrm{j}^{n+1} a_{1nm} \boldsymbol{F}_{1nm}(\Theta, \Phi) + \mathrm{j}^{n} a_{2nm} \boldsymbol{F}_{2nm}(\Theta, \Phi) \right] \tag{6.24}$$

By substituting (6.24) into (6.23), it results:

$$P_{rad} = \frac{1}{2\zeta} \sum_{n=1}^{N_{\max}} \sum_{m=-n}^{n} \sum_{\bar{n}=1}^{N_{\max}} \sum_{\bar{m}=-\bar{n}}^{\bar{n}} \int_0^{\pi} \int_{-\pi}^{\pi} \left[\mathrm{j}^{n+1} a_{1nm} \boldsymbol{F}_{1nm}(\Theta, \Phi) + \mathrm{j}^{n} a_{2nm} \boldsymbol{F}_{2nm}(\Theta, \Phi) \right]$$

$$\cdot \left[\mathrm{j}^{\bar{n}+1} a_{1\bar{n}\bar{m}} \boldsymbol{F}_{1\bar{n}\bar{m}}(\Theta, \Phi) + \mathrm{j}^{\bar{n}} a_{2\bar{n}\bar{m}} \boldsymbol{F}_{2\bar{n}\bar{m}}(\Theta, \Phi) \right]^* \sin\Theta \, \mathrm{d}\Phi \, \mathrm{d}\Theta \tag{6.25}$$

From this last, by taking into account (6A.40), (6A.41), (6A.43), and (6A.44), it follows:

$$P_{rad} = \frac{1}{2\zeta} \sum_{n=1}^{N_{\max}} \sum_{m=-n}^{n} \left[|a_{1nm}|^2 + |a_{2nm}|^2 \right] \tag{6.26}$$

which shows that, when the spherical vector wave functions $\boldsymbol{F}_{1nm}$ and $\boldsymbol{F}_{2nm}$ are defined according to relations (6A.44), the spherical wave functions result to be power normalized.

In the following, a numerical example, which assesses the effectiveness of the classical spherical NF–FF transformation without probe compensation as modified in [93] and of the two-dimensional OSI expansion (6.21), is shown. The

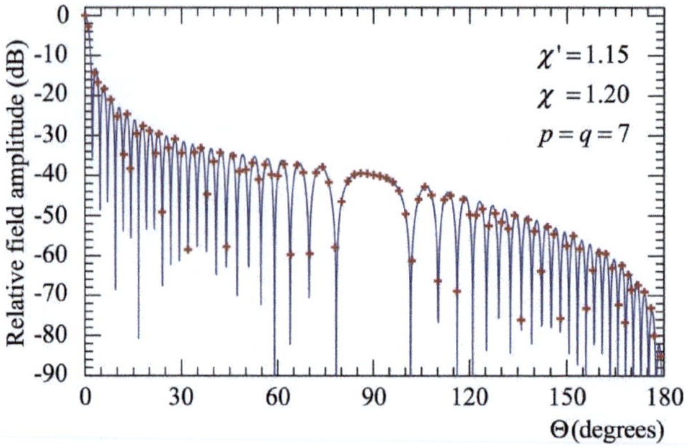

Figure 6.6 *E-plane pattern. Solid line: exact field. Crosses: reconstructed from NF data.*

simulations are relevant to the field radiated by a uniform planar square array lying in the plane $z = 0$ and having side $L = 24\,\lambda$. Its elements, elementary Huygens sources linearly polarized along the y axis, are $\lambda/2$ spaced on both the x and y directions. The radius of the scanning sphere is 25 λ and an FFT algorithm using a number of terms power of 2, 3, and 5 has been employed both in the evaluation of the SWE coefficients and in that of the FF samples. The antenna FF pattern in the E-plane, reconstructed from the NF samples by applying the abovementioned NF–FF transformation, is compared with the exact one in Figure 6.6, whereas the reconstructions of the FF components in the cut plane at $\Phi = 60°$ are shown in Figure 6.7. As can be seen, all the FF reconstructions result to be everywhere very accurate, thus assessing the effectiveness of the NF–FF transformation technique and of the FF OSI expansion. It is useful to note that an enlargement bandwidth factor $\chi' = 1.15$ has been adopted to fix the highest spherical wave considered in the SWE and to determine the sampling points in the two-dimensional OSI expansion (6.21).

It can be interesting to compare the number of the used NF samples (21058) with that (29040) needed by the classical spherical NF–FF transformation.

6.3 Classical spherical NF–FF transformation with probe compensation

It can be easily realized that it is not possible to measure, using a real (not ideal) probe, the tangential components E_ϑ and E_φ of the NF radiated by the AUT on the scanning sphere. In fact, the probe sees each AUT portion under different directions. As a consequence, according to the discussion made at Sect. 4.3, the response (measured voltage) of a real probe is not proportional to the tangential components of the electric field radiated by the AUT. For this reason, the far field radiated by the

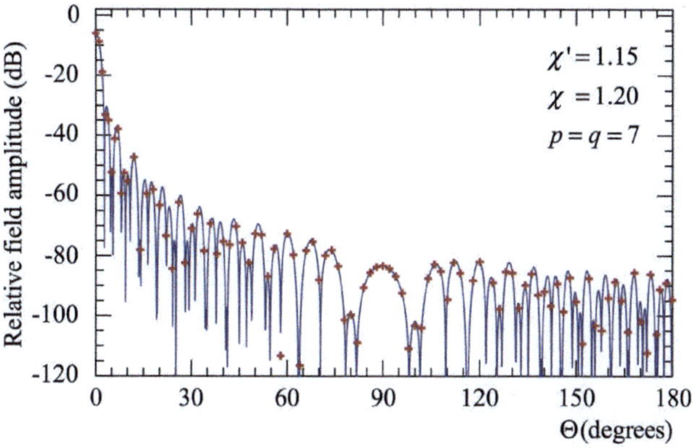

(a)

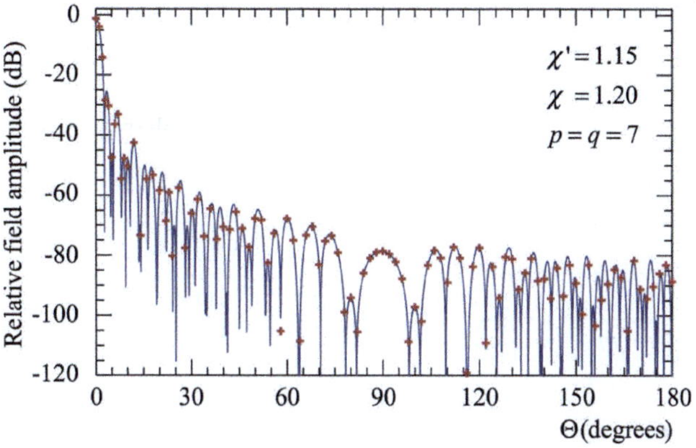

(b)

Figure 6.7 Amplitude of the FF spherical components in the cut plane at $\Phi = 60°$. Solid line: exact field. Crosses: reconstructed from NF data. (a) Amplitude of E_Φ, (b) Amplitude of E_Θ.

AUT cannot be precisely reconstructed from the acquired NF data by applying the described uncompensated spherical NF–FF transformation.

The probe-compensated spherical NF–FF transformation [76], wherein the directional effects introduced by a real probe are properly taken into account, has been developed by exploiting the source scattering matrix formulation. According to [76], if the probe exhibits only a first-order azimuthal dependence in its radiated far field[†], i.e., is such that its FF components are given by (4A.54), the SWE coefficients

[†]Such a dependence is exhibited by an open-ended circular cross-section waveguide excited by a TE_{11} mode and, in a very good approximation [303], by an open-ended rectangular waveguide excited by a TE_{10} mode (see Appendix 4A.3).

a_{1nm} and a_{2nm} can be determined from the voltages V_p and V_r measured by the probe and the rotated probe by means of the following expressions:

$$a_{1nm} = \frac{2n+1}{16\pi} \frac{\int_0^\pi \int_{-\pi}^\pi \left[I_1(\vartheta) V_p - \mathrm{j} I_2(\vartheta) V_r\right] \mathrm{e}^{-\mathrm{j}m\varphi} \sin\vartheta \, \mathrm{d}\varphi \, \mathrm{d}\vartheta}{\sum_{\upsilon=1}^{\upsilon_{\max}} \left[a'_{1\upsilon 1} A^n_{\upsilon 1}(\beta d) - a'_{2\upsilon 1} B^n_{\upsilon 1}(\beta d)\right]} \tag{6.27a}$$

$$a_{2nm} = \frac{2n+1}{16\pi} \frac{\int_0^\pi \int_{-\pi}^\pi \left[I_2(\vartheta) V_p - \mathrm{j} I_1(\vartheta) V_r\right] \mathrm{e}^{-\mathrm{j}m\varphi} \sin\vartheta \, \mathrm{d}\varphi \, \mathrm{d}\vartheta}{\sum_{\upsilon=1}^{\upsilon_{\max}} \left[a'_{1\upsilon 1} B^n_{\upsilon 1}(\beta d) - a'_{2\upsilon 1} A^n_{\upsilon 1}(\beta d)\right]} \tag{6.27b}$$

wherein $\upsilon_{\max}$ is the order of the highest spherical wave to be considered in the SWE of the field radiated by the probe or the rotated probe when they work as transmitting antennas, $a'_{1,2\upsilon 1}$ are the related SWE coefficients, and

$$I_1(\vartheta) = \left[d^n_{1m}(\vartheta) - d^n_{-1m}(\vartheta)\right]; \qquad I_2(\vartheta) = \left[d^n_{1m}(\vartheta) + d^n_{-1m}(\vartheta)\right] \tag{6.28}$$

with

$$d^n_{\mu m}(\vartheta) = \sqrt{\frac{(n+\mu)!\,(n-\mu)!}{(n+m)!\,(n-m)!}} \sum_\sigma \binom{n+m}{n-\mu-\sigma} \binom{n-m}{\sigma} (-1)^{n-\mu-\sigma} \times \left(\cos\frac{\vartheta}{2}\right)^{2\sigma+\mu+m} \left(\sin\frac{\vartheta}{2}\right)^{2n-2\sigma-\mu-m} \tag{6.29}$$

being the rotation coefficients [76, 311]. In this last relation, the symbol

$$\binom{n}{m} = \frac{n!}{(n-m)!\,m!} \tag{6.30}$$

is the binomial coefficient and the summation over σ is extended to all terms which do not involve negative arguments for the factorials. In expressions (6.27),

$$A^n_{\upsilon\mu}(\beta d) = \sqrt{\frac{(2n+1)(2\upsilon+1)}{n(n+1)\,\upsilon(\upsilon+1)}} \sqrt{\frac{(\upsilon+\mu)!\,(n-\mu)!}{(\upsilon-\mu)!\,(n+\mu)!}} (-1)^\mu \frac{1}{2} \mathrm{j}^{n-\upsilon} \times \sum_{p=|n-\upsilon|}^{n+\upsilon} \left\{\mathrm{j}^{-p}\left[n(n+1) + \upsilon(\upsilon+1) - p(p+1)\right] \tau(\mu, n, -\mu, \upsilon, p)\, h_p^{(2)}(\beta d)\right\} \tag{6.31a}$$

$$B^n_{\upsilon\mu}(\beta d) = \sqrt{\frac{(2n+1)(2\upsilon+1)}{n(n+1)\,\upsilon(\upsilon+1)}} \sqrt{\frac{(\upsilon+\mu)!\,(n-\mu)!}{(\upsilon-\mu)!\,(n+\mu)!}} (-1)^\mu \frac{1}{2} \mathrm{j}^{n-\upsilon} \times \sum_{p=|n-\upsilon|}^{n+\upsilon} \left[\mathrm{j}^{-p}(2\mathrm{j}\mu\beta d)\, \tau(\mu, n, -\mu, \upsilon, p)\, h_p^{(2)}(\beta d)\right] \tag{6.31b}$$

are the translation coefficients [76, 312], $\tau\left(\mu, n, -\mu, \upsilon, p\right)$ being the linearization coefficients identified by the expansion of the product of two associated Legendre functions (see (A.50))

$$P_n^{\mu}(x)P_{\upsilon}^{-\mu}(x) = \sum_{p=|n-\upsilon|}^{n+\upsilon} \tau\left(\mu, n, -\mu, \upsilon, p\right) P_p(x) \tag{6.32}$$

The SWE coefficients $a'_{1\upsilon 1}$ and $a'_{2\upsilon 1}$ of the probe and rotated probe to be used in (6.27) can be determined, by means of relation (6.9), from the knowledge of the tangential electric field radiated by them on a scanning sphere when working as transmitting antennas. Namely, these coefficients can be evaluated by means of the spherical uncompensated NF–FF transformation from the simulated measurements of the field radiated by the probe and rotated probe.

The highest spherical wave to be considered in the SWE is again fixed in the classical approach [76] according to the empirical minimum sphere rule (6.6), whereas, in the modified approach [94, 95], it is given by (6.7), being rigorously determined by the spatial band-limitation properties of radiated EM fields [102]. The sampling spacings $\Delta\vartheta$ and $\Delta\varphi$ relevant to the classical probe-compensated NF–FF transformation with spherical scanning [76] and to its rearranged version in [94, 95] and, hence, the number of scanning parallels and of samples acquired on each of them are the same as those reported in the previous section for the corresponding uncompensated NF–FF transformation techniques.

The efficient evaluation of the integrals in (6.27) can be carried out [76] in a quite similar way to those in (6.9). The integration over φ can be performed by expanding, on each scanning parallel, the voltages measured by the probe and rotated probe in Fourier series with respect to φ via the FFT algorithm. The integration over ϑ can be efficiently executed [76] by extending the aforementioned Fourier series coefficients from the range $[0, \pi]$ to the $[-\pi, \pi]$ one and then expanding them in Fourier series with respect to ϑ via the FFT. The rotation coefficients are they also expanded in Fourier series in ϑ [76]. In such a way, the double integration is so transformed in a double summation, of indexes ℓ and i, which involves the abovementioned Fourier series coefficients and the integrals:

$$\int_0^{\pi} \mathrm{e}^{\mathrm{j}(i-\ell)\vartheta} \sin\vartheta \, \mathrm{d}\vartheta$$

Also in this case, to take advantage from the numerical efficiency of the FFT, the number of the scanning parallels to be considered and the number of NF samples on each of them must be the first power of 2, or power of 2, 3, and 5 (in dependence of the available FFT algorithm) greater or equal to $N_{\max}$ and $2M$, respectively. Once the SWE coefficients a_{1nm} and a_{2nm} have been evaluated, the FF spherical components of the electric field radiated by the AUT can be attained by substituting them in the expression (6.14) of the SWE valid in the FF region.

Since the SWE is independent on the way the modal coefficients have been determined, the expression (6.20) can be again exploited to efficiently evaluate the antenna far field at the considered zenithal angle Θ by carrying out the summation using the FFT. Moreover, the OSI formula (6.21) can be applied also in this case to accurately determine the FF spherical components of the electric field radiated by the AUT from the FF samples reconstructed via the NF–FF transformation.

In the following, a numerical example pointing out the need to apply a probe-compensated NF–FF transformation to accurately reconstruct the antenna FF pattern from the NF data gathered by a real probe is shown. The simulations refer to the field radiated by a uniform planar array of 0.5 λ spaced elementary Huygens sources polarized along the y axis and covering a circle of radius $a = 17\ \lambda$, which lies in the plane $z = 0$. The NF samples radiated by such an array at the working frequency of 10 GHz are acquired on a sphere having radius $d = 23\ \lambda$ by the probe, an open-ended WR-90 rectangular waveguide. Figure 6.8 shows the amplitudes on the meridian at $\varphi = 90°$ of the rotated probe voltage V_r and of the NF ϑ-component E_ϑ. As can be seen, the plots exhibit some differences, thus confirming that the voltage measured by a real probe is not proportional to the tangential component of the electric field. However, these differences are much smaller than those present in the analogous Figure 4.6 and Figure 5.3, relevant to the plane-rectangular and cylindrical scans, respectively, since the probe is always pointed towards the AUT centre in the spherical scanning. The antenna FF pattern in the E-plane reconstructed from the voltage data by applying the classical spherical NF–FF transformation without probe compensation as modified in [93] and its probe-compensated version [95] are reported in Figures 6.9 and 6.10, respectively. As can be seen, some discrepancies are evident in the reconstruction got by applying the uncompensated NF–FF transformation. For the abovementioned reason, the discrepancies are smaller than those occurring in the cases of the plane-rectangular and cylindrical NF–FF transformations. On the other hand, the pattern reconstructed via the probe-compensated NF–FF transformation [95] and the exact one are practically indistinguishable.

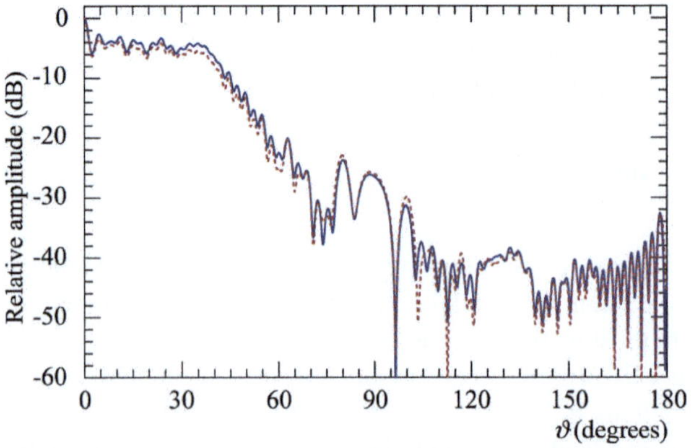

Figure 6.8 Amplitude of the rotated probe voltage V_r (blue solid line) and of the NF component E_ϑ (red dashes) on the meridian at φ=90°.

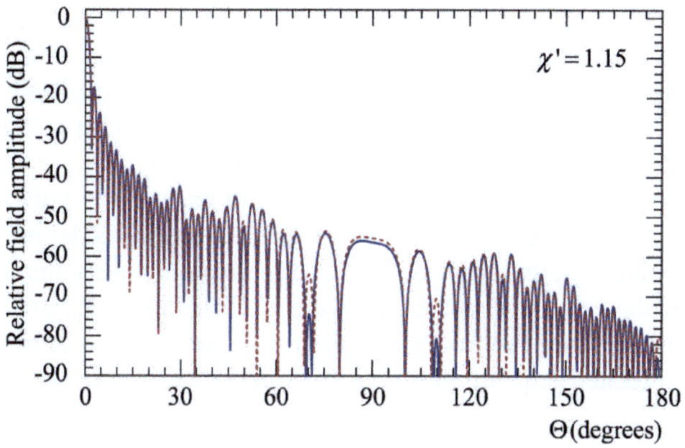

Figure 6.9 *E-plane pattern. Blue solid line: exact field. Red dashes: reconstructed from voltage data by the uncompensated NF–FF transformation.*

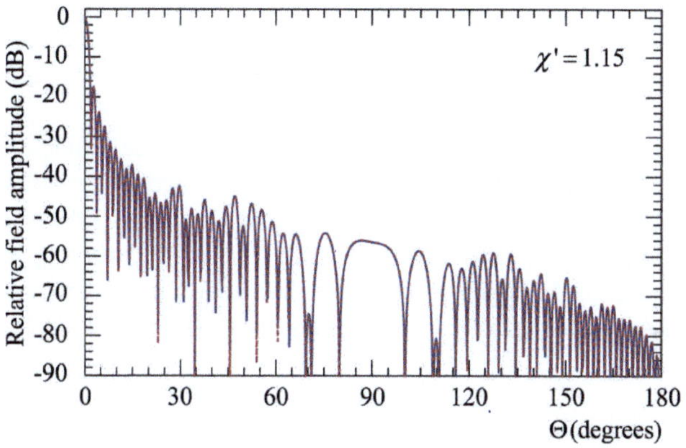

Figure 6.10 *E-plane pattern. Blue solid line: exact field. Red dashes: reconstructed from voltage data by the probe-compensated NF–FF transformation.*

It is noteworthy that the number of employed NF data is (21098) and, therefore, significantly smaller than the one (29040) which would be required when applying the classical NF–FF transformation [76].

6.4 Non-redundant NF–FF transformations with spherical scanning

In this section, NR NF–FF transformations with spherical scanning for non-volumetric AUTs, i.e., antennas having one or two predominant dimensions, will be presented. The description of the NF–FF transformations [93, 94, 97], using the spheroidal source modellings (an oblate spheroid for quasi-planar antennas and a prolate spheroid for the elongated ones), is reported in the subsection 6.4.1, whereas the NF–FF transformations [95, 96], employing the flexible modellings (a rounded cylinder for elongated antennas and a double bowl for the quasi-planar ones), are described in the subsequent subsection 6.4.2.

Although these NR NF–FF transformations are particularly tailored for elongated or quasi-planar antennas in dependence of the adopted modelling, they result to be valid for volumetric antennas too, since all these modellings contain the spherical one as particular case. In any case, when the AUT geometry departs significantly from the spherical one, they allow a remarkable reduction in the number of required NF samples and related acquisition time with respect to those employing the spherical modelling.

6.4.1 Non-redundant spherical NF–FF transformations: spheroidal AUT modellings

6.4.1.1 The oblate modelling case

Let a non-directive probe, having small dimensions with respect to the wavelength λ, be acquiring the NF data radiated on a sphere of radius d by a quasi-planar AUT contained in an oblate spheroid Σ with major and minor semi-axes equal to a and b (see Figure 6.11).

According to [41], the effective spatial bandwidth of the voltage measured by such a kind of probe is almost the same of the field radiated by the AUT, so that the NR sampling representations [104, 105] and the corresponding OSI expansions can be applied to the probe reduced voltage $\tilde{V}(\xi) = V(\xi)\, e^{j\gamma(\xi)}$, where ξ is an optimal parameter to be adopted for describing the observation curve and $\gamma(\xi)$ a proper phase function. As shown in Sect. 2.2, the bandwidth W_ξ, the phase function γ and

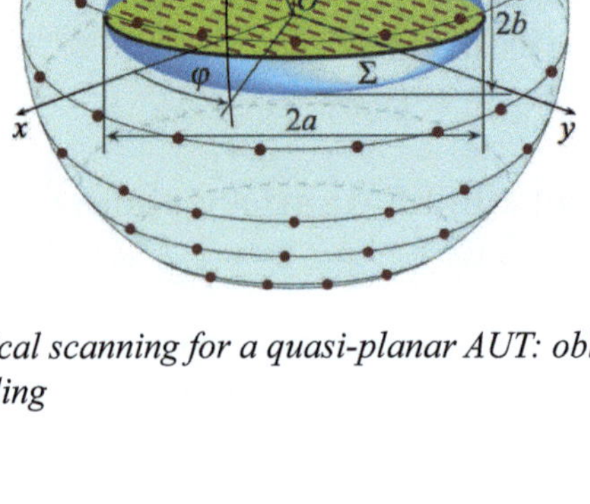

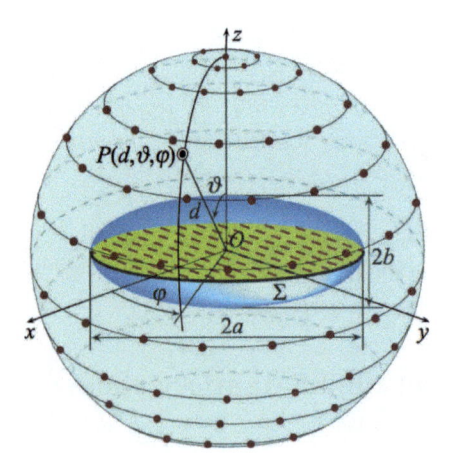

Figure 6.11 Spherical scanning for a quasi-planar AUT: oblate spheroidal modelling

the optimal parameter ξ relevant to a meridian of the scanning sphere are given by (2.25), (2.27) and (2.28b), whereas the azimuthal bandwidth[‡] W_φ is given by relation (2.29b). The probe and rotated probe voltages at any point on the sphere and, hence, the NF data needed by the classical spherical NF–FF transformation can be accurately recovered by exploiting the two-dimensional OSI expansion (2.71), here reported for reader's convenience:

$$V\left(\xi(\vartheta),\varphi\right) = \mathrm{e}^{-\mathrm{j}\gamma(\xi)}$$

$$\times \sum_{n=n_0-q+1}^{n_0+q} \left\{ SF\left(\xi,\xi_n,\bar{\xi},N,N''\right) \sum_{m=m_0-p+1}^{m_0+p} \tilde{V}\left(\xi_n,\varphi_{m,n}\right) SF\left(\varphi,\varphi_{m,n},\bar{\varphi}_n,M,M_n''\right) \right\} \tag{6.33}$$

all symbols having the same meanings as in (2.68). It must be noted that $\vartheta_\infty = \sin^{-1} u$ (see (2.30b)) takes the place of the angle $\vartheta\,(\xi_n)$ in the expression (2.67) of the azimuthal enlargement bandwidth factor χ^*.

Some numerical examples assessing the effectiveness of this NR spherical NF–FF transformation technique are shown in the following. The simulations are relevant to the same uniform planar circular array, probe, and scanning sphere already considered in Sect. 6.3. The AUT has been modelled by an oblate spheroid with major and minor semi-axes equal to 17 λ and 4.5 λ, respectively. A representative reconstruction example of the amplitude and phase of the voltage V_r on the meridian at $\varphi = 90°$ is shown in Figure 6.12a and b. As can be seen, the agreement between the recovered voltage and the exact one results to be very good. The stability of the algorithm, that is to say its robustness with respect to the errors altering the samples, has been investigated by corrupting the exact samples with uniformly distributed random errors. In particular, a background noise (bounded to Δa dB in amplitude and with arbitrary phase) and uncertainties on the data of $\pm\Delta a_r$ dB in amplitude and $\pm\Delta\phi$ degrees in phase have been simulated. As testified by Figure 6.13, the algorithm results to be stable. The plots of the maximum and mean-square reconstruction errors, shown in Figure 6.14a and b, assess the accuracy of the two-dimensional OSI algorithm in a more quantitative way. The errors are normalized to the voltage maximum value over the scanning sphere and have been determined by comparing the interpolated values of V_r with those directly evaluated on a close spherical grid. As can be noticed, they decrease up to very low values on increasing the oversampling factor and/or the number of retained samples. At last, the two-dimensional OSI expansion (6.33) has been applied to recover the NF data needed for the classical spherical NF–FF transformation as modified in [95]. The reconstruction of the antenna FF pattern in the E-plane is shown in Figure 6.15. As can be seen, the exact and recovered patterns are practically indistinguishable, thus fully validating the effectiveness of the described NR spherical NF–FF transformation for quasi-planar antennas.

‡The optimal parameter to be used to describe a parallel of the scanning sphere is the azimuthal angle φ.

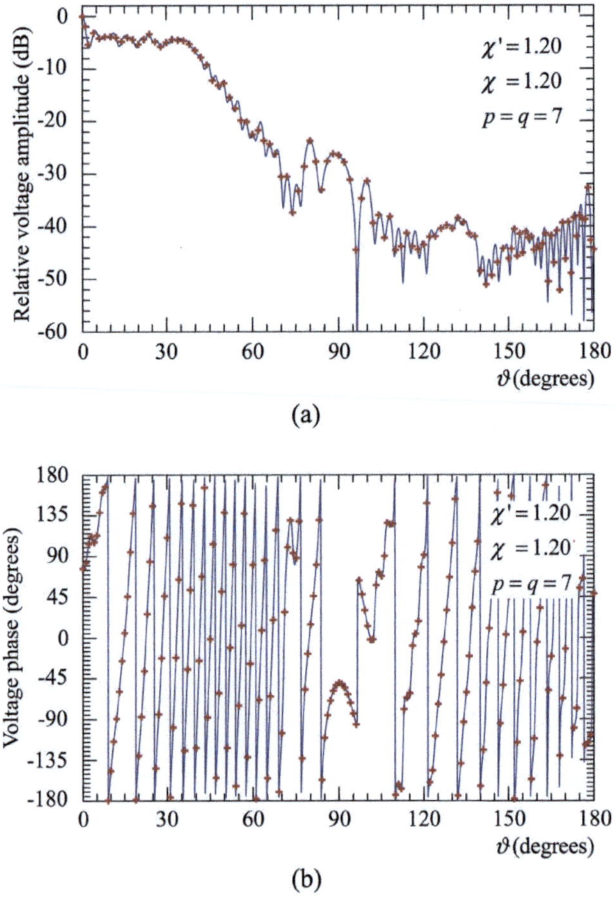

Figure 6.12 *Voltage V_r on the meridian at $\varphi = 90°$. Solid line: exact. Crosses: obtained by interpolating the NR spherical NF samples. (a) Amplitude, (b) Phase.*

It must be stressed that the number of employed NF data is (19210) and, accordingly, remarkably smaller than the one (29040) which would be needed when applying the classical NF–FF transformation [76].

The interested reader can find in [97] experimental results, which assess the effectiveness of this NF–FF transformation also from a practical viewpoint.

6.4.1.2 The prolate modelling case

Let an electrically small, non-directive probe, be gathering the NF data radiated on a sphere of radius d by a long AUT modelled by a prolate spheroid Σ with major and minor semi-axes equal to a and b (see Figure 6.16). As already stressed in the previous subsection, the NR sampling representations [104] can be applied to the probe reduced voltage $\tilde{V}(\xi) = V(\xi)\, e^{j\gamma(\xi)}$.

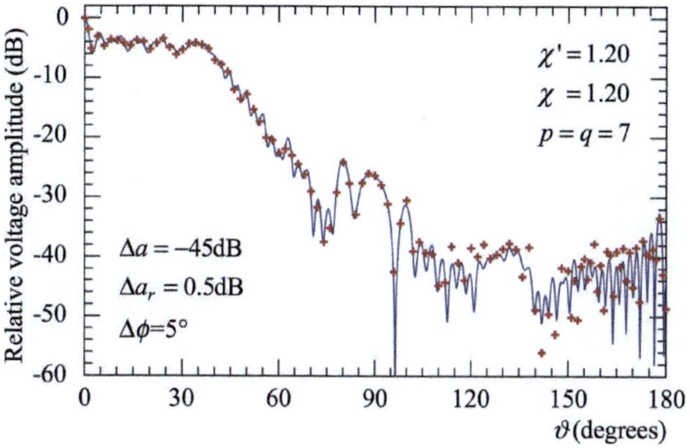

Figure 6.13 *Amplitude of the voltage V_r on the meridian at $\varphi = 90°$. Solid line: exact. Crosses: interpolated from error affected NR spherical NF samples.*

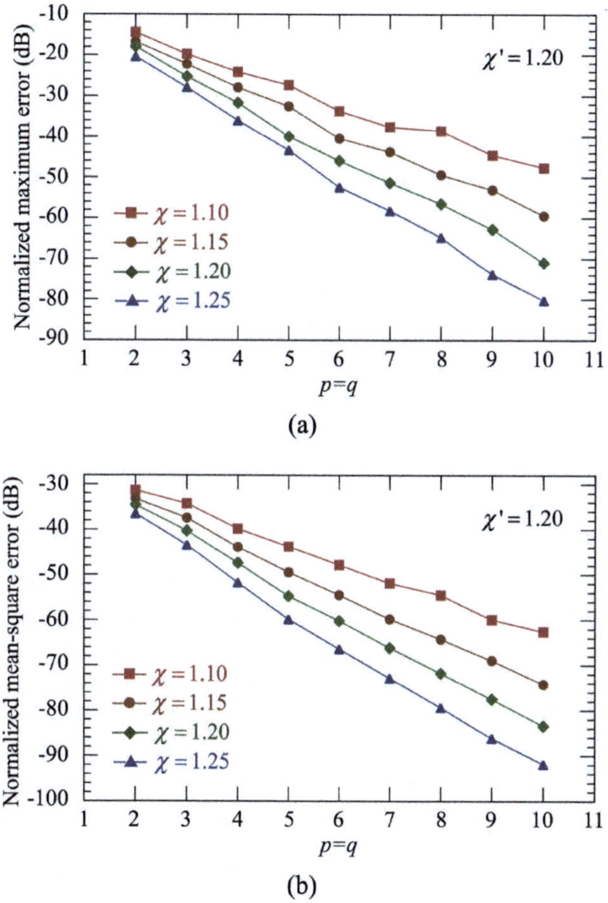

Figure 6.14 *Normalized errors in the reconstruction of the voltage V_r. (a) Maximum error, (b) Mean-square error.*

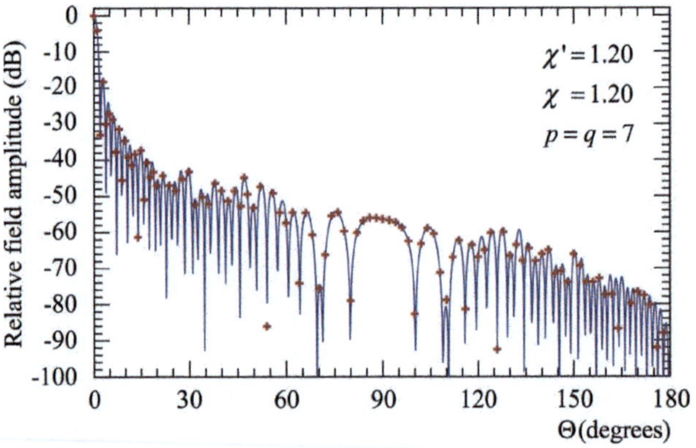

Figure 6.15 *E-plane pattern. Solid line: exact field. Crosses: reconstructed from the NR spherical NF samples.*

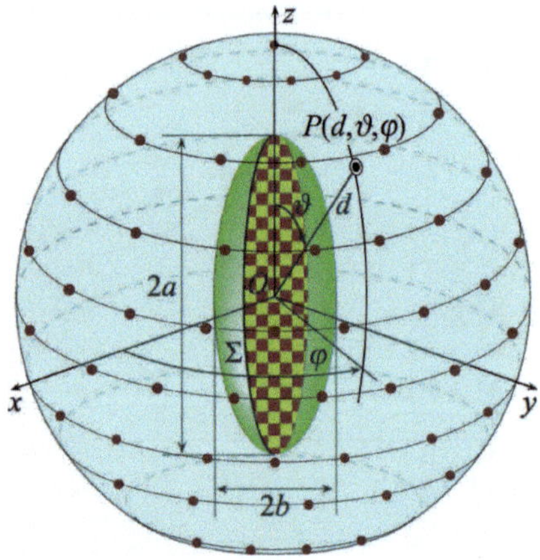

Figure 6.16 *Spherical scanning for an elongated AUT: prolate spheroidal modelling*

In accordance with the results in Sect. 2.2, the bandwidth W_ξ and the phase function γ relevant to a meridian of the scanning sphere are still given by the relations (2.25) and (2.27), whereas the corresponding optimal parameter ξ and the azimuthal bandwidth W_φ are now given by expressions (2.28a) and (2.29a).

The OSI expansion (6.33) allows again the efficient reconstruction of the probe and rotated probe voltages at any point on the scanning sphere and, therefore, can be used to accurately determine the values of V_p and V_r at the points required by the classical NF–FF transformation with spherical scanning in its original [76] or modified [95] version.

The described NR NF–FF transformation has been experimentally validated through the flexible NF measurement facility available in the laboratory of antenna characterization of the University of Salerno, arranged in such a way to perform a spherical scanning (see Sect. 1.2).

The probe and rotated probe voltages V_p and V_r have been acquired on sphere of radius 45.2 cm by an open-ended WR90 rectangular waveguide having its end tapered for minimizing the diffraction effects. The AUT is an X-band resonant slotted waveguide array working at 10.4 GHz (see Figure 6.17). It has been manufactured by PROCOM A/S by making 12 round-ended slots on both the broad walls of a WR90 rectangular waveguide and welding two cylinders on its narrow walls. This AUT has been mounted in such a way that its axis is coincident with the z one and its broad walls are parallel to the plane $y = 0$. According to the presented sampling representation, it has been modelled by a prolate spheroid having its major semi-axis $a = 18.17$ cm and the minor semi-axis $b = 3.75$ cm.

In order to validate the accuracy of the two-dimensional OSI expansion, the amplitude and phase of the reconstructed probe voltage V_p (the most significant one) relevant to the meridian at $\varphi = 0°$ are compared in Figure 6.18a and b with the directly measured ones. The reconstruction of the amplitude of the voltage V_p on the meridian at $\varphi = 90°$ is also shown in Figure 6.19. As can be seen, the agreement between the directly measured voltages (solid line) and those interpolated from the NR samples (crosses), by using $\chi' = 1.30$, $\chi = 1.20$, and $p = q = 7$, is very good except where the voltage values are very low, where the error is due to the noise and the residual environmental reflections. The overall effectiveness of this NF–FF transformation is assessed by comparing (see Figure 6.20a and b) the FF patterns in the principal planes E and H reconstructed from the acquired NF samples with those (references) got, by using the software package MI-3000 (implementing the classical spherical NF–FF transformation), from the data directly measured on the classical spherical grid. For completeness, the reconstruction of the AUT FF pattern in the cut plane at $\Phi = 90°$ is also shown in Figure 6.21. It is noteworthy that the same software package has been used to

Figure 6.17 Photo of the X-band slotted waveguide antenna. The photo was taken before covering the AUT support with the absorbers.

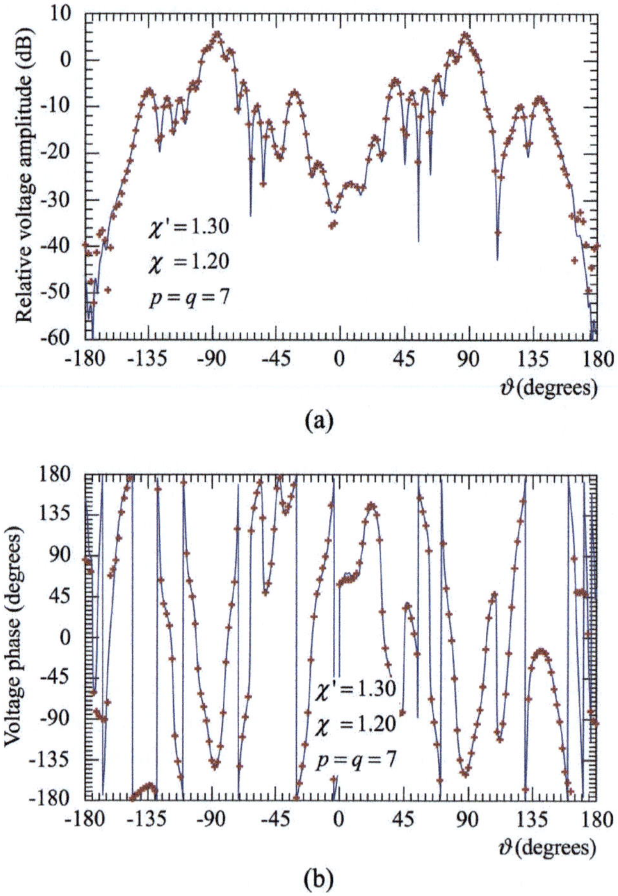

Figure 6.18 *Probe voltage V_p on the meridian at $\varphi = 0°$. Solid line: measured. Crosses: obtained by interpolating the NR spherical NF samples. (a) Amplitude, (b) Phase.*

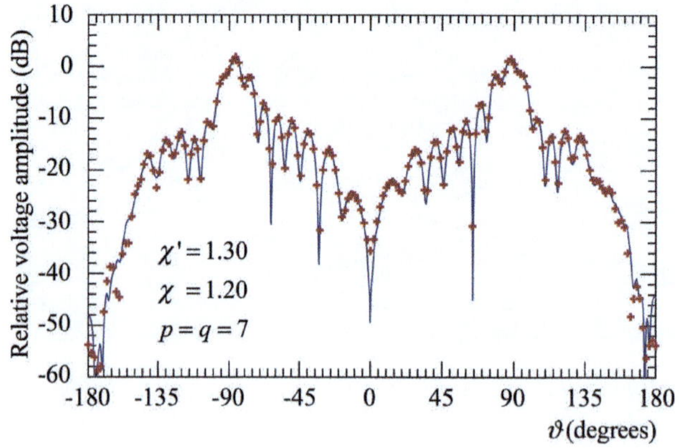

Figure 6.19 *Amplitude of the probe voltage V_p on the meridian at $\varphi = 90°$. Solid line: measured. Crosses: got by interpolating the NR spherical NF samples.*

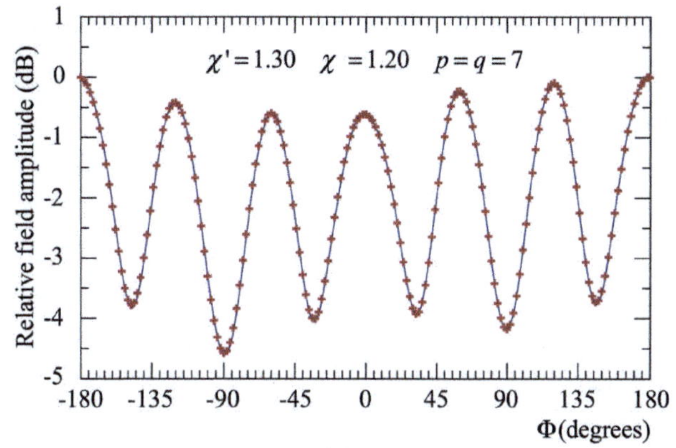

(a)

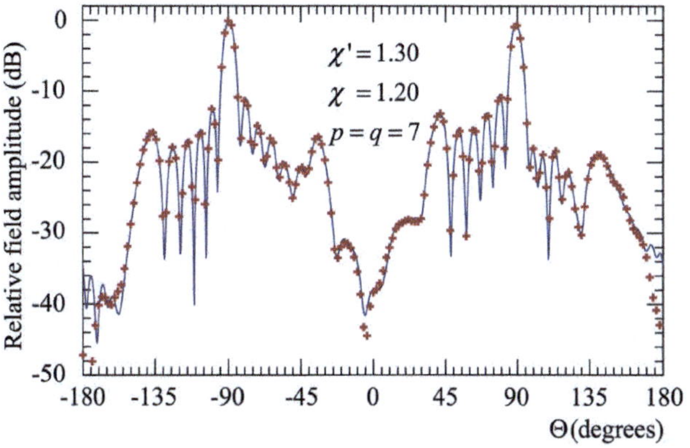

(b)

Figure 6.20 *FF patterns in the principal planes. Solid line: reference. Crosses: reconstructed from the acquired NR spherical NF samples. (a) E-plane (Θ = 90°), (b) H-plane (Φ = 0°).*

obtain the FF reconstruction from the NR samples. To this end, the OSI algorithm has been exploited to recover the massive spherical data needed to carry out the NF–FF transformation. As can be seen, there is a very good agreement, thus confirming the effectiveness of the approach.

It must be underlined that the number of employed NF samples is 1032 and, therefore, remarkably smaller than that (5100) required by the MI software package.

6.4.2 Non-redundant spherical NF–FF transformations: flexible AUT modellings

6.4.2.1 The rounded cylinder modelling case

Let a non-directive probe, whose sizes are small with respect to the wavelength λ, be acquiring the NF data radiated on a sphere of radius d by an elongated AUT

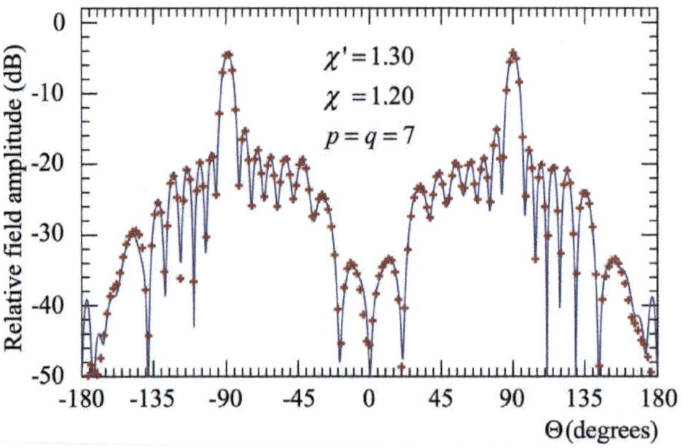

Figure 6.21 FF pattern in the cut plane $\Phi = 90°$. Solid line: reference. Crosses: reconstructed from the acquired NR spherical NF samples.

contained in a rounded cylinder Σ, i.e., a cylinder of height h' ended by two hemispherical caps of radius a' (see Figures 6.22 and 6.23). As emphasized in the previous section, the NR sampling representations [104] can be applied to the probe reduced voltage $\tilde{V}(\xi) = V(\xi)\, e^{j\gamma(\xi)}$.

In accordance with the results in Sect. 2.3, the bandwidth W_ξ, the optimal parameter ξ to describe a meridian of the scanning sphere and the corresponding phase function γ are given by relations (2.10), (2.11), and (2.8), here reported for reader's convenience:

$$W_\xi = \beta\ell'/2\pi; \quad \xi = \frac{\pi}{\ell'}\left[R_1 - R_2 + s_1' + s_2'\right]; \quad \gamma = \frac{\beta}{2}\left[R_1 + R_2 + s_1' - s_2'\right] \tag{6.34}$$

wherein $\ell' = 2\,(h' + \pi a')$ is the length of the curve C' intersection of Σ with a meridian plane. The explicit expressions of γ and ξ can be attained [95, 96] by substituting the appropriate values of the distances $R_{1,2}$ and curvilinear abscissae $s'_{1,2}$ in the above relations. The expressions of $R_{1,2}$ and of $s'_{1,2}$ change depending on the position of the tangency points $P_{1,2}$, which, in turn, depends on the observation point location. As shown in Sect. 2.3, where these expressions are reported§, three cases must be considered when the angle ϑ covers the range $[0, \pi]$.

As regards the evaluation of the azimuthal bandwidth W_φ, according to the results presented in Sect. 2.3, the maximum in (2.14), which allows its determination, is obtained in correspondence of the z' value given by (2.38).

The OSI expansion (6.33), wherein the azimuthal enlargement bandwidth factor χ^* is given by (2.67), allows again the effective reconstruction of the probe and rotated probe voltages at any point on the scanning sphere and, therefore, can be exploited to accurately determine their values at the points required by the classical spherical NF–FF transformation.

§Obviously, in these expressions $r = d$.

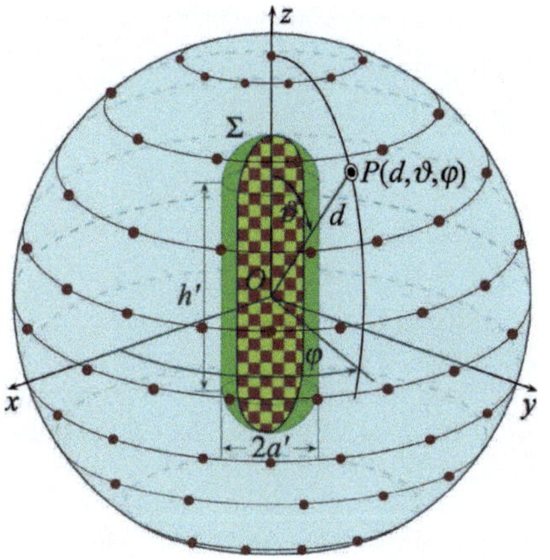

Figure 6.22 NR spherical scanning: rounded cylinder modelling

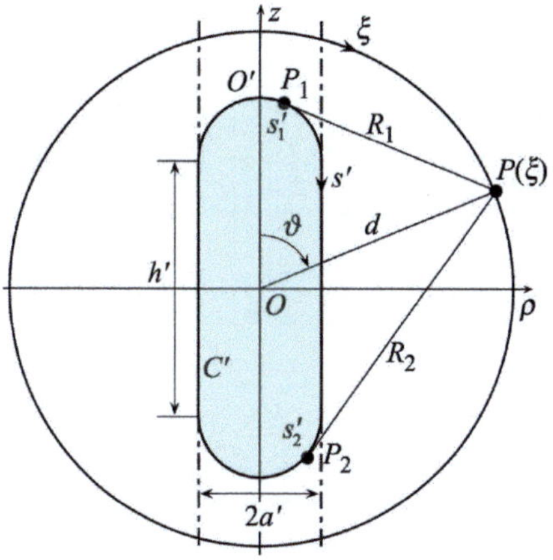

Figure 6.23 Rounded cylinder modelling: relevant to the evaluation of ξ

In the following, some results of numerical simulations, which prove the effectiveness of such a NR spherical NF–FF transformation technique, are shown. These simulations are relevant to the field radiated by a uniform planar array of $\lambda/2$ spaced elementary Huygens sources polarized along the z axis and covering a zone in the plane $y = 0$ formed by a $14\ \lambda \times 40\ \lambda$ rectangle ended in two half-circles. Accordingly, such an AUT has been considered as enclosed in a rounded cylinder with $h' = 40\ \lambda$ and $a' = 7\ \lambda$. An open-ended WR-90 rectangular waveguide scanning a sphere having radius $d = 33\lambda$ has been chosen as probe. Figure 6.24a and b show a representative reconstruction example of the amplitude and phase of the voltage V_r on the meridian at $\varphi = 90°$. As can be seen, the agreement between the exact voltage and that obtained by interpolating the NR samples is very

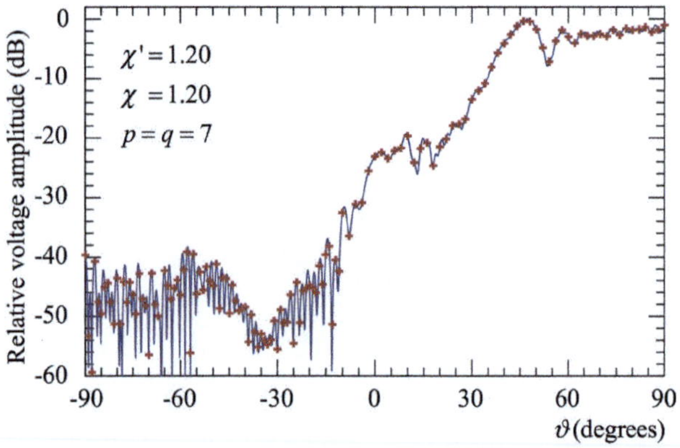

(a)

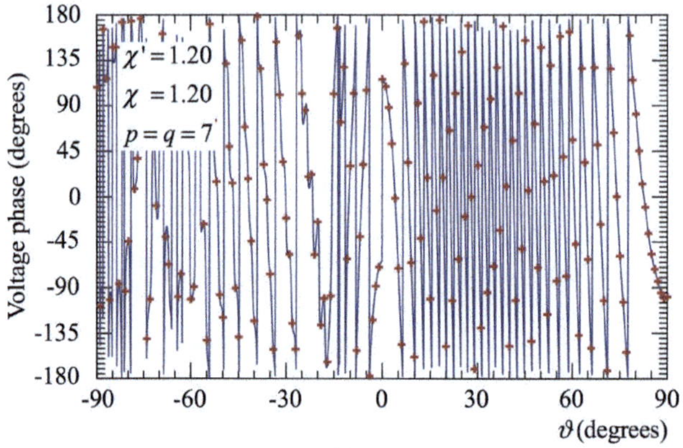

(b)

Figure 6.24 *Voltage V_r on the meridian at $\varphi = 90°$. Solid line: exact. Crosses: obtained by interpolating the NR spherical NF samples. (a) Amplitude, (b) Phase.*

good. The precision of the two-dimensional OSI expansion is assessed in a more quantitative way by the plots of the maximum and mean-square reconstruction errors, shown in Figure 6.25a and b. These errors have been got by comparing the interpolated values of V_r with those directly evaluated on a close grid of the scanning sphere and then normalizing the attained errors to the maximum value of this voltage on the sphere. As can be clearly observed, they decrease up to very low values on increasing the retained samples number and/or the oversampling factor. The algorithm stability with respect to errors affecting the samples has been investigated by corrupting the exact samples with uniformly distributed random errors, which simulate a background noise (bounded to Δa dB in amplitude and with arbitrary phase) and uncertainties on the data of $\pm\Delta a_r$ dB in amplitude and $\pm\Delta\phi$ degrees in phase. As can be seen

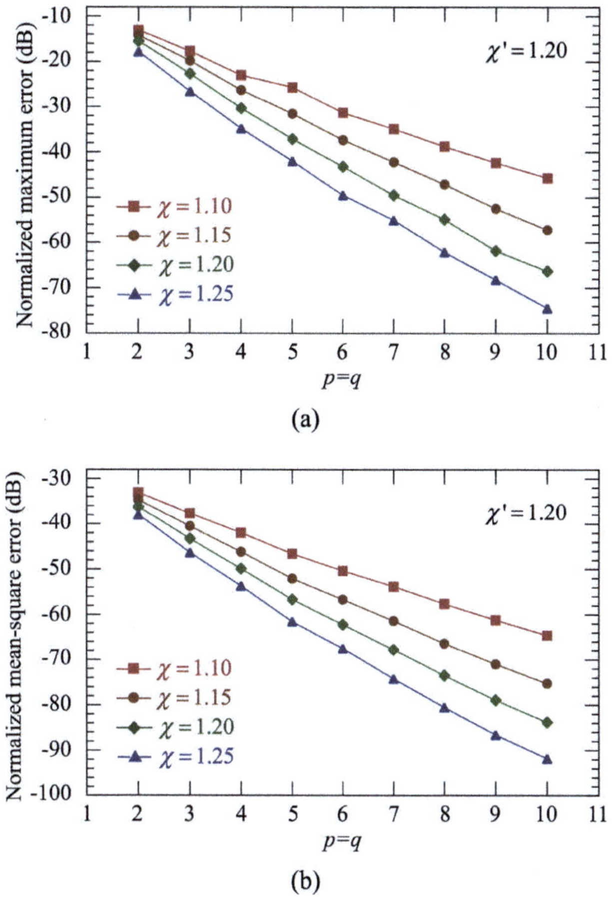

Figure 6.25 *Normalized errors in the reconstruction of the voltage V_r. (a) Maximum error, (b) Mean-square error.*

from Figure 6.26, the algorithm is stable, since it does not amplify the errors present on the input samples. At last, the overall effectiveness of the described NR spherical NF–FF transformation for long antennas has been assessed by applying the two-dimensional OSI expansion (6.33) for recovering the NF data required to execute the classical spherical NF–FF transformation as modified in [95]. The reconstructions of the antenna FF pattern in the principal planes are reported in Figure 6.27a and b. As can be seen, the exact and recovered patterns are in perfect agreement, thus fully proving the accuracy of such a NR spherical NF–FF transformation.

It can be interesting to compare the number of used NF samples (21292) with that (65150) which would be required by applying the classical spherical NF–FF transformation [76]. In conclusion, the described NR spherical NF–FF

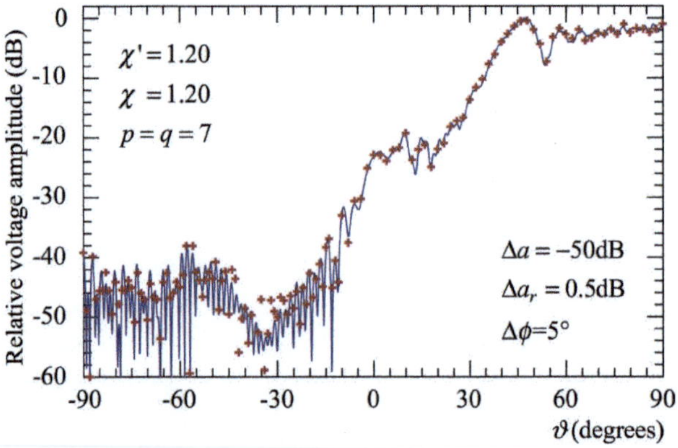

Figure 6.26 *Amplitude of the voltage V_r on the meridian at $\varphi = 90°$. Solid line: exact. Crosses: interpolated from error affected NR spherical NF samples.*

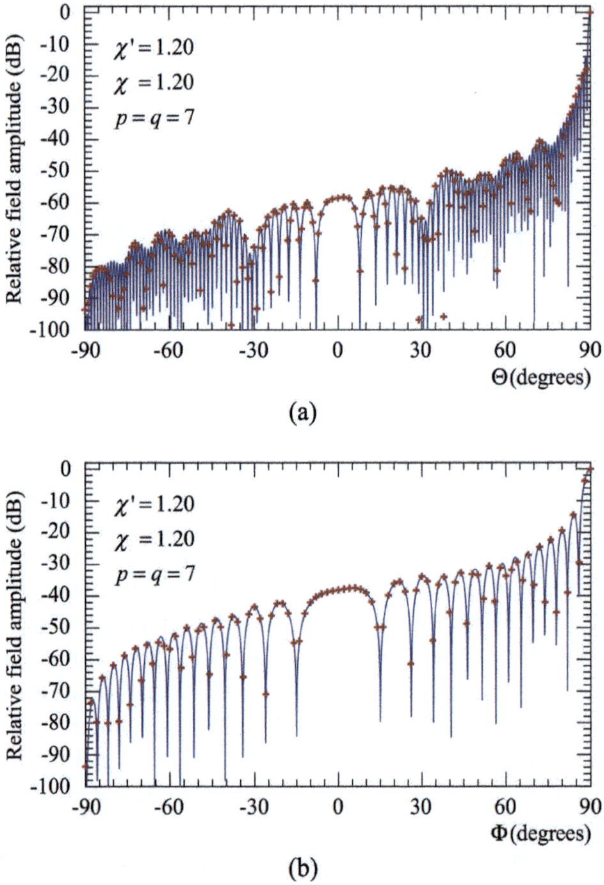

Figure 6.27 *FF patterns in the principal planes. Solid line: exact. Crosses: reconstructed from the acquired NR spherical NF samples. (a) E-plane, (b) H-plane.*

transformation allows a remarkable reduction of the needed NF data and related acquisition time without any loss of accuracy in the reconstruction of the AUT far field.

Experimental results assessing the effectiveness of such a NF–FF transformation also from a practical viewpoint can be found in [96].

6.4.2.2 The double bowl modelling case

Let a non-directive probe, having small dimensions with respect to the working wavelength λ, be collecting the NF data radiated on a sphere of radius d by a quasi-planar AUT enclosed in a double bowl Σ, i.e., a surface formed by two circular bowls sharing their apertures of radius a. Their lateral surfaces are got by rotating two arcs of a quarter of a circumference, whose radii c and c' can be different to fit better the shape of the considered AUT (see Figures 6.28 and 6.29). As in the previous subsection, the NR sampling representations [104] can be applied to the probe reduced voltage $\tilde{V}(\xi) = V(\xi)\, \mathrm{e}^{\mathrm{j}\gamma(\xi)}$.

According to the results presented in Sect. 2.3, the bandwidth W_ξ, the optimal parameter ξ for describing a meridian of the scanning sphere and the related phase function γ are given by relations (6.34), where the length ℓ' of the curve C', intersection of Σ with a meridian plane, is now given by

$$\ell' = 2\left[b + b' + \left(c + c'\right)\pi/2\right] \tag{6.35}$$

wherein $b = a - c$ and $b' = a - c'$. The explicit expressions of γ and ξ can be determined [95, 96] by substituting in these relations the proper values of $R_{1,2}$ and $s'_{1,2}$, which change in dependence of the observation point location. As shown in Sect. 2.3, where these expressions are reported¶, five cases have to be considered for ϑ ranging in $[0, \pi]$.

The optimal parameter to describe a parallel is again the azimuthal angle φ and the maximum in the relation (2.14), which allows one to evaluate the corresponding bandwidth W_φ, can be determined [95, 96] as detailed in Sect. 2.3.

The accurate reconstruction of the probe and rotated probe voltages at any point on the scanning sphere and, in particular, at the points required by the classical spherical NF–FF transformation in its original [76] or modified [95] version can be again performed by applying the two-dimensional OSI expansion (6.33), wherein the azimuthal enlargement bandwidth factor χ^* is given by (2.67).

Some results of experimental tests assessing the effectiveness of the described NR spherical NF–FF transformation for quasi-planar antennas are shown in the following. These tests have been carried out by using the flexible NF measurement facility available in the UNISA Antenna Characterization Lab, arranged in the spherical scanning configuration (see Sect. 1.2).

The AUT is an X-band flat plate slot array (see Figure 6.30) having a radius of about 23 cm, located on the plane $z = 0$, and working at 9.4 GHz. It has been manufactured by Rantec Microwave Systems Inc. and has been modelled by a double bowl with $a = 23.46$ cm, $c = 6.38$ cm, and $c' = 4.79$ cm. The radius d of the scanning

¶Obviously, in these expressions $r = d$.

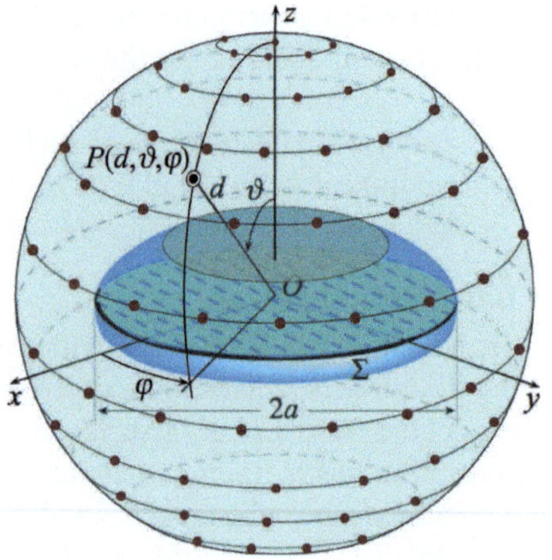

Figure 6.28 NR spherical scanning: double bowl modelling

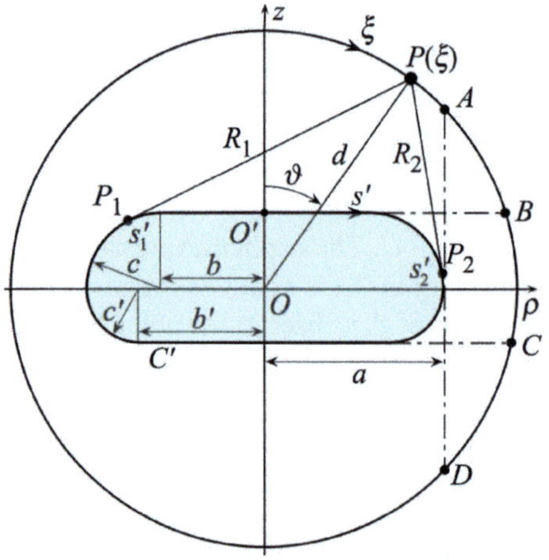

Figure 6.29 Double bowl modelling: relevant to the evaluation of ξ

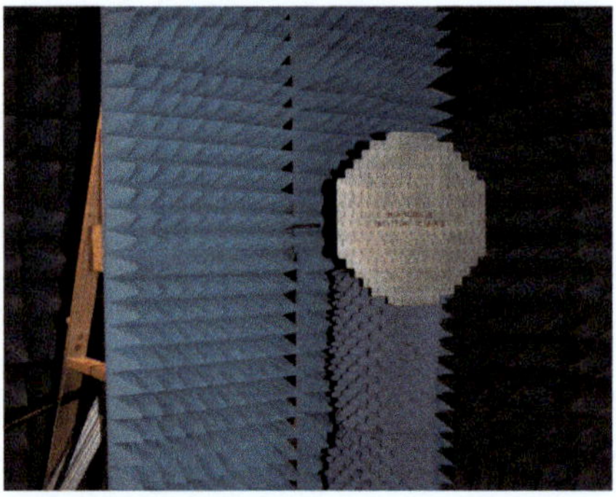

Figure 6.30 Photo of the X-band flat plate slot array

sphere is 42.0 cm and the employed probe is an open-ended WR90 rectangular waveguide, whose end is tapered to minimize the diffraction effects.

In Figure 6.31a and b, the amplitude and phase of the reconstructed voltage V_r on the meridian at $\varphi = 0°$ are compared with the directly measured ones. The reconstruction of the amplitude of the voltage V_p on the meridian at $\varphi = 90°$ is also shown in Figure 6.32. As can be observed, the agreement between the reconstructed voltages and the directly measured ones is very good, thus assessing the accuracy of the two-dimensional OSI expansion (6.33). At last, the E- and H-plane FF patterns reconstructed from the acquired NR spherical samples (3951) are compared in Figure 6.33a and b with those (references) attained, by using the software package MI-3000, from the NF data (7320) directly acquired

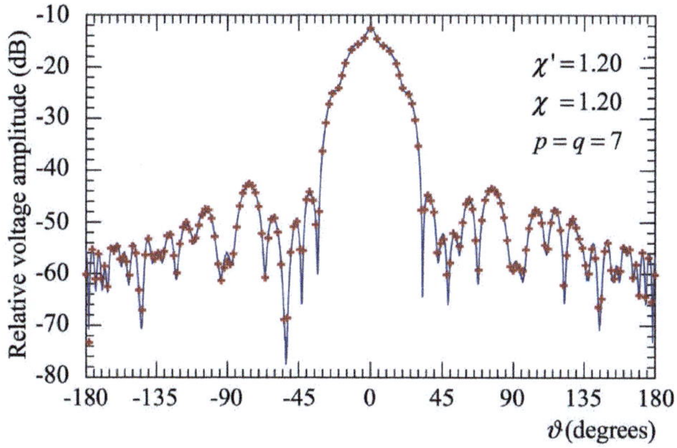

(a)

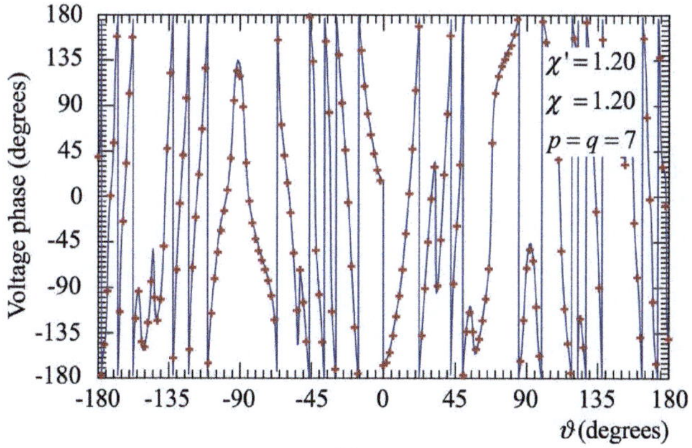

(b)

Figure 6.31 *Rotated probe voltage V_r on the meridian at $\varphi = 0°$. Solid line: measured. Crosses: obtained by interpolating the NR spherical NF samples. (a) Amplitude, (b) Phase.*

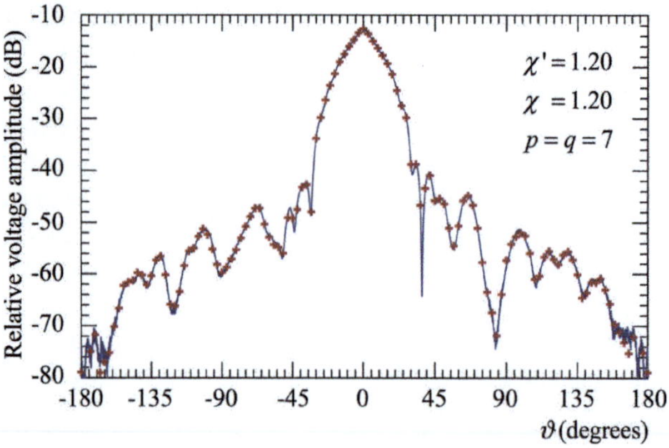

Figure 6.32 *Amplitude of the probe voltage V_p on the meridian at $\varphi = 90°$. Solid line: measured. Crosses: obtained by interpolating the NR spherical NF samples.*

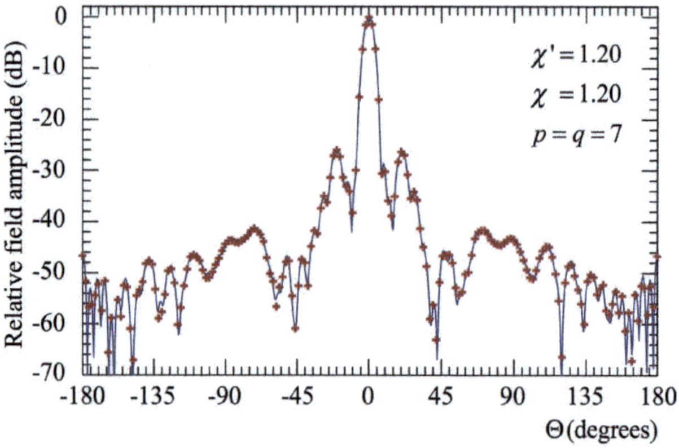

(a)

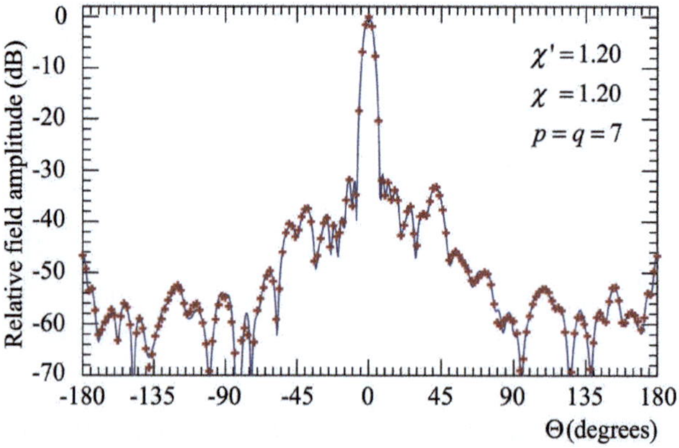

(b)

Figure 6.33 *FF patterns in the principal planes. Solid line: reference. Crosses: reconstructed from the acquired NR spherical NF samples. (a) E-plane, (b) H-plane.*

on the classical spherical grid. As can be seen, also in such a case there is a very good agreement, thus confirming the effectiveness of the described NR spherical NF–FF transformation. It is noteworthy that the FF reconstruction from the NR samples has been achieved by recovering, via the OSI expansion (6.33), the massive NF data required by this software package and then applying it to determine the AUT far field.

6.5 Non-redundant NF–FF transformations with spherical spiral scanning

The NR NF–FF transformations with spherical spiral scanning for non-volumetric AUTs will be presented in this section. They allow a further considerable measurement time saving with respect to the NR spherical NF–FF transformations described in the previous section, since the NR samples are acquired on fly and continuous and synchronized motions of the two rotating tables arranged in the roll (φ axis) over azimuth (ϑ axis) configuration allowing the spherical scan are adopted (see Sect. 1.2), thus realizing a spherical spiral scanning. The NF–FF transformations using the spheroidal source modellings (a prolate spheroid [133, 138] for elongated antennas and an oblate spheroid [133, 136] for the quasi-planar ones) are described in the subsection 6.5.1, whereas the description of NF–FF transformations employing the flexible modellings (a rounded cylinder [126, 134, 137] for elongated antennas and a double bowl [134, 135] for the quasi-planar ones) is reported in the subsequent subsection 6.5.2.

It is useful to note that, even if these NR NF–FF transformations are suitable to quasi-planar or long antennas depending on the employed AUT model, they are also valid for volumetric antennas, since all these models contain the spherical one as a particular case. In any case, when the AUT geometry differs significantly from the spherical one, they allow a considerable reduction in the amount of needed NF samples and related measurement time with respect to those using the spherical modelling [131, 132].

6.5.1 Non-redundant NF–FF transformations with spherical spiral scanning: spheroidal AUT modellings

6.5.1.1 The prolate modelling case

Let a long AUT be considered as contained in a prolate spheroid Σ with minor and major semi-axes equal to b and a, respectively, and let the voltage be acquired by a non-directive, electrically small probe, which is scanning a proper spiral, wrapping a sphere of radius d (see Figure 6.34). An effective sampling representation [133, 138], which allows the accurate recovery of the NF data, needed by the classical probe-compensated NF–FF transformation with spherical scanning, from the NR samples collected along the spiral, can be developed by properly exploiting (see Sect. 3.2) the unified theory of spiral scans for non-spherical antennas [124].

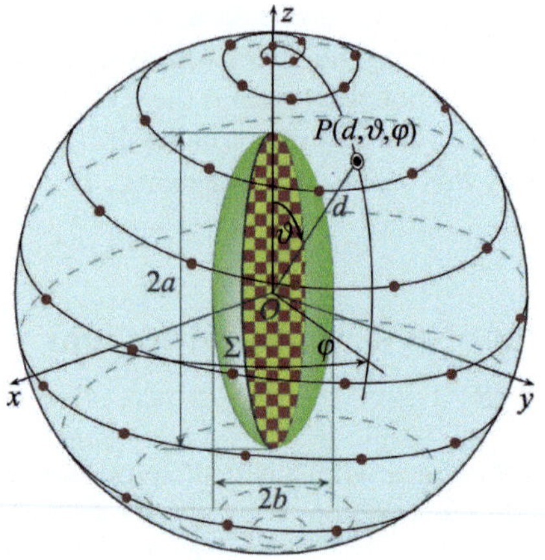

Figure 6.34 Spherical spiral scanning for a long AUT: prolate spheroidal modelling

The spiral equations (3.23), by taking into account that $r(\theta) = d$ in the case of a spiral wrapping a sphere of radius d, become:

$$\begin{cases} x = d \sin\theta\,(\xi) \cos\phi \\ y = d \sin\theta\,(\xi) \sin\phi \\ z = d \cos\theta\,(\xi) \end{cases} \tag{6.36}$$

wherein $\xi = k\,\phi$, with ϕ being the angular parameter which describes the spiral. The spiral step $2\pi k$, determined by two consecutive intersections at the points $Q(\phi)$ and $Q(\phi + 2\pi)$ of the spiral with a meridian, must be equal to the sampling spacing $\Delta\xi$ required for the interpolation along a meridian, which, in accordance with (2.59), is given by $\Delta\xi = 2\pi/(2N'' + 1)$. As a consequence, it results $k = 1/(2N'' + 1)$.

It is useful to note that θ becomes negative when the angular parameter ϕ is less than zero, whereas the zenithal angle ϑ is always a positive quantity. Moreover, ϕ is always continuous, while the azimuthal angle φ presents a discontinuity jump of π in correspondence of the poles, as can be easily recognized from (6.36).

It is worthy to note that the spiral can be viewed as attained by projecting on the measurement sphere, by means of the curves at ξ = constant (see Figure 2.6), the spiral which wraps, with the same step, the prolate spheroid Σ.

According to the unified theory of spiral scannings for non-spherical antennas [124] and as pointed out in Chapter 3, the NR sampling representation of the probe voltage from its samples acquired on the spiral, whose step is equal to the sample spacing required for the interpolation along a meridian, can be obtained by developing a NR sampling voltage representation along such a spiral. In fact, the voltage at any point P on the sphere can be recovered from the samples acquired on the spiral by first evaluating, through an interpolation along the spiral, the intermediate reduced voltage samples (those at the intersection points between the spiral and the meridian through P) and then reconstructing the voltage at P by interpolating, along the meridian, the intermediate samples.

The bandwidth W_ξ, the phase function γ, and the optimal parameter ξ for describing a meridian are given by the expressions (2.25), (2.27), and (2.28a).

The NR sampling representation along the spiral can be determined by exploiting the heuristic procedure [124] described in Sect. 3.2. In accordance with such a procedure, the parameter η to describe the spiral is β/W_η times the curvilinear abscissa of the projecting point lying on the spiral wrapping the prolate spheroid Σ, the corresponding phase function ψ coincides with that γ relevant to a meridian, and W_η is β/π times the length of the spiral which wraps Σ from the north to the south pole.

The probe and rotated probe voltages at any point on the sphere and, accordingly, at the points wherein the NF data to perform the classical spherical NF–FF transformation, in its original [76] or modified [95] version, are required can be accurately reconstructed from the NR samples acquired along the spiral by applying the OSI expansion (3.24), here reported for reader's convenience:

$$V\big(\xi(\vartheta), \varphi\big) = \mathrm{e}^{-\mathrm{j}\gamma(\xi)}$$

$$\times \sum_{n=n_0-q+1}^{n_0+q} \left[SF\left(\xi, \xi_n, \bar{\xi}, N, N''\right) \sum_{m=m_0-p+1}^{m_0+p} \tilde{V}\left(\eta_m\right) SF\left(\eta\left(\xi_n\right), \eta_m, \bar{\eta}, M, M''\right) \right] \tag{6.37}$$

where all symbols have the same meanings as in (3.24). It must be remembered that the enlargement bandwidth factor χ' relevant to the representation along the spiral must be suitably increased nearby the poles, to avoid a significant growth of the bandlimitation error in these zones [124, 133, 138].

Some numerical tests, which prove the validity of this NR NF–FF transformation with spherical spiral scanning, are reported in the following. The simulations are relevant to the field radiated by a uniform planar array of 0.6 λ spaced elementary Huygens sources, polarized along the z axis and covering an elliptical zone in the xz plane with minor and major semi-axes equal to 6 λ and 24 λ. An open-ended WR-90 rectangular waveguide, which is scanning a spiral wrapping a sphere of radius $d =$ 30 λ, has been chosen as probe. The reconstructions of the amplitude and phase of the voltage V_r on the meridian at $\varphi = 90°$ are shown in Figure 6.35a and b. As can be observed, there is an excellent agreement between the reconstructed voltage and the exact one. It is useful to note that, in the zones of the spiral specified by the 30 samples surrounding each pole, the control of the aliasing error has been achieved by suitably increasing the enlargement bandwidth factor χ' in such a way that the sample spacing is reduced by a factor 7. The stability of the algorithm with respect to errors on the samples values has been assessed (see Figure 6.36) by corrupting the exact samples with uniformly distributed random errors, which simulate a background noise (bounded to Δa dB in amplitude and with arbitrary phase) and uncertainties on the data of $\pm\Delta a_r$ dB in amplitude and $\pm\Delta\phi$ degrees in phase. The plot of the mean-square reconstruction error shown in Figure 6.37 assesses the precision

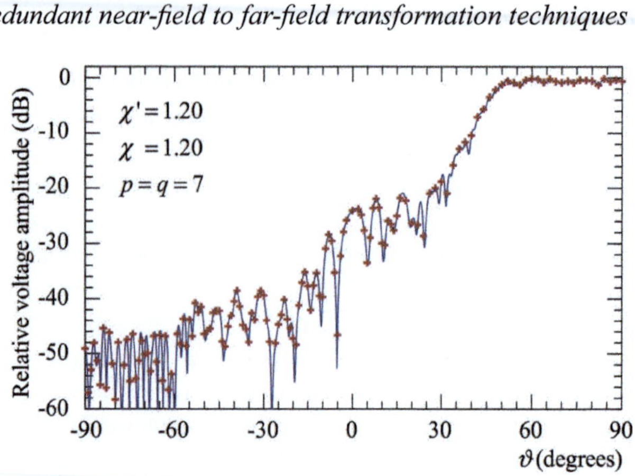

(a)

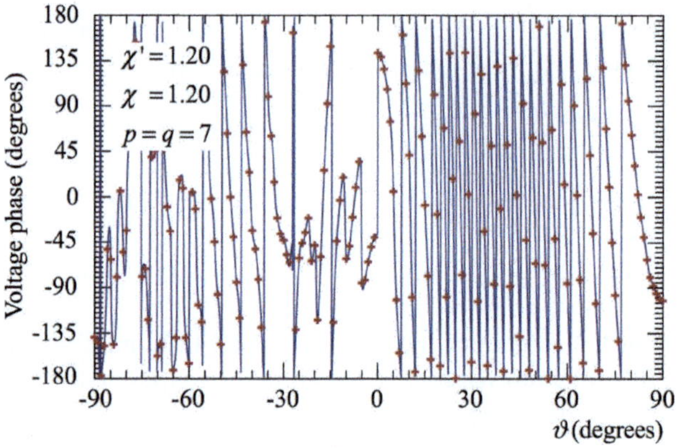

(b)

Figure 6.35 *Voltage V_r on the meridian at $\varphi = 90°$. Solid line: exact. Crosses: obtained by interpolating the NF samples acquired along the spiral. (a) Amplitude, (b) Phase.*

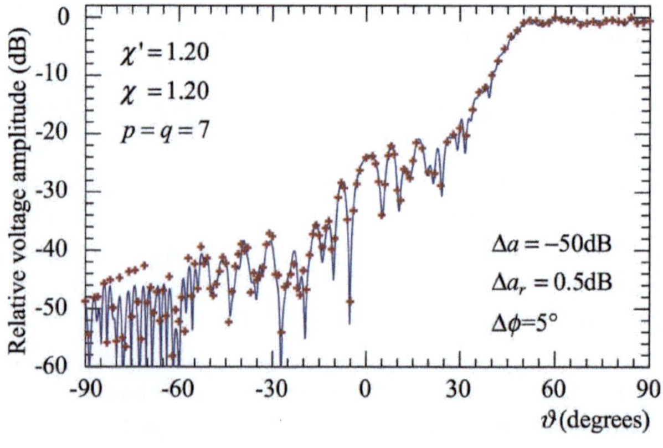

Figure 6.36 *Amplitude of the voltage V_r on the meridian at $\varphi = 90°$. Solid line: exact. Crosses: interpolated from error affected samples.*

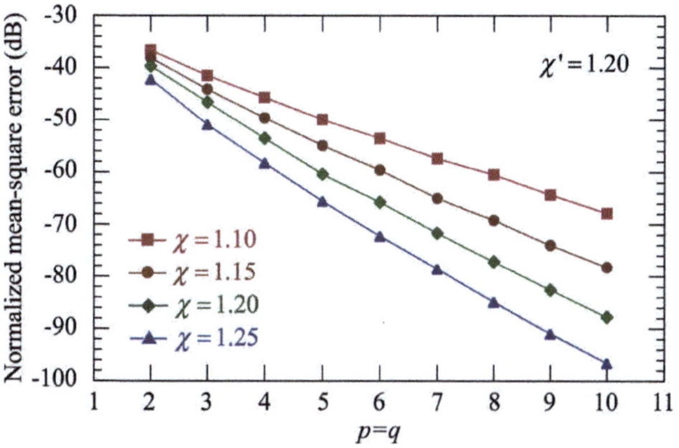

Figure 6.37 *Normalized mean-square error in the reconstruction of the rotated probe voltage* V_r

of the two-dimensional OSI expansion (6.37) in a more quantitative way. The evaluation has been performed by comparing, on a close grid of the scanning sphere, the interpolated values of V_r with the exact ones and then normalizing the obtained result to the maximum value of this voltage on the sphere. As usual, the mean-square error decreases up to very low values when increasing the oversampling factor and/or the number of retained samples. The curves relevant to the maximum error are not reported here for space saving and exhibit a quite analogous behaviour running about 20 dB above with respect to the mean-square error ones. The reconstructions of the E-plane and H-plane FF patterns, obtained by applying the classical spherical NF–FF transformation as modified in [95] to the NF data reconstructed via the OSI expansion (6.37) from the NR samples collected along the spiral, are shown in Figure 6.38a and b. As can be seen, the recovered and exact patterns are practically indistinguishable, thus fully validating the effectiveness of the described NR NF–FF transformation.

It must be stressed that the number of used NF data is (12842 including the 360 extra samples at reduced spacing) and, accordingly, remarkably smaller than the one (51520) required when applying the classical NF–FF transformation [76].

The interested reader can find in [138] experimental results assessing the effectiveness of this NF–FF transformation also from a practical point of view.

6.5.1.2 The oblate modelling case

Let an electrically small, non-directive probe, be acquiring the NF data radiated on a proper spiral, wrapping a sphere of radius d, by a quasi-planar AUT modelled by an oblate spheroid Σ with major and minor semi-axes equal to a and b (see Figure 6.39). As shown in the previous subsection, an efficient sampling representation, which allows the accurate recovering of the input NF data for the classical spherical NF–FF transformation from the NR samples gathered along the spiral, can be developed [133, 136] by properly making use of the unified theory of spiral scans for non-spherical antennas [124].

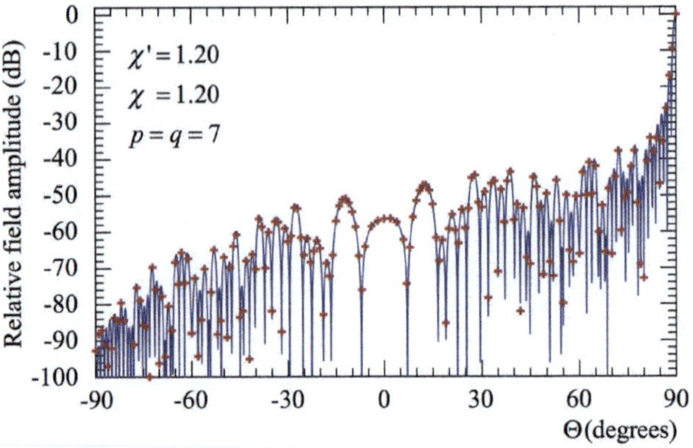

(a)

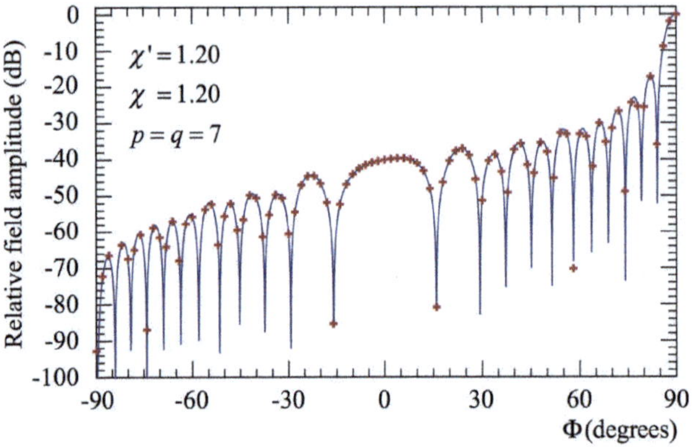

(b)

Figure 6.38 FF patterns in the principal planes. Solid line: exact. Crosses: reconstructed from the NF samples acquired along the spiral. (a) E-plane, (b) H-plane.

Since, when adopting the oblate spheroidal modelling, the development of the NR sampling representation of the probe voltage on the sphere from its samples collected on the spiral is quite similar to that previously described for the prolate modelling case, in the following, for space saving, only the differences will be pointed out. The projecting curves at ξ = constant are now the hyperbolas shown in Figure 2.7. The bandwidth W_ξ and the phase function γ are still given by (2.25) and (2.27), but the optimal parameter ξ to describe a meridian must be now evaluated according to the expression (2.28b).

The two-dimensional OSI expansion (6.37) can be again applied to accurately reconstruct the probe and rotated probe voltages at any point on the sphere and, in particular, at those needed by the classical NF–FF transformation with spherical scanning. It is useful to stress that, also in this case, the enlargement bandwidth

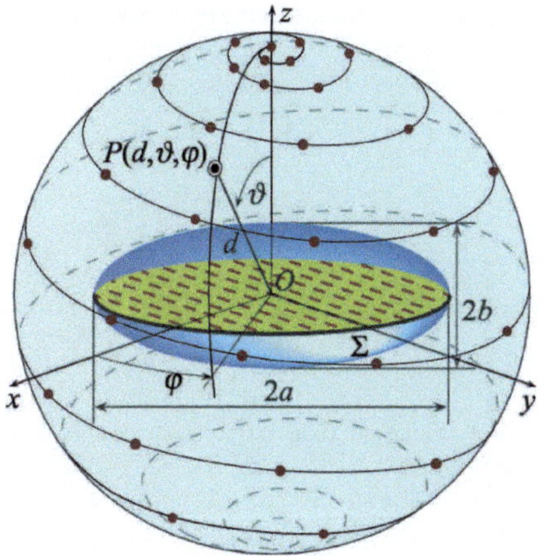

Figure 6.39 Spherical spiral scanning for a quasi-planar AUT: oblate spheroidal modelling

factor χ', relevant to the representation along the spiral, must be properly augmented in the neighbourhoods of the poles to prevent a significant increase of the bandlimitation error in these zones [124, 133, 136].

The experimental assessment of the described NF–FF transformation with spherical spiral scanning has been performed using the flexible NF measurement facility existing in the Antenna Characterization Lab of the University of Salerno, arranged in the spherical scanning configuration (see Sect. 1.2). The AUT is the X-band flat plate slot array manufactured by Rantec Microwave Systems Inc. already considered in the subsection 6.4.2.2, but it is now working at 9.3 GHz. According to the described sampling representation, it has been modelled by an oblate spheroid with semi-axes a = 23.9 cm and b = 7.4 cm. The samples of the probe and rotated probe voltages have been acquired by an open-ended WR-90 rectangular waveguide on a spiral wrapping a spherical surface with radius d = 45.2 cm.

The amplitude and phase of the reconstructed voltage V_r on the meridian at $\varphi = 90°$ are compared in Figure 6.40a and b with those directly measured at greater resolution. The reconstruction of the amplitude of the voltage V_p on the meridian at $\varphi = 0°$ is also shown in Figure 6.41. As can be clearly seen, the agreement between the reconstructed voltages and the measured ones is very good, thus assessing the accuracy of the OSI expansion (6.37). All these reconstructions have been attained by using $\chi = 1.20$, $p = q = 7$, and an enlargement bandwidth factor $\chi' = 1.20$, save for the zones of the spiral determined by the 18 samples around each pole, where, to avoid the growth of the bandlimitation error, its value has been suitably increased in such a way that the sample spacing is exactly reduced by a factor 5. The overall effectiveness of this NF–FF transformation with spherical spiral scanning is assessed by comparing in Figure 6.42a and b the E- and H-plane FF patterns reconstructed from the NR samples acquired along the spiral (4044 including the 144 extra samples at

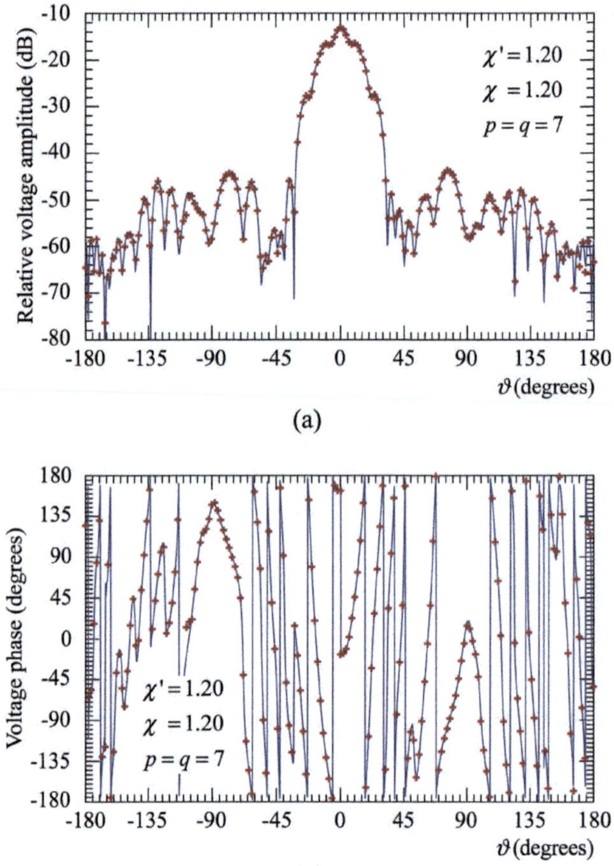

Figure 6.40 *Voltage V_r on the meridian at $\varphi = 90°$. Solid line: measured. Crosses: obtained by interpolating the NF samples acquired along the spiral. (a) Amplitude, (b) Phase.*

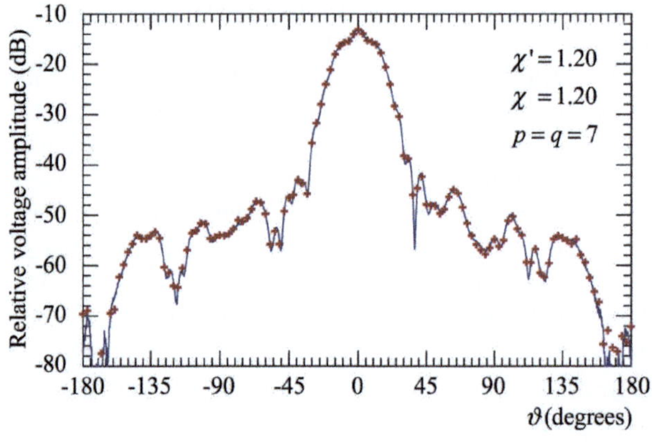

Figure 6.41 *Amplitude of V_p on the meridian at $\varphi = 0°$. Solid line: measured. Crosses: obtained by interpolating the NF samples acquired along the spiral.*

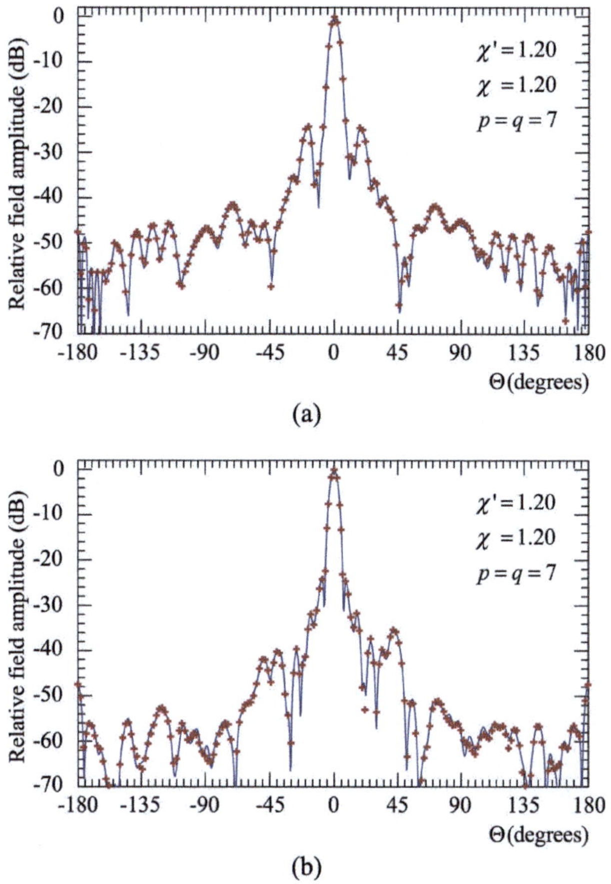

Figure 6.42 *FF patterns in the principal planes. Solid line: reference. Crosses: reconstructed from the NF samples acquired along the spiral. (a) E-plane, (b) H-plane.*

reduced spacing) with those (references) obtained, by using the software package MI-3000, from the NF data (7320) directly acquired on the classical spherical grid. It is useful to note that FF reconstruction from the NR samples collected on the spiral has been achieved by recovering, through the OSI expansion (6.37), the NF data needed by such a software package and then applying it to determine the AUT far field.

6.5.2 Non-redundant NF–FF transformations with spherical spiral scanning: flexible AUT modellings

6.5.2.1 The double bowl modelling case

Let a non-directive probe, whose dimensions are small with respect to the wavelength λ, be gathering, on a proper spiral wrapping a sphere of radius d, the NF data radiated by a quasi-planar AUT enclosed in a double bowl Σ (see Figure 6.43). As

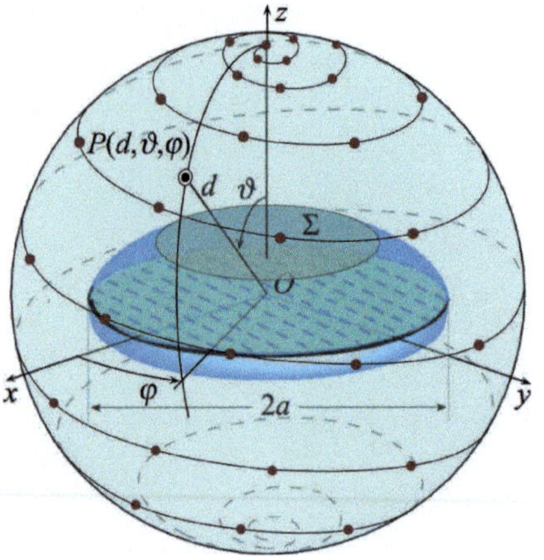

Figure 6.43 Spherical spiral scanning for a quasi-planar AUT: double bowl modelling

already done in the previous subsections, the NR sampling representations [104] can be applied to the probe reduced voltage $\tilde{V}(\xi) = V(\xi)\,\mathrm{e}^{\mathrm{j}\gamma(\xi)}$ and the unified theory of spiral scans for non-spherical antennas [124] can be conveniently exploited to develop [134, 135] an effective sampling representation, which allows the precise reconstruction of the input NF data for the classical spherical NF–FF transformation from the NR samples acquired on the spiral.

According to such a theory, the voltage at any point P on the sphere can be recovered from the samples acquired on the spiral by first evaluating, through an interpolation along the spiral, the intermediate reduced voltage samples and then reconstructing the voltage at P by interpolating, along the meridian, the intermediate samples. Hence, the spiral step must be equal to the sample spacing required for the interpolation along a meridian. Because the development of the sampling representation is quite similar to that described in the previous subsections, only the differences will be pointed out in the following. The curves at ξ = constant, which project the spiral wrapping the double bowl Σ onto the scan sphere, are now no longer hyperbolas, but those shown in Figure 4.53. The bandwidth W_ξ, the optimal parameter ξ for describing a meridian of the scanning sphere and the related phase function γ are given by relations (6.34), where the length ℓ' of the curve C' is given by (6.35). The explicit expressions of γ and ξ can be determined [134, 135] by substituting (see Sect. 2.3) the appropriate values of $R_{1,2}$ and $s'_{1,2}$ in the relations (6.34).

The probe and rotated probe voltages at any point on the scan sphere and, in particular, at those required by the classical spherical NF–FF transformation can be still accurately recovered by exploiting the OSI expansion (6.37) and again the factor χ', relevant to the representation along the spiral, must be suitably increased nearby the poles to avoid a significant growth of the bandlimitation error in these zones [124, 134, 135].

The validity of the described NF–FF transformation with spherical spiral scanning is proved by some numerical results reported in the following. The simulations are relevant to the same uniform planar circular array, probe, and scanning sphere already considered in Sect. 6.3, but now the NR samples of the probe and rotated probe voltages have been simulated as acquired along a spiral wrapping the sphere. Such an array has been considered as enclosed in a double bowl with $a = 17\,\lambda$, $c = c' = 4.5\,\lambda$.

In order to assess the accuracy of the two-dimensional OSI expansion (6.37), the amplitude and phase of the rotated probe voltage V_r, reconstructed from the NR spiral samples on the meridian at $\varphi = 90°$, are compared in Figure 6.44a and b with the exact ones. As can be observed, there is a perfect agreement between the recovered voltage and the exact one. It is worth noting that, in order to prevent

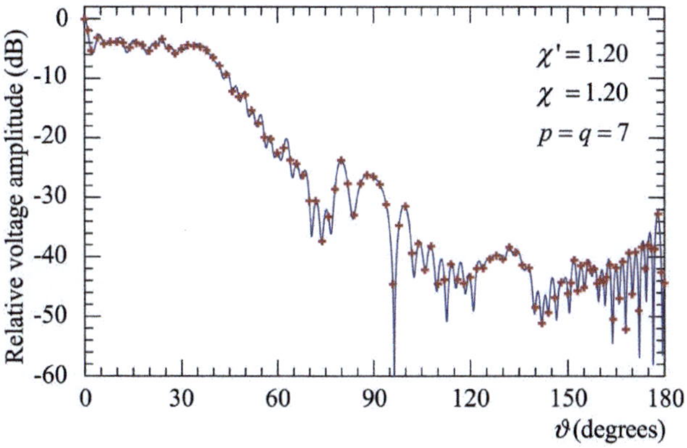

(a)

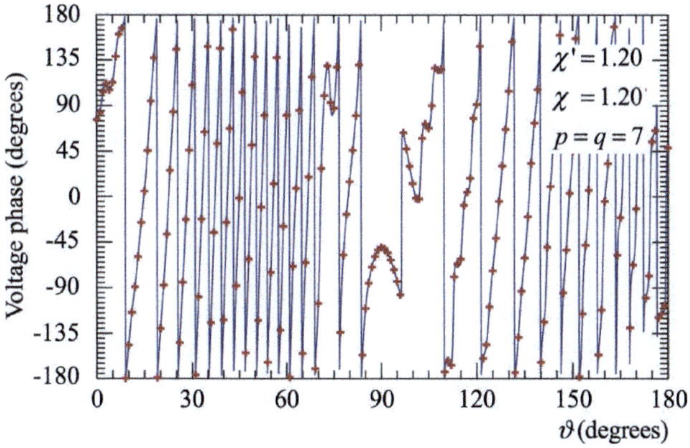

(b)

Figure 6.44 *Voltage V_r on the meridian at $\varphi = 90°$. Solid line: exact. Crosses: obtained by interpolating the NF samples acquired along the spiral. (a) Amplitude, (b) Phase.*

the growth of the bandlimitation error nearby the poles, the factor χ', relevant to the representation along the spiral, has been locally augmented in such a way that the sample spacing is reduced by a factor 7 in the zones of the spiral corresponding to the 30 samples surrounding each pole. The algorithm stability with respect to errors affecting the samples has been investigated by corrupting the exact samples with random errors, which simulate a background noise (bounded to Δa dB in amplitude and with arbitrary phase) and uncertainties on the data of $\pm\Delta a_r$ dB in amplitude and $\pm\Delta\phi$ degrees in phase. As can be seen from Figure 6.45, the algorithm results to be stable, since it does not amplify the errors which affect the input samples. The accuracy of the two-dimensional OSI expansion (6.37) is assessed in a more quantitative way by the plot of the mean-square reconstruction errors shown in Figure 6.46. These errors have been evaluated by

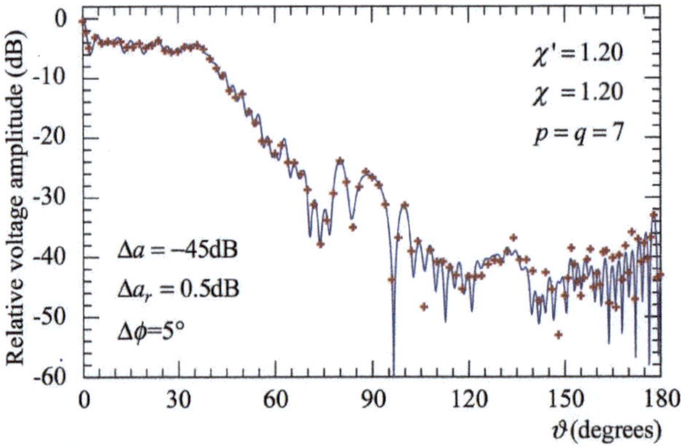

Figure 6.45 Amplitude of the voltage V_r on the meridian at $\varphi = 90°$. Solid line: exact. Crosses: interpolated from error affected samples.

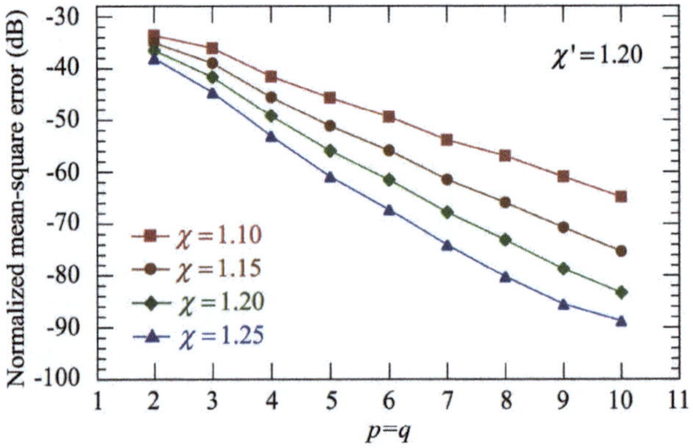

Figure 6.46 Normalized mean-square error in the reconstruction of the rotated probe voltage V_r

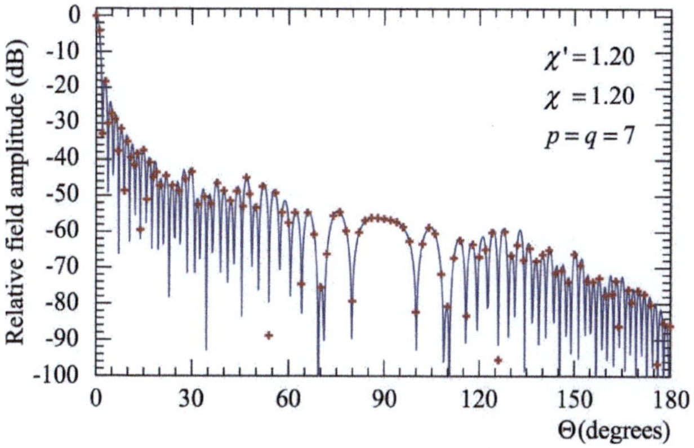

Figure 6.47 *E-plane pattern. Solid line: exact field. Crosses: reconstructed from the NF samples acquired along the spiral.*

comparing, on a close grid of the scanning sphere, the interpolated values of V_r with the exact ones and then normalizing the obtained results to the maximum value of this voltage on the sphere. As can be observed, the mean-square error decreases up to very low values on increasing the number of retained samples and/or the value of the oversampling factor. The curves relevant to the maximum errors, not reported here to save space, exhibit a quite analogous behaviour running about 20 dB above with respect to the mean-square error ones. At last, the OSI expansion (6.37) has been employed to accurately reconstruct the input NF data for the classical spherical NF–FF transformation as modified in [95]. The so reconstructed antenna FF pattern in the E-plane is compared with exact one in Figure 6.47. As can be seen, the exact and recovered patterns are practically indistinguishable, thus further validating the effectiveness of the described NF–FF transformation for quasi-planar antennas.

It can be interesting to compare the number of employed NF samples (20282 including the 360 extra samples at reduced spacing) with that (29040) which would be required when applying the classical NF–FF transformation [76].

Experimental results proving the effectiveness of this NF–FF transformation technique also from a practical point of view can be found in [135].

6.5.2.2 The rounded cylinder modelling case

Let an elongated AUT be modelled by a rounded cylinder Σ, formed by a cylinder of height h' terminated by two hemi-spheres of radius a', and let the voltage be acquired by a non-directive, electrically small probe, which is scanning a proper spiral, wrapping a sphere of radius d (see Figure 6.48). As in the previous subsections, the unified theory of spiral scans for non-spherical antennas [124] can be applied to accurately recover the NF data needed by the classical probe-compensated NF–FF transformation with spherical scanning, thus allowing the development of an efficient NF–FF transformation with spherical spiral scanning for long AUTs [134, 137].

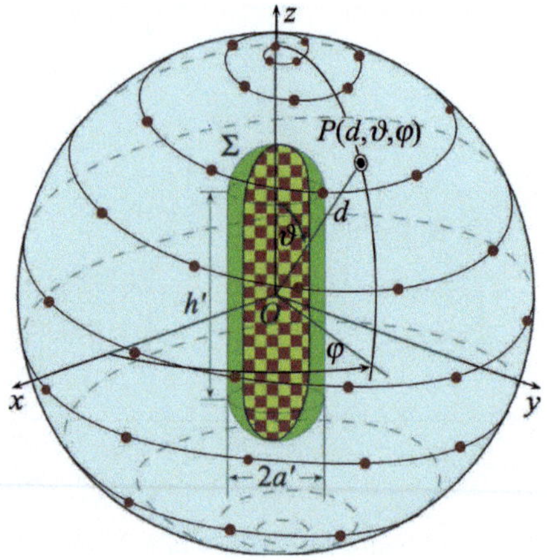

Figure 6.48 Spherical spiral scanning for a long AUT: rounded cylinder modelling

To save space, since the development of the sampling representation is very similar to those of the previous subsections, only the differences will be underlined in the following. The projecting curves at ξ = constant are now those displayed in Figure 5.32. The bandwidth W_ξ, the optimal parameter ξ to describe a meridian, and the corresponding phase function γ are given by relations (6.34), wherein the involved parameters can be determined as described in Sect. 6.4.2.1. Also in such a case, the two-dimensional OSI expansion (6.37) can be employed to accurately reconstruct the NF data required by the classical spherical NF–FF transformation in its original [76] or modified [95] version.

Some experimental results assessing the effectiveness of this NR NF–FF transformation with spherical spiral scanning are reported in the following. The experimental proofs have been performed through the spherical NF facility available in the Antenna Characterization Lab of the University of Salerno (see Sect. 1.2).

The samples of the probe and rotated probe voltages have been acquired by an open-ended WR-90 rectangular waveguide on a spiral wrapping a sphere with radius $d = 15.07\,\lambda$. The considered AUT is an X-band resonant slotted waveguide array, placed on the plane $y = 0$. It has been realized by making 8 round-ended slots on both the broad walls of a tapered rectangular waveguide and is fed by means of a coaxial to rectangular waveguide transition. Such an AUT has been modelled by a rounded cylinder with $h' = 10\,\lambda$ and $a' = 0.85\,\lambda$. To assess the effectiveness of the two-dimensional OSI algorithm, the amplitude and phase of the recovered voltage V_p relevant to the meridian at $\varphi = 0°$ are compared in Figure 6.49a and b with the directly measured ones at greater resolution. The reconstruction of the amplitude of the voltage V_p relevant to the meridian at $\varphi = 90°$ is also shown in Figure 6.50. A good agreement between the retrieved voltages and those directly acquired results save for the zones characterized by a very low voltage level, where the error is due to the residual reflections from the anechoic chamber walls and to the noise. It must be observed that, in the zones of the spiral specified by the 20 samples surrounding

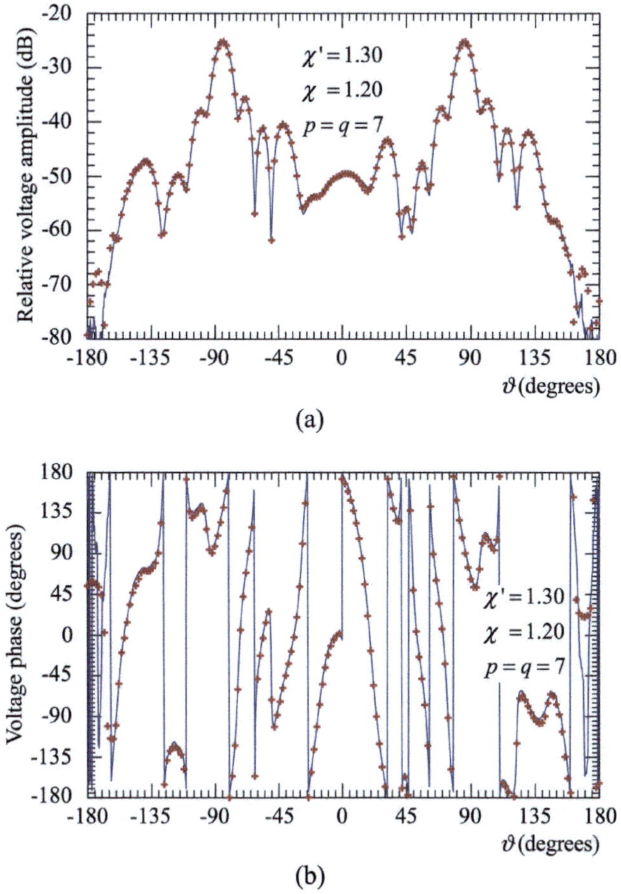

Figure 6.49 *Voltage V_p on the meridian at $\varphi = 0°$. Solid line: measured. Crosses: obtained by interpolating the NF samples acquired along the spiral. (a) Amplitude, (b) Phase.*

each pole, the control of the aliasing error has been achieved by enlarging the factor χ', relevant to the sampling representation along the spiral, in such a way that the sample spacing is reduced by a factor 5.

The effectiveness of the described NF–FF transformation is definitively validated by the very good agreement observable in the comparison of the FF patterns in the principal planes E and H recovered from the NF data collected on the spiral (Figure 6.51a and b) with those (references) attained from the NF data directly acquired on the classical spherical grid. For completeness, the reconstruction of the AUT FF pattern in the cut plane at $\Phi = 0°$ is also shown in Figure 6.52. It is noteworthy that the software package MI-3000 has been used to reconstruct the AUT FF pattern from the NF data of the classical spherical grid, either directly acquired, or recovered via the OSI expansion (6.37) from the samples collected along the spiral.

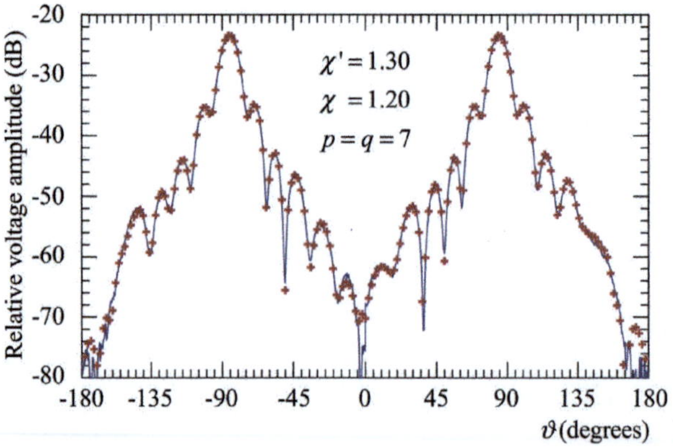

Figure 6.50 *Amplitude of the probe voltage V_p on the meridian at $\varphi = 90°$. Solid line: measured. Crosses: obtained by interpolating the NF samples acquired along the spiral.*

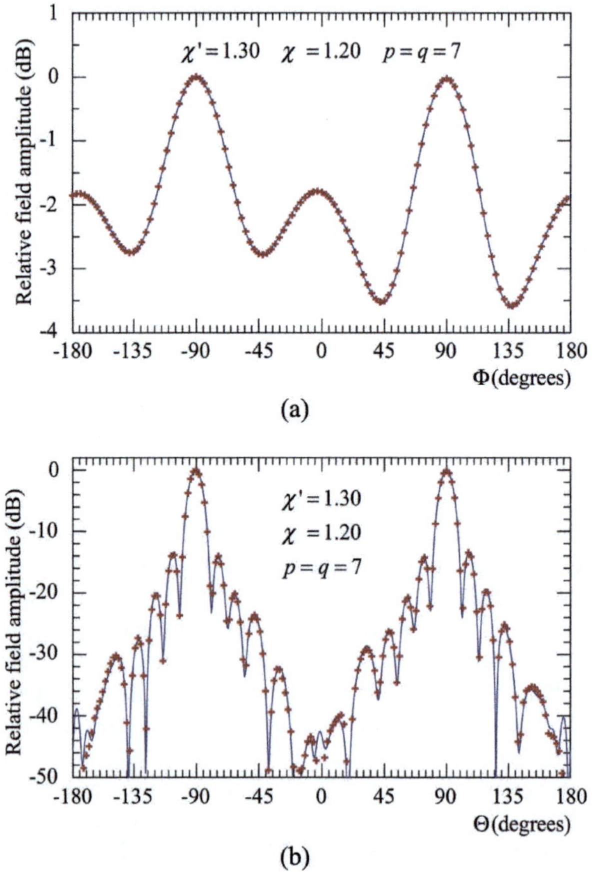

Figure 6.51 *FF patterns in the principal planes. Solid line: reference. Crosses: reconstructed from the NF samples acquired along the spiral. (a) E-plane ($\Theta = 90°$), (b) H-plane ($\Phi = 90°$).*

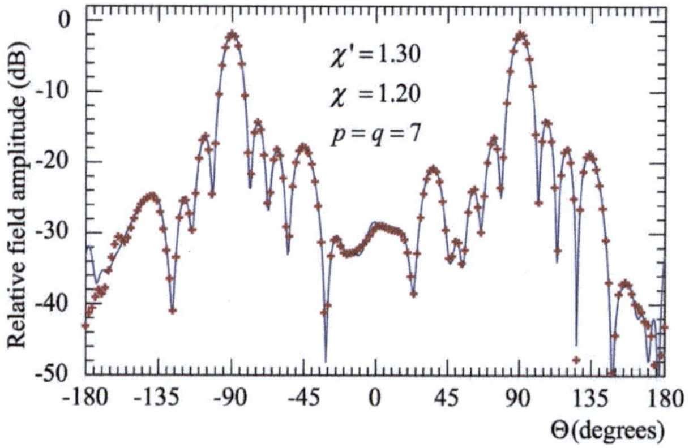

Figure 6.52 FF pattern in the cut plane Φ = 0°. Solid line: reference. Crosses: reconstructed from the NF samples acquired along the spiral.

It must be stressed that the number of used samples (comprehensive of the 160 extra samples at reduced spacing) is 983 and, hence, remarkably smaller than that (5100) required by the MI software package.

Appendix

6A.1 *Spherical wave expansion*

In this appendix, the SWE will be summarized. In such an expansion the EM field is expressed as superposition of transverse electric (TE) and transverse magnetic (TM) spherical modes, elementary solutions of the homogeneous wave equation in the spherical coordinates (r, ϑ, φ). In fact, in a homogeneous source-free region, an arbitrary EM field can be always represented as sum of a TE field and a TM field [308]. The expressions of these modes in a spherical coordinate system can be conveniently derived [17] by employing the vector potentials $\boldsymbol{A}$ and $\boldsymbol{F}$. In particular, the field expressions of the modes TE with respect to r, i.e., having the electric field $\boldsymbol{E}$ transverse to the radial direction $\hat{r}$, are determined [17] by putting

$$\boldsymbol{F} = \hat{r} F(r, \vartheta, \varphi) \tag{6A.1a}$$

$$\boldsymbol{A} = 0 \tag{6A.1b}$$

whereas those relevant to the TM to r modes are derived by assuming

$$\boldsymbol{A} = \hat{r} A(r, \vartheta, \varphi) \tag{6A.2a}$$

$$\boldsymbol{F} = 0 \tag{6A.2b}$$

However, unlike the cylindrical case, the vector potentials $\boldsymbol{A}$ and $\boldsymbol{F}$ can no longer be determined by means of the vector Helmholtz equations:

$$\nabla^2 \boldsymbol{A} + \beta^2 \boldsymbol{A} = 0 \tag{6A.3a}$$

$$\nabla^2 \boldsymbol{F} + \beta^2 \boldsymbol{F} = 0 \tag{6A.3b}$$

valid in a source-free region when assuming that the vector potentials $\boldsymbol{A}$, $\boldsymbol{F}$ and the scalar ones Φ, Ψ satisfy the Lorentz conditions:

$$\nabla \cdot \boldsymbol{A} + \mathrm{j}\omega\varepsilon\mu\Phi = 0 \tag{6A.4a}$$

$$\nabla \cdot \boldsymbol{F} + \mathrm{j}\omega\varepsilon\mu\Psi = 0 \tag{6A.4b}$$

As a matter of fact, since $\nabla^2(\hat{r}\ S) \neq \hat{r}\ \nabla^2 S$, the vector Helmholtz equations (6A.3) no longer reduce to the corresponding scalar ones. Obviously, in such a case, also the expressions (A.17) and (A.18), which relate the fields $\boldsymbol{E}$ and $\boldsymbol{H}$ to the potentials, are not valid. Different expressions for the fields and potentials, which are no longer based on the Lorentz conditions, must be derived.

To this end, let us consider the Maxwell's equations in a source-free region

$$\nabla \times \boldsymbol{E} = -\mathrm{j}\omega\mu\boldsymbol{H} \tag{6A.5a}$$

$$\nabla \times \boldsymbol{H} = \mathrm{j}\omega\varepsilon\boldsymbol{E} \tag{6A.5b}$$

$$\nabla \cdot \varepsilon\boldsymbol{E} = 0 \tag{6A.5c}$$

$$\nabla \cdot \mu\boldsymbol{H} = 0 \tag{6A.5d}$$

When $\boldsymbol{A} \neq 0$, $\boldsymbol{F} = 0$, it can be easily recognized that

$$\boldsymbol{H} = \frac{1}{\mu}\nabla \times \boldsymbol{A} \tag{6A.6}$$

whereas from the second of the Maxwell's equations (6A.5b), it follows that:

$$\boldsymbol{E} = \frac{1}{\mathrm{j}\omega\varepsilon}\nabla \times \boldsymbol{H} = \frac{1}{\mathrm{j}\omega\varepsilon\mu}\nabla \times \nabla \times \boldsymbol{A} \tag{6A.7}$$

By substituting (6A.6) in the first of the Maxwell's equations (6A.5a), it results:

$$\nabla \times (\boldsymbol{E} + \mathrm{j}\omega\boldsymbol{A}) = 0 \tag{6A.8}$$

and, accordingly,

$$\boldsymbol{E} = -\mathrm{j}\omega\boldsymbol{A} - \nabla\Phi \tag{6A.9}$$

By substituting this last into (6A.7), it follows that:

$$\nabla \times \nabla \times \boldsymbol{A} = \mathrm{j}\omega\varepsilon\mu\,(-\mathrm{j}\omega\boldsymbol{A} - \nabla\Phi) = \beta^2\boldsymbol{A} - \mathrm{j}\omega\varepsilon\mu\nabla\Phi \tag{6A.10}$$

A similar procedure can be applied when $\boldsymbol{A} = 0$, $\boldsymbol{F} \neq 0$, thus obtaining:

$$\boldsymbol{E} = -\frac{1}{\varepsilon}\nabla \times \boldsymbol{F} \tag{6A.11}$$

$$\boldsymbol{H} = \frac{1}{\mathrm{j}\omega\varepsilon\mu}\nabla \times \nabla \times \boldsymbol{F} \tag{6A.12}$$

$$\nabla \times \nabla \times \boldsymbol{F} = \beta^2\boldsymbol{F} - \mathrm{j}\omega\varepsilon\mu\nabla\Psi \tag{6A.13}$$

As already stated and clearly evident from (6A.11), TE fields can be constructed by assuming $\boldsymbol{A} = 0$, $\boldsymbol{F} = \hat{r}\ F(r, \vartheta, \varphi)$ and determining $\boldsymbol{F}$ by means of equations

(6A.13). Expanding equations (6A.13) in spherical coordinates and taking into account that

$$\nabla \times \boldsymbol{F} = \nabla \times \left(\hat{r}\ F\right) = \hat{\vartheta}\ \frac{1}{r\sin\vartheta}\frac{\partial F}{\partial \varphi} - \hat{\varphi}\ \frac{1}{r}\frac{\partial F}{\partial \vartheta} \tag{6A.14a}$$

$$\begin{aligned}\nabla \times \nabla \times \boldsymbol{F} &= \hat{r}\,\frac{1}{r\sin\vartheta}\left[\frac{\partial}{\partial \vartheta}\left(-\frac{\sin\vartheta}{r}\frac{\partial F}{\partial \vartheta}\right) - \frac{\partial}{\partial \varphi}\left(\frac{1}{r\sin\vartheta}\frac{\partial F}{\partial \varphi}\right)\right] \\ &\quad + \hat{\vartheta}\ \frac{1}{r}\frac{\partial^2 F}{\partial r\,\partial \vartheta} + \hat{\varphi}\ \frac{1}{r\sin\vartheta}\frac{\partial^2 F}{\partial r\,\partial \varphi}\end{aligned} \tag{6A.14b}$$

$$\nabla \Psi = \hat{r}\,\frac{\partial \Psi}{\partial r} + \hat{\vartheta}\ \frac{1}{r}\frac{\partial \Psi}{\partial \vartheta} + \hat{\varphi}\ \frac{1}{r\sin\vartheta}\frac{\partial \Psi}{\partial \varphi} \tag{6A.14c}$$

the following expressions for the r, ϑ, and φ components of the vector equations (6A.13) are obtained [17]:

$$\frac{1}{r\sin\vartheta}\left[\frac{\partial}{\partial \vartheta}\left(-\frac{\sin\vartheta}{r}\frac{\partial F}{\partial \vartheta}\right) - \frac{\partial}{\partial \varphi}\left(\frac{1}{r\sin\vartheta}\frac{\partial F}{\partial \varphi}\right)\right] - \beta^2 F = -\mathrm{j}\omega\varepsilon\mu\frac{\partial \Psi}{\partial r} \tag{6A.15a}$$

$$\frac{1}{r}\frac{\partial^2 F}{\partial r\,\partial \vartheta} = -\frac{\mathrm{j}\omega\varepsilon\mu}{r}\frac{\partial \Psi}{\partial \vartheta} \tag{6A.15b}$$

$$\frac{1}{r\sin\vartheta}\frac{\partial^2 F}{\partial r\,\partial \varphi} = -\frac{\mathrm{j}\omega\varepsilon\mu}{r\sin\vartheta}\frac{\partial \Psi}{\partial \varphi} \tag{6A.15c}$$

The last two relations can be rewritten as:

$$\frac{\partial^2 F}{\partial r\,\partial \vartheta} = \frac{\partial}{\partial \vartheta}\left(\frac{\partial F}{\partial r}\right) = \frac{\partial}{\partial \vartheta}\left(-\mathrm{j}\omega\varepsilon\mu\Psi\right) \tag{6A.16a}$$

$$\frac{\partial^2 F}{\partial r\,\partial \varphi} = \frac{\partial}{\partial \varphi}\left(\frac{\partial F}{\partial r}\right) = \frac{\partial}{\partial \varphi}\left(-\mathrm{j}\omega\varepsilon\mu\Psi\right) \tag{6A.16b}$$

and are simultaneously satisfied if

$$\Psi = -\frac{1}{\mathrm{j}\omega\varepsilon\mu}\frac{\partial F}{\partial r} \tag{6A.17}$$

By substituting (6A.17) into (6A.15a), the following differential equation, which allows one to determine F, is finally obtained:

$$\frac{\partial^2 F}{\partial r^2} + \frac{1}{r^2\sin\vartheta}\frac{\partial}{\partial \vartheta}\left(\sin\vartheta\,\frac{\partial F}{\partial \vartheta}\right) + \frac{1}{r^2\sin^2\vartheta}\frac{\partial^2 F}{\partial \varphi^2} + \beta^2 F = 0 \tag{6A.18}$$

It can be easily verified that this last can be rewritten in the compact form

$$\nabla^2\left(\frac{F}{r}\right) + \beta^2\left(\frac{F}{r}\right) = 0 \tag{6A.19}$$

Accordingly, F can be determined by solving the scalar Helmholtz equation

$$\nabla^2 \psi + \beta^2 \psi = 0 \tag{6A.20}$$

where

$$\psi\left(r, \vartheta, \varphi\right) = F\left(r, \vartheta, \varphi\right)/r \tag{6A.21}$$

The elementary solution to (6A.20) can be attained [17] by means of the separation of variables method and, in the considered case of an observation region external to the smallest sphere enclosing the antenna, is given by (see Appendix A.3):

$$\psi\left(r, \vartheta, \varphi\right) = h_n^{(2)}\left(\beta r\right) \bar{P}_n^{|m|}\left(\cos\vartheta\right) \mathrm{e}^{\mathrm{j}m\varphi} \tag{6A.22}$$

where n and m are integer, $h_n^{(2)}(\beta r)$ is the spherical Hankel function of second kind and order n, and $\bar{P}_n^m(\cos\vartheta)$ is the normalized associated Legendre function (see (A.56)) as defined by Belousov [310].

From (6A.11), by taking into account (6A.14a), it results:

$$E_r = 0 \tag{6A.23a}$$

$$E_\vartheta = -\frac{1}{\varepsilon r \sin\vartheta}\frac{\partial F}{\partial \varphi} \tag{6A.23b}$$

$$E_\varphi = \frac{1}{\varepsilon r}\frac{\partial F}{\partial \vartheta} \tag{6A.23c}$$

By taking into account (6A.21), (6A.22), and (6A.23), it can be verified that the electric field related to an elementary spherical TE wave is given by:

$$\boldsymbol{E} = -\frac{1}{\varepsilon}\left(\frac{\mathrm{j}m}{\sin\vartheta}\bar{P}_n^{|m|}\left(\cos\vartheta\right)\hat{\vartheta} - \frac{\mathrm{d}}{\mathrm{d}\vartheta}\bar{P}_n^{|m|}\left(\cos\vartheta\right)\hat{\varphi}\right) h_n^{(2)}\left(\beta r\right) \mathrm{e}^{\mathrm{j}m\varphi} \tag{6A.24}$$

As seen, TM fields can be constructed by assuming $\boldsymbol{A} = \hat{r}\, A\left(r,\ \vartheta,\ \varphi\right)$, $\boldsymbol{F} = 0$ and determining $\boldsymbol{A}$ by means of equation (6A.10). Expanding equation (6A.10) in spherical coordinates and imposing that

$$\Phi = -\frac{1}{\mathrm{j}\omega\varepsilon\mu}\frac{\partial A}{\partial r} \tag{6A.25}$$

it can be shown, by following the same procedure adopted for F, that A can still be determined by solving the scalar Helmholtz equation (6A.20), where now

$$\psi\left(r, \vartheta, \varphi\right) = A\left(r, \vartheta, \varphi\right)/r \tag{6A.26}$$

The elementary solution $\psi\,(r, \vartheta, \varphi)$ to the scalar Helmholtz equation is, obviously, again given by (6A.22).

From (6A.7), by taking into account (6A.14b), it results:

$$E_r = \frac{-1}{\mathrm{j}\omega\varepsilon\mu r^2 \sin\vartheta}\left[\frac{\partial}{\partial\vartheta}\left(\sin\vartheta\frac{\partial A}{\partial\vartheta}\right) + \frac{1}{\sin\vartheta}\frac{\partial^2 A}{\partial\varphi^2}\right] \tag{6A.27a}$$

$$E_\vartheta = \frac{1}{\mathrm{j}\omega\varepsilon\mu r}\frac{\partial^2 A}{\partial r\,\partial\vartheta} \tag{6A.27b}$$

$$E_\varphi = \frac{1}{\mathrm{j}\omega\varepsilon\mu r\sin\vartheta}\frac{\partial^2 A}{\partial r\,\partial\varphi} \tag{6A.27c}$$

Since a relation quite analogous to (6A.18) holds also for A, it is possible to rewrite (6A.27a) in the form:

$$E_r = \frac{1}{\mathrm{j}\omega\varepsilon\mu}\left[\frac{\partial^2 A}{\partial r^2} + \beta^2 A\right] \tag{6A.27d}$$

By taking into account (6A.26), (6A.22), and (6A.27), it can be verified that the electric field related to an elementary spherical TM wave is given by:

$$\begin{aligned}\boldsymbol{E} = \frac{\beta}{\mathrm{j}\omega\varepsilon\mu}&\left[\left((\beta r)\,h_n^{(2)}(\beta r) + \frac{\mathrm{d}^2}{\mathrm{d}(\beta r)^2}\left(\beta r\,h_n^{(2)}(\beta r)\right)\right)\bar{P}_n^{|m|}(\cos\vartheta)\,\hat{r}\,+\right.\\ &\left.+\left(\frac{\mathrm{d}}{\mathrm{d}\vartheta}\bar{P}_n^{|m|}(\cos\vartheta)\,\hat{\vartheta} + \frac{\mathrm{j}m}{\sin\vartheta}\bar{P}_n^{|m|}(\cos\vartheta)\,\hat{\varphi}\right)\frac{1}{\beta r}\frac{\mathrm{d}}{\mathrm{d}(\beta r)}\left(\beta r\,h_n^{(2)}(\beta r)\right)\right]\mathrm{e}^{\mathrm{j}m\varphi}\end{aligned} \tag{6A.28}$$

By using the following recurrence relations for the spherical Bessel functions [313]

$$\frac{\mathrm{d}}{\mathrm{d}x}h_n^{(2)}(x) = h_{n-1}^{(2)}(x) - \frac{n+1}{x}h_n^{(2)}(x) \tag{6A.29a}$$

$$\frac{\mathrm{d}}{\mathrm{d}x}h_{n-1}^{(2)}(x) = \frac{n-1}{x}h_{n-1}^{(2)}(x) - h_n^{(2)}(x) \tag{6A.29b}$$

it can be shown that

$$x\,h_n^{(2)}(x) + \frac{\mathrm{d}}{\mathrm{d}x}\left[\frac{\mathrm{d}}{\mathrm{d}x}x\,h_n^{(2)}(x)\right] = n\,(n+1)\,\frac{h_n^{(2)}(x)}{x} \tag{6A.30}$$

and, accordingly, (6A.28) can be rewritten as:

$$\begin{aligned}\boldsymbol{E} = \frac{\beta}{\mathrm{j}\omega\varepsilon\mu}&\left[n(n+1)\frac{h_n^{(2)}(\beta r)}{\beta r}\bar{P}_n^{|m|}(\cos\vartheta)\,\hat{r}\right.\\ &\left.+\left(\frac{\mathrm{d}}{\mathrm{d}\vartheta}\bar{P}_n^{|m|}(\cos\vartheta)\,\hat{\vartheta} + \frac{\mathrm{j}m}{\sin\vartheta}\bar{P}_n^{|m|}(\cos\vartheta)\,\hat{\varphi}\right)\frac{1}{\beta r}\frac{\mathrm{d}}{\mathrm{d}(\beta r)}\left(\beta r h_n^{(2)}(\beta r)\right)\right]\mathrm{e}^{\mathrm{j}m\varphi}\end{aligned} \tag{6A.31}$$

According to the previous relations, the transverse electric field $\boldsymbol{E}_t$ related to an elementary spherical TE wave is given by:

$$\boldsymbol{E}_t = -\frac{1}{\varepsilon}\,g_{1n}(\beta r)\,\tilde{\boldsymbol{F}}_{1nm}(\vartheta,\varphi) = -\frac{1}{\varepsilon}\,g_{1n}(\beta r)\,\tilde{\boldsymbol{f}}_{1nm}(\vartheta)\,\mathrm{e}^{\mathrm{j}m\varphi} \tag{6A.32}$$

where

$$g_{1n}(\beta r) = h_n^{(2)}(\beta r) \tag{6A.33a}$$

$$\tilde{\boldsymbol{f}}_{1nm}(\vartheta) = \frac{\mathrm{j}m}{\sin\vartheta}\bar{P}_n^{|m|}(\cos\vartheta)\hat{\vartheta} - \frac{\mathrm{d}}{\mathrm{d}\vartheta}\bar{P}_n^{|m|}(\cos\vartheta)\hat{\varphi} \tag{6A.33b}$$

whereas that relevant to an elementary spherical TM wave is:

$$\boldsymbol{E}_t = \frac{\beta}{\mathrm{j}\omega\varepsilon\mu} g_{2n}(\beta r)\tilde{\boldsymbol{F}}_{2nm}(\vartheta,\varphi) = \frac{\beta}{\mathrm{j}\omega\varepsilon\mu} g_{2n}(\beta r)\tilde{\boldsymbol{f}}_{2nm}(\vartheta)\,\mathrm{e}^{\mathrm{j}m\varphi} \tag{6A.34}$$

where

$$g_{2n}(\beta r) = \frac{1}{\beta r}\frac{\mathrm{d}}{\mathrm{d}(\beta r)}\left(\beta r h_n^{(2)}(\beta r)\right) \tag{6A.35a}$$

$$\tilde{\boldsymbol{f}}_{2nm}(\vartheta) = \frac{\mathrm{d}}{\mathrm{d}\vartheta}\bar{P}_n^{|m|}(\cos\vartheta)\,\hat{\vartheta} + \frac{\mathrm{j}m}{\sin\vartheta}\bar{P}_n^{|m|}(\cos\vartheta)\,\hat{\varphi} \tag{6A.35b}$$

According to [308], the EM field in a homogeneous source-free region can be expressed as superposition of TE and TM elementary spherical waves (6A.32) and (6A.34). Accordingly, the transverse electric field $\boldsymbol{E}_t$ on a sphere of radius r enclosing the AUT can be expressed as:

$$\boldsymbol{E}_t = \sum_{n=1}^{N_{\max}}\sum_{m=-n}^{n}\left[\tilde{a}_{1nm}\,g_{1n}(\beta r)\,\tilde{\boldsymbol{F}}_{1nm}(\vartheta,\varphi) + \tilde{a}_{2nm}\,g_{2n}(\beta r)\,\tilde{\boldsymbol{F}}_{2nm}(\vartheta,\varphi)\right] \tag{6A.36}$$

wherein N_{max} is the index of the highest spherical wave to be considered and the constants $-1/\varepsilon$ and $\beta/\mathrm{j}\omega\varepsilon\mu$ have been incorporated in the coefficients $\tilde{a}_{1nm}$ and $\tilde{a}_{2nm}$.

It is useful to note that the summation on m extends from $-n$ to n, since $\bar{P}_n^{|m|}(\cos\vartheta)$ is zero for $|m| > n$ (see Appendix A.3). Moreover, the summation on n starts from 1, since $\tilde{\boldsymbol{F}}_{100}$ and $\tilde{\boldsymbol{F}}_{200}$ are both null.

It can be easily verified that the spherical wave functions $\tilde{\boldsymbol{F}}_{1nm}$ and $\tilde{\boldsymbol{F}}_{2\bar{n}\bar{m}}$ are orthogonal on the sphere. As a matter of fact,

$$\begin{aligned}\left\langle\tilde{\boldsymbol{F}}_{1nm},\tilde{\boldsymbol{F}}_{2\bar{n}\bar{m}}\right\rangle &= \int_0^{\pi}\int_{-\pi}^{\pi}\tilde{\boldsymbol{F}}_{1nm}(\vartheta,\varphi)\cdot\tilde{\boldsymbol{F}}^*_{2\bar{n}\bar{m}}(\vartheta,\varphi)\sin\vartheta\,\mathrm{d}\varphi\,\mathrm{d}\vartheta \\ &= \int_{-\pi}^{\pi}\mathrm{e}^{\mathrm{j}(m-\bar{m})\varphi}\mathrm{d}\varphi\int_0^{\pi}\tilde{\boldsymbol{f}}_{1nm}(\vartheta)\cdot\tilde{\boldsymbol{f}}^*_{2\bar{n}\bar{m}}(\vartheta)\sin\vartheta\,\mathrm{d}\vartheta = 2\pi\,\delta_{m\bar{m}}\,\mathrm{j}m \\ &\times\int_{-1}^{1}\left[\frac{\bar{P}_n^{|m|}(\cos\vartheta)}{\sin\vartheta}\frac{\mathrm{d}}{\mathrm{d}\vartheta}\bar{P}_{\bar{n}}^{|\bar{m}|}(\cos\vartheta)\right. \\ &\left.+\frac{\bar{P}_{\bar{n}}^{|\bar{m}|}(\cos\vartheta)}{\sin\vartheta}\frac{\mathrm{d}}{\mathrm{d}\vartheta}\bar{P}_n^{|m|}(\cos\vartheta)\right]\mathrm{d}(\cos\vartheta)\end{aligned} \tag{6A.37}$$

wherein

$$\delta_{m\bar{m}} = \begin{cases} 1 & \text{for } \bar{m} = m \\ 0 & \text{for } \bar{m} \neq m \end{cases} \tag{6A.38}$$

is the Kronecker delta, the asterisk (*) indicates the complex conjugation, and the symbol (·) denotes the scalar product. Since [76]

$$\int_{-1}^{1}\left[\frac{\bar{P}_n^{|m|}(\cos\vartheta)}{\sin\vartheta}\frac{\mathrm{d}}{\mathrm{d}\vartheta}\bar{P}_{\bar{n}}^{|m|}(\cos\vartheta)+\frac{\bar{P}_{\bar{n}}^{|m|}(\cos\vartheta)}{\sin\vartheta}\frac{\mathrm{d}}{\mathrm{d}\vartheta}\bar{P}_n^{|m|}(\cos\vartheta)\right]\mathrm{d}(\cos\vartheta)=0 \qquad (6A.39)$$

it results:

$$\langle\tilde{\boldsymbol{F}}_{1nm},\tilde{\boldsymbol{F}}_{2\bar{n}\bar{m}}\rangle=0 \qquad (6A.40)$$

that is to say, $\tilde{\boldsymbol{F}}_{1nm}$ and $\tilde{\boldsymbol{F}}_{2\bar{n}\bar{m}}$ are always orthogonal even when $\bar{n}=n$ and $\bar{m}=m$.

Also the spherical wave functions $\tilde{\boldsymbol{F}}_{1nm}$ and $\tilde{\boldsymbol{F}}_{1\bar{n}\bar{m}}$ are orthogonal. In fact,

$$\begin{aligned}\langle\tilde{\boldsymbol{F}}_{1nm},\tilde{\boldsymbol{F}}_{1\bar{n}\bar{m}}\rangle &= \int_0^{\pi}\int_{-\pi}^{\pi}\tilde{\boldsymbol{F}}_{1nm}(\vartheta,\varphi)\cdot\tilde{\boldsymbol{F}}^*_{1\bar{n}\bar{m}}(\vartheta,\varphi)\sin\vartheta\,\mathrm{d}\varphi\,\mathrm{d}\vartheta\\
&=\int_{-\pi}^{\pi}\mathrm{e}^{\mathrm{j}(m-\bar{m})\varphi}\mathrm{d}\varphi\int_0^{\pi}\tilde{\boldsymbol{f}}_{1nm}(\vartheta)\cdot\tilde{\boldsymbol{f}}^*_{1\bar{n}\bar{m}}(\vartheta)\sin\vartheta\,\mathrm{d}\vartheta=2\pi\,\delta_{m\bar{m}}\\
&\times\int_{-1}^{1}\left[\frac{m^2\bar{P}_n^{|m|}(\cos\vartheta)\,\bar{P}_{\bar{n}}^{|\bar{m}|}(\cos\vartheta)}{\sin^2\vartheta}\right.\\
&\left.+\frac{\mathrm{d}}{\mathrm{d}\vartheta}\bar{P}_n^{|m|}(\cos\vartheta)\frac{\mathrm{d}}{\mathrm{d}\vartheta}\bar{P}_{\bar{n}}^{|\bar{m}|}(\cos\vartheta)\right]\mathrm{d}(\cos\vartheta)\\
&=2\pi\,n(n+1)\delta_{m\bar{m}}\,\delta_{n\bar{n}}\end{aligned} \qquad (6A.41)$$

since [76]

$$\int_{-1}^{1}\left[\frac{m^2\bar{P}_n^{|m|}(\cos\vartheta)\,\bar{P}_{\bar{n}}^{|m|}(\cos\vartheta)}{\sin^2\vartheta}+\frac{\mathrm{d}}{\mathrm{d}\vartheta}\bar{P}_n^{|m|}(\cos\vartheta)\,\frac{\mathrm{d}}{\mathrm{d}\vartheta}\bar{P}_{\bar{n}}^{|m|}(\cos\vartheta)\right]\mathrm{d}(\cos\vartheta)=n(n+1)\delta_{n\bar{n}} \qquad (6A.42)$$

In a similar way, it can be shown that

$$\langle\tilde{\boldsymbol{F}}_{2nm},\tilde{\boldsymbol{F}}_{2\bar{n}\bar{m}}\rangle=2\pi n\,(n+1)\,\delta_{m\bar{m}}\,\delta_{n\bar{n}} \qquad (6A.43)$$

It is convenient to introduce the new functions

$$\boldsymbol{F}_{1nm}(\vartheta,\varphi)=\left(\frac{-m}{|m|}\right)^m\frac{1}{\sqrt{2\pi n(n+1)}}\tilde{\boldsymbol{F}}_{1nm}(\vartheta,\varphi) \qquad (6A.44a)$$

$$\boldsymbol{F}_{2nm}(\vartheta,\varphi)=\left(\frac{-m}{|m|}\right)^m\frac{1}{\sqrt{2\pi n(n+1)}}\tilde{\boldsymbol{F}}_{2nm}(\vartheta,\varphi) \qquad (6A.44b)$$

in order to have orthonormal spherical wave functions and power-normalized spherical waves (see Sect. 6.2). It is noteworthy that the factor $(-m/|m|)^m$ has been introduced to ensure that the phase of the modes follows the phase of the spherical harmonics as defined by Edmonds [311].

Accordingly, the expression of the transverse electric field $\boldsymbol{E}_t$ radiated by the AUT on a sphere of radius r enclosing it can be rewritten in the form:

$$\boldsymbol{E}_t = \beta \sum_{n=1}^{N_{\max}} \sum_{m=-n}^{n} \left[a_{1nm}\, g_{1n}(\beta r)\, \boldsymbol{F}_{1nm}(\vartheta, \varphi) + a_{2nm}\, g_{2n}(\beta r)\, \boldsymbol{F}_{2nm}(\vartheta, \varphi) \right] \tag{6A.45}$$

Since, according to the Sommerfeld radiation conditions, in the FF region the longitudinal component E_r of the electric field must vanish, the expression of the electric field in the antenna FF region can be easily derived by replacing in the above relation the spherical Hankel function and its first derivative by their asymptotic expansions

$$h_n^{(2)}(\beta r) \underset{\beta r \to \infty}{\approx} \mathrm{j}^{n+1} \frac{\mathrm{e}^{-\mathrm{j}\beta r}}{\beta r} \tag{6A.46a}$$

$$\frac{1}{\beta r} \frac{\mathrm{d}}{\mathrm{d}(\beta r)} \left(\beta r h_n^{(2)}(\beta r) \right) \underset{\beta r \to \infty}{\approx} \mathrm{j}^{n} \frac{\mathrm{e}^{-\mathrm{j}\beta r}}{\beta r} \tag{6A.46b}$$

thus getting:

$$\boldsymbol{E}_t(r \to \infty, \vartheta, \varphi) = \frac{\mathrm{e}^{-\mathrm{j}\beta r}}{r} \sum_{n=1}^{N_{\max}} \sum_{m=-n}^{n} \left[\mathrm{j}^{n+1} a_{1nm}\, \boldsymbol{f}_{1nm}(\vartheta) + \mathrm{j}^{n} a_{2nm}\, \boldsymbol{f}_{2nm}(\vartheta) \right] \mathrm{e}^{\mathrm{j}m\varphi} \tag{6A.47}$$

wherein

$$\boldsymbol{f}_{1,2nm}(\vartheta) = \left(\frac{-m}{|m|} \right)^m \frac{1}{\sqrt{2\pi\, n(n+1)}} \tilde{\boldsymbol{f}}_{1,2nm}(\vartheta) \tag{6A.48}$$

On the other hand, the longitudinal component of the electric field of each elementary TE spherical wave is null (see (6A.24)), whereas that relevant to a TM wave decreases asymptotically as $1/r^2$, as can be easily verified from (6A.31) and (6A.46a).

Appendices

A.1 Radiation and auxiliary vector potentials

In the study of the radiation from given sources, it is convenient to introduce auxiliary vector potentials as aids to achieve the solution. The most commonly used ones are the magnetic vector potential $\boldsymbol{A}$ and the electric vector potential $\boldsymbol{F}$.

The case of the radiation from electric sources characterized by a charge density ρ and a current density $\boldsymbol{J}$ is first considered. In such a case, by assuming and suppressing a time dependence $\mathrm{e}^{\mathrm{j}\omega t}$, the electromagnetic (EM) field radiated by finite size sources in an unbounded homogeneous medium (having permittivity ε and permeability μ) must be solution of the Maxwell's equations

$$\nabla \times \boldsymbol{E} = -\mathrm{j}\,\omega\,\mu \boldsymbol{H} \tag{A.1a}$$

$$\nabla \times \boldsymbol{H} = \mathrm{j}\,\omega\,\varepsilon\,\boldsymbol{E} + \boldsymbol{J} \tag{A.1b}$$

$$\nabla \cdot \varepsilon \boldsymbol{E} = \rho \tag{A.1c}$$

$$\nabla \cdot \mu \boldsymbol{H} = 0 \tag{A.1d}$$

and must satisfy the Sommerfeld's radiation conditions:

$$\lim_{r\to\infty} r\left|\boldsymbol{E}\right| < q; \qquad \lim_{r\to\infty} r\left|\boldsymbol{H}\right| < q \tag{A.2a}$$

$$\lim_{r\to\infty} r\left[\boldsymbol{E} - \zeta \boldsymbol{H} \times \hat{r}\right] = 0; \qquad \lim_{r\to\infty} r\left[\boldsymbol{H} - \left(\hat{r} \times \boldsymbol{E}\right)/\zeta\right] = 0 \tag{A.2b}$$

where q is a positive real number and ζ is the intrinsic impedance of the medium.

Since $\nabla \cdot \mu\boldsymbol{H} = 0$, the vector $\mu\boldsymbol{H}$ can be represented as the curl of another vector $\boldsymbol{A}$. In fact, for any arbitrary vector $\boldsymbol{A}$, it results $\nabla \cdot \nabla \times \boldsymbol{A} = 0$. Accordingly,

$$\boldsymbol{H} = \frac{1}{\mu}\nabla \times \boldsymbol{A} \tag{A.3}$$

where $\boldsymbol{A}$ is called magnetic vector potential. By substituting (A.3) into the Maxwell's equation (A.1a), it results

$$\nabla \times \left(\boldsymbol{E} + \mathrm{j}\omega \boldsymbol{A}\right) = 0 \tag{A.4}$$

Since any vector function with zero curl can be expressed as the gradient of a scalar function, from (A.4), it follows that,

$$\boldsymbol{E} = -\mathrm{j}\omega \boldsymbol{A} - \nabla\Phi \tag{A.5}$$

where Φ is the electric scalar potential. By substituting (A.3) and (A.5) into the Maxwell's equation (A.1b), it results

$$\nabla \times \nabla \times \boldsymbol{A} = \mathrm{j}\omega\varepsilon\mu\left(-\mathrm{j}\omega\boldsymbol{A} - \nabla\Phi\right) + \mu\boldsymbol{J} = \beta^2\boldsymbol{A} - \mathrm{j}\omega\varepsilon\mu\nabla\Phi + \mu\boldsymbol{J} \tag{A.6}$$

which, by using the vector identity $\nabla^2 \boldsymbol{A} = \nabla\nabla\cdot\boldsymbol{A} - \nabla\times\nabla\times\boldsymbol{A}$, leads to

$$\nabla^2 \boldsymbol{A} + \beta^2 \boldsymbol{A} = -\mu \boldsymbol{J} + \nabla\left(\nabla\cdot\boldsymbol{A} + \mathrm{j}\omega\varepsilon\mu\Phi\right) \tag{A.7}$$

On the other hand, by substituting (A.5) into the Maxwell's equation (A.1c), it results

$$\mathrm{j}\omega\nabla\cdot\boldsymbol{A} + \nabla^2\Phi = -\frac{\rho}{\varepsilon} \tag{A.8}$$

which can be rewritten in the form

$$\nabla^2\Phi + \beta^2\Phi + \mathrm{j}\omega\left(\nabla\cdot\boldsymbol{A} + \mathrm{j}\omega\varepsilon\mu\Phi\right) = -\frac{\rho}{\varepsilon} \tag{A.9}$$

Equations (A.7) and (A.9) relate the potentials $\boldsymbol{A}$ and Φ to the sources. Once these potentials have been determined, the fields $\boldsymbol{E}$ and $\boldsymbol{H}$ can be found from (A.5) and (A.3), respectively.

In any case, the choice of the potentials is largely arbitrary. In fact, as can be easily recognized, the introduction of two new potentials $\boldsymbol{A}' = \boldsymbol{A} - \nabla\psi$ and $\Phi' = \Phi + \mathrm{j}\omega\psi$ leaves the fields unchanged. This suggests to choose the potentials so that they satisfy proper conditions. A convenient one is the Lorentz condition.

$$\nabla\cdot\boldsymbol{A} + \mathrm{j}\omega\varepsilon\mu\Phi = 0 \tag{A.10}$$

By imposing such a condition, equations (A.7) and (A.9) reduce to:

$$\nabla^2\boldsymbol{A} + \beta^2\boldsymbol{A} = -\mu\boldsymbol{J} \tag{A.11}$$

$$\nabla^2\Phi + \beta^2\Phi = -\frac{\rho}{\varepsilon} \tag{A.12}$$

which are no more coupled. It is worthy to note that only the Helmholtz equation (A.11) must be solved in such a case, since the scalar potential Φ is related to vector potential $\boldsymbol{A}$ through the Lorentz condition. As a consequence, once equation (A.11) has been solved, the magnetic field $\boldsymbol{H}$ can be found from (A.3), whereas the electric field $\boldsymbol{E}$ can be determined by means of the relation

$$\boldsymbol{E} = -\mathrm{j}\omega\boldsymbol{A} + \frac{\nabla\nabla\cdot\boldsymbol{A}}{\mathrm{j}\omega\varepsilon\mu} \tag{A.13}$$

According to the duality theorem [17], the solution for the EM field radiated by magnetic sources ρ_m and $\boldsymbol{J}_m$ can be easily derived from the previous one by applying the following transformations: $\boldsymbol{E}\rightarrow\boldsymbol{H}$, $\boldsymbol{H}\rightarrow-\boldsymbol{E}$, $\varepsilon\rightleftarrows\mu$, $\rho\rightarrow\rho_m$, $\boldsymbol{J}\rightarrow\boldsymbol{J}_m$, $\boldsymbol{A}\rightarrow\boldsymbol{F}$. Accordingly from (A.3) and (A.13), it results

$$\boldsymbol{E} = -\frac{1}{\varepsilon}\nabla\times\boldsymbol{F} \tag{A.14}$$

$$\boldsymbol{H} = -\mathrm{j}\omega\boldsymbol{F} + \frac{\nabla\nabla\cdot\boldsymbol{F}}{\mathrm{j}\omega\varepsilon\mu} \tag{A.15}$$

wherein the electric vector potential $\boldsymbol{F}$ can be determined by solving the equation:

$$\nabla^2\boldsymbol{F} + \beta^2\boldsymbol{F} = -\varepsilon\boldsymbol{J}_m \tag{A.16}$$

When both electric and magnetic sources are simultaneously present, the fields $\boldsymbol{E}$ and $\boldsymbol{H}$ can be obtained by the superposition of the individual fields due to the electric and magnetic sources. Accordingly, it results:

$$\boldsymbol{E} = -\mathrm{j}\omega \boldsymbol{A} + \frac{\nabla\nabla \cdot \boldsymbol{A}}{\mathrm{j}\omega\varepsilon\mu} - \frac{1}{\varepsilon}\nabla \times \boldsymbol{F} \tag{A.17}$$

$$\boldsymbol{H} = \frac{1}{\mu}\nabla \times \boldsymbol{A} - \mathrm{j}\omega \boldsymbol{F} + \frac{\nabla\nabla \cdot \boldsymbol{F}}{\mathrm{j}\omega\varepsilon\mu} \tag{A.18}$$

wherein the magnetic and electric vector potentials $\boldsymbol{A}$ and $\boldsymbol{F}$ can be determined by solving the equations (A.11) and (A.16), respectively.

A.2 Solution of the scalar Helmholtz equation in cylindrical coordinates

The scalar Helmholtz equation

$$\nabla^2 A + \beta^2 A = 0 \tag{A.19}$$

in the cylindrical coordinates (ρ, φ, z) becomes:

$$\frac{\partial^2 A}{\partial \rho^2} + \frac{1}{\rho}\frac{\partial A}{\partial \rho} + \frac{1}{\rho^2}\frac{\partial^2 A}{\partial \varphi^2} + \frac{\partial^2 A}{\partial z^2} + \beta^2 A = 0 \tag{A.20}$$

According to the separation of variables method, we seek to find solutions of the form:

$$A(\rho, \varphi, z) = R(\rho)\,\Phi(\varphi)\,Z(z) \tag{A.21}$$

By substituting (A.21) into (A.20) and dividing by A, it results:

$$\frac{1}{R}\frac{\mathrm{d}^2 R}{\mathrm{d}\rho^2} + \frac{1}{R}\frac{1}{\rho}\frac{\mathrm{d}R}{\mathrm{d}\rho} + \frac{1}{\Phi}\frac{1}{\rho^2}\frac{\mathrm{d}^2\Phi}{\mathrm{d}\varphi^2} + \frac{1}{Z}\frac{\mathrm{d}^2 Z}{\mathrm{d}z^2} + \beta^2 = 0 \tag{A.22}$$

The fourth term is explicitly independent of ρ and φ and it must also be independent of z since the other terms do not depend on z. Accordingly:

$$\frac{1}{Z}\frac{\mathrm{d}^2 Z}{\mathrm{d}z^2} = -\eta^2 \tag{A.23}$$

where η is a constant.

By substituting this last relation into (A.22) and multiplying by ρ^2, it results:

$$\frac{\rho^2}{R}\frac{\mathrm{d}^2 R}{\mathrm{d}\rho^2} + \frac{\rho}{R}\frac{\mathrm{d}R}{\mathrm{d}\rho} + \frac{1}{\Phi}\frac{\mathrm{d}^2\Phi}{\mathrm{d}\varphi^2} + \left(\beta^2 - \eta^2\right)\rho^2 = 0 \tag{A.24}$$

Now, the third term of (A.24) is independent of ρ and z, and the other terms are independent of φ. Hence:

$$\frac{1}{\Phi}\frac{\mathrm{d}^2\Phi}{\mathrm{d}\varphi^2} = -\nu^2 \tag{A.25}$$

where ν is a constant. Then, equation (A.24) becomes

$$\frac{\rho^2}{R}\frac{\mathrm{d}^2R}{\mathrm{d}\rho^2} + \frac{\rho}{R}\frac{\mathrm{d}R}{\mathrm{d}\rho} - \nu^2 + \left(\beta^2 - \eta^2\right)\rho^2 = 0 \tag{A.26}$$

which is a differential equation in ρ only.

The wave equation is now separated. By defining

$$\Lambda^2 = \beta^2 - \eta^2 \tag{A.27}$$

and multiplying equation (A.26) by R, it results

$$\rho^2\frac{\mathrm{d}^2R}{\mathrm{d}\rho^2} + \rho\frac{\mathrm{d}R}{\mathrm{d}\rho} + \left[(\Lambda\rho)^2 - \nu^2\right]R = 0 \tag{A.28}$$

As well-known, the solutions to the harmonic equations (A.23) and (A.25) take the form, respectively, of

$$Z(z) = B_1 \mathrm{e}^{-\mathrm{j}\eta z} + B_2 \mathrm{e}^{\,\mathrm{j}\eta z} \tag{A.29a}$$

or

$$Z(z) = B_3 \cos(\eta z) + B_4 \sin(\eta z) \tag{A.29b}$$

and

$$\Phi(\varphi) = C_1 \mathrm{e}^{-\mathrm{j}\nu\varphi} + C_2 \mathrm{e}^{\,\mathrm{j}\nu\varphi} \tag{A.30a}$$

or

$$\Phi(\varphi) = C_3 \cos(\nu\varphi) + C_4 \sin(\nu\varphi) \tag{A.30b}$$

where B_1, B_2, B_3, B_4, C_1, C_2, C_3, and C_4 are integration constants.

Equation (A.28) is the classic Bessel's differential equation of order ν, whose solution can be written in the form

$$R(\rho) = D_1 H_\nu^{(2)}(\Lambda\rho) + D_2 H_\nu^{(1)}(\Lambda\rho) \tag{A.31a}$$

or

$$R(\rho) = D_3 J_\nu(\Lambda\rho) + D_4 Y_\nu(\Lambda\rho) \tag{A.31b}$$

where D_1, D_2, D_3, and D_4 are integration constants, $J_\nu(\Lambda\rho)$ is the Bessel function of first kind and order ν, $Y_\nu(\Lambda\rho)$ is the Bessel function of second kind and order ν, $H_\nu^{(1)}(\Lambda\rho)$ is the Hankel function of first kind and order ν, and $H_\nu^{(2)}(\Lambda\rho)$ is the Hankel function of second kind and order ν.

It is worthy to note that the φ coordinate represents an angle and, as such, it implies some restrictions on the values of ν. For instance, if the desired solution has to be valid in a whole 2π-range, then it must be periodic and, accordingly, ν must be an integer.

The forms (A.29a), (A.30a), and (A.31a) are more suitable to represent cylindrical travelling waves, whereas (A.29b), (A.30b), and (A.31b) are tailored for characterizing a standing wave behaviour.

In the considered case, the desired solution must be valid in the region external to the smallest cylinder enclosing the AUT, accordingly, ν must be an integer and the forms (A.29a), (A.30a), and (A.31a) must be adopted. Moreover, since there are no sources at infinity, only outgoing cylindrical waves have to be contemplated. Accordingly, the most general solution of (A.19) is

$$A = \sum_{\nu=-\infty}^{\infty} \int_{-\infty}^{\infty} D_1 H_\nu^{(2)}(\Lambda\rho) \left(B_1 \mathrm{e}^{-\mathrm{j}\eta z} + B_2 \mathrm{e}^{\mathrm{j}\eta z}\right)\left(C_1 \mathrm{e}^{-\mathrm{j}\nu\varphi} + C_2 \mathrm{e}^{\mathrm{j}\nu\varphi}\right) \mathrm{d}\eta \quad \text{(A.32)}$$

As a matter of fact, only the Hankel function of second kind has the correct behaviour to represent an outgoing cylindrical wave.

By properly taking into account (A.27) and the following relation

$$H_{-\nu}^{(2)}(\Lambda\rho) = (-1)^\nu H_\nu^{(2)}(\Lambda\rho) \quad \text{(A.33)}$$

it can be easily recognized that (A.32) reduces to

$$A(\rho,\varphi,z) = \sum_{\nu=-\infty}^{\infty} \int_{-\infty}^{\infty} C_\nu(\eta) H_\nu^{(2)}(\Lambda\rho) \mathrm{e}^{-\mathrm{j}\eta z} \mathrm{e}^{\mathrm{j}\nu\varphi} \mathrm{d}\eta \quad \text{(A.34)}$$

A.3 Solution of the scalar Helmholtz equation in spherical coordinates

The scalar Helmholtz equation

$$\nabla^2 A + \beta^2 A = 0 \quad \text{(A.35)}$$

in the spherical coordinates (r, ϑ, φ) becomes:

$$\frac{1}{r^2}\frac{\partial}{\partial r}\left(r^2 \frac{\partial A}{\partial r}\right) + \frac{1}{r^2 \sin\vartheta}\frac{\partial}{\partial \vartheta}\left(\sin\vartheta \frac{\partial A}{\partial \vartheta}\right) + \frac{1}{r^2 \sin^2\vartheta}\frac{\partial^2 A}{\partial \varphi^2} + \beta^2 A = 0 \quad \text{(A.36)}$$

According to the separation of variables method, we seek to find solutions of the form:

$$A(r,\vartheta,\varphi) = R(r)\,\Theta(\vartheta)\,\Phi(\varphi) \quad \text{(A.37)}$$

By substituting (A.37) into (A.36), dividing by A, and multiplying by $r^2 \sin^2\vartheta$, it results:

$$\frac{\sin^2\vartheta}{R}\frac{\mathrm{d}}{\mathrm{d}r}\left(r^2 \frac{\mathrm{d}R}{\mathrm{d}r}\right) + \frac{\sin\vartheta}{\Theta}\frac{\mathrm{d}}{\mathrm{d}\vartheta}\left(\sin\vartheta \frac{\mathrm{d}\Theta}{\mathrm{d}\vartheta}\right) + \frac{1}{\Phi}\frac{\mathrm{d}^2\Phi}{\mathrm{d}\varphi^2} + (\beta r \sin\vartheta)^2 = 0 \quad \text{(A.38)}$$

The third term is explicitly independent of r and ϑ and it must also be independent of φ, since the other terms do not depend on φ. Accordingly:

$$\frac{1}{\Phi}\frac{\mathrm{d}^2\Phi}{\mathrm{d}\varphi^2} = -m^2 \quad \text{(A.39)}$$

where m is a constant.

By substituting this last relation into (A.38) and dividing by $\sin^2\vartheta$, it results:

$$\frac{1}{R}\frac{\mathrm{d}}{\mathrm{d}r}\left(r^2\frac{\mathrm{d}R}{\mathrm{d}r}\right)+(\beta r)^2+\frac{1}{\Theta\sin\vartheta}\frac{\mathrm{d}}{\mathrm{d}\vartheta}\left(\sin\vartheta\frac{\mathrm{d}\Theta}{\mathrm{d}\vartheta}\right)-\left(\frac{m}{\sin\vartheta}\right)^2=0 \qquad \text{(A.40)}$$

The last two terms of (A.40) depend only on ϑ, whereas the other two do not depend on ϑ, so that they must be constant. Hence,

$$\frac{1}{\Theta\sin\vartheta}\frac{\mathrm{d}}{\mathrm{d}\vartheta}\left(\sin\vartheta\frac{\mathrm{d}\Theta}{\mathrm{d}\vartheta}\right)-\left(\frac{m}{\sin\vartheta}\right)^2=-n\,(n+1) \qquad \text{(A.41)}$$

where n is usually an integer. Then, equation (A.40) becomes

$$\frac{1}{R}\frac{\mathrm{d}}{\mathrm{d}r}\left(r^2\frac{\mathrm{d}R}{\mathrm{d}r}\right)+(\beta r)^2-n\,(n+1)=0 \qquad \text{(A.42)}$$

In summary, by applying the separation of variables method to the scalar wave equation (A.36), the following three separated differential equations have been found

$$\frac{\mathrm{d}^2\Phi}{\mathrm{d}\varphi^2}+m^2\Phi=0 \qquad \text{(A.43a)}$$

$$\frac{1}{\sin\vartheta}\frac{\mathrm{d}}{\mathrm{d}\vartheta}\left(\sin\vartheta\frac{\mathrm{d}\Theta}{\mathrm{d}\vartheta}\right)+\left[n\,(n+1)-\left(\frac{m}{\sin\vartheta}\right)^2\right]\Theta=0 \qquad \text{(A.43b)}$$

$$\frac{\mathrm{d}}{\mathrm{d}r}\left(r^2\frac{\mathrm{d}R}{\mathrm{d}r}\right)+\left[(\beta r)^2-n\,(n+1)\right]R=0 \qquad \text{(A.43c)}$$

As well-known, the solutions to the harmonic equation (A.43a) take the form

$$\Phi(\varphi)=B_1\,\mathrm{e}^{-\mathrm{j}m\varphi}+B_2\,\mathrm{e}^{\,\mathrm{j}m\varphi} \qquad \text{(A.44a)}$$

or

$$\Phi(\varphi)=B_3\cos(m\varphi)+B_4\sin(m\varphi) \qquad \text{(A.44b)}$$

where B_1, B_2, B_3, and B_4 are integration constants.

It is worth noting that the φ coordinate represents an angle and, accordingly, if the solution has to be valid in a whole 2π-range, then it must be periodic and, as a consequence, m must be an integer.

Equation (A.43b) is the associated Legendre differential equation, whose solutions are

$$\Theta(\vartheta)=C_1\,P_n^m(\cos\vartheta)+C_2\,Q_n^m(\cos\vartheta) \qquad \text{(A.45a)}$$

when n is a non-negative integer, or

$$\Theta(\vartheta)=C_3\,P_n^m(\cos\vartheta)+C_4\,P_n^m(-\cos\vartheta) \qquad \text{(A.45b)}$$

when n is not an integer, $P_n^m(\cos\vartheta)$ and $Q_n^m(\cos\vartheta)$ are the associated Legendre functions of first and second kind, respectively.

It must be stressed that all solutions to the associated Legendre differential equation have singularities at the poles $\vartheta=0$ or $\vartheta=\pi$ except $P_n^m(\cos\vartheta)$ with a non-negative integer value of n [17, 308, 309]. Accordingly, if the desired solution has to be finite in the whole ϑ-range $[0, \pi]$, the correct solution to the differential equation (A.43b) reduces to

$$\Theta(\vartheta) = C_1 P_n^m(\cos\vartheta) \tag{A.46}$$

Equation (A.43c) is strictly related to the Bessel's differential equation. Its solutions can be written in the form

$$R(r) = D_1 h_n^{(2)}(\beta r) + D_2 h_n^{(1)}(\beta r) \tag{A.47a}$$

or

$$R(r) = D_3 j_n(\beta r) + D_4 y_n(\beta r) \tag{A.47b}$$

where $j_n(\beta r)$ and $y_n(\beta r)$ are the spherical Bessel function of order n of the first and second kind, respectively, $h_n^{(1)}(\beta r)$ and $h_n^{(2)}(\beta r)$ are the spherical Hankel functions of order n of the first and second kind.

The spherical Bessel functions $j_n(\beta r)$ and $y_n(\beta r)$ are used to represent radial standing waves and are related to the ordinary Bessel functions by

$$j_n(\beta r) = \sqrt{\frac{\pi}{2\beta r}}\, J_{n+1/2}(\beta r) \tag{A.48a}$$

$$y_n(\beta r) = \sqrt{\frac{\pi}{2\beta r}}\, Y_{n+1/2}(\beta r) \tag{A.48b}$$

The spherical Hankel functions $h_n^{(2)}(\beta r)$ and $h_n^{(1)}(\beta r)$ are employed to represent outward and inward radial travelling waves, respectively. They are related to the ordinary Hankel functions by

$$h_n^{(2)}(\beta r) = \sqrt{\frac{\pi}{2\beta r}}\, H_{n+1/2}^{(2)}(\beta r) \tag{A.49a}$$

$$h_n^{(1)}(\beta r) = \sqrt{\frac{\pi}{2\beta r}}\, H_{n+1/2}^{(1)}(\beta r) \tag{A.49b}$$

Before writing the most general solution to the scalar Helmholtz equation (A.35) in the considered case of an observation region external to the smallest sphere enclosing the AUT, it is worthy to recall some properties of the associated Legendre functions $P_n^m(\cos\vartheta)$. They are solutions of the associated Legendre differential equation (A.43b) and are related to the Legendre polynomials $P_n(\cos\vartheta)$ by the following equation

$$P_n^m(x) = \left(1 - x^2\right)^{m/2} \frac{\mathrm{d}^m P_n(x)}{\mathrm{d}x^m} \tag{A.50}$$

wherein $x = \cos\vartheta$. Such a relation is valid for $m \geq 0$. The Legendre polynomials are solutions of the Legendre differential equation, which is easily obtained by putting $m = 0$ in the equation (A.43b). They are, obviously, polynomials and a convenient expression for them is the Rodrigues' formula

$$P_n(x) = \frac{1}{2^n n!} \frac{\mathrm{d}^n}{\mathrm{d}x^n} \left(x^2 - 1\right)^n \tag{A.51}$$

The Legendre polynomials of lower degree are $P_0(x) = 1$, $P_1(x) = x$, and $P_2(x) = 0.5\,(3x^2 - 1)$.

Since $P_n(x)$ is a polynomial of degree n, from (A.50) it follows that $P_n^m(x) = 0$ for $m > n$.

By substituting (A.51) into (A.50), it results:

$$P_n^m(x) = \frac{1}{2^n n!}\left(1 - x^2\right)^{m/2} \frac{\mathrm{d}^{m+n}}{\mathrm{d}x^{m+n}}\left(x^2 - 1\right)^n \tag{A.52}$$

This equation allows the extension of associated Legendre functions also to negative values of m. It can be shown that the associated Legendre functions for negative values of m are linked to those with positive values of m via the relation:

$$P_n^{-|m|}(x) = (-1)^m \frac{(n - |m|)!}{(n + |m|)!} P_n^{|m|}(x) \tag{A.53}$$

It is worth noting that, according to (A.50), the associated Legendre function is a polynomial for m even, whereas for m odd, it is not a polynomial, due to the presence of the factor $(1 - x^2)^{1/2}$.

The associated Legendre functions satisfy the orthogonality relation [76]

$$\int_{-1}^{1} P_n^m(x)\, P_k^m(x)\, \mathrm{d}x = \frac{2}{2n+1} \frac{(n+m)!}{(n-m)!} \delta_{nk} \tag{A.54}$$

wherein

$$\delta_{nk} = \begin{cases} 1 & \text{for } k = n \\ 0 & \text{for } k \neq n \end{cases} \tag{A.55}$$

is the Kronecker delta.

Accordingly, it is convenient to introduce the normalized associated Legendre functions as defined by Belousov [310]:

$$\bar{P}_n^m(x) = \sqrt{\frac{2n+1}{2} \frac{(n-m)!}{(n+m)!}}\, P_n^m(x) \tag{A.56}$$

so that

$$\int_{-1}^{1} \bar{P}_n^m(x) \bar{P}_k^m(x)\, \mathrm{d}x = \delta_{nk} \tag{A.57}$$

In the light of the previous discussion, the most general solution to the scalar Helmholtz equation in the region external to the smallest sphere enclosing the AUT is:

$$A(r, \vartheta, \varphi) = \sum_{n=0}^{\infty} \sum_{m=-n}^{n} \tilde{C}_{nm}\, h_n^{(2)}(\beta r)\, \bar{P}_n^m(\cos\vartheta)\, \mathrm{e}^{\,\mathrm{j}m\varphi} \tag{A.58}$$

By taking into account that the normalized associated Legendre functions for negative values of m are proportional (see (A.53) and (A.56)) to those with positive values, relation (A.58) reduces to

$$A(r,\vartheta,\varphi)=\sum_{n=0}^{\infty}\sum_{m=-n}^{n}C_{nm}\,h_n^{(2)}(\beta r)\,\bar{P}_n^{|m|}(\cos\vartheta)\,\mathrm{e}^{\,\mathrm{j}m\varphi} \tag{A.59}$$

where C_{nm} are proper integration constants.

A.4 Method of the stationary phase

The asymptotic evaluation of single and double integrals by means of the stationary phase method is briefly described in this appendix.

A.4.1 Single integrals

Let

$$I(\Omega)=\int_{-\infty}^{\infty}F(\tau)\,\mathrm{e}^{\,\mathrm{j}\Omega f(\tau)}\,\mathrm{d}\tau \tag{A.60}$$

be a single integral, where Ω is a large positive parameter, $F(\tau)$ is a regular real or complex function of the real variable τ, and $f(\tau)$ is a regular real function. In the hypothesis that $F(\tau)$ is a slowly varying function of τ and that $f(\tau)$ has an isolated first order stationary phase point at τ_s, i.e., $f'(\tau_s)=0$, but $f''(\tau_s)\neq 0$, such an integral can be asymptotically evaluated by means of the method of the stationary phase [306].

It can be easily recognized that, for $\Omega\to\infty$, only the integration region in the neighbourhood of the stationary phase point gives a significant contribution to the integral. In fact, outside such a region, the real and imaginary part of the integrand become the product of a slowly varying amplitude function for another one which oscillates very rapidly between +1 and –1, so that the related contributions tend to cancel each other. Accordingly, $I(\Omega)$ can be approximated as:

$$I(\Omega)\underset{\Omega\to\infty}{\approx}F(\tau_s)\,\mathrm{e}^{\,\mathrm{j}\Omega f(\tau_s)}\int_{-\infty}^{\infty}\mathrm{e}^{\,\mathrm{j}\Omega[f(\tau)-f(\tau_s)]}\,\mathrm{d}\tau=F(\tau_s)\,\mathrm{e}^{\,\mathrm{j}\Omega f(\tau_s)}\tilde{I}(\Omega) \tag{A.61}$$

where

$$\tilde{I}(\Omega)=\int_{-\infty}^{\infty}\mathrm{e}^{\,\mathrm{j}\Omega[f(\tau)-f(\tau_s)]}\,\mathrm{d}\tau \tag{A.62}$$

To evaluate this last integral, it is convenient, in the neighbourhood of the stationary phase point τ_s, to approximate $f(\tau)$ by a Taylor series truncated at the second order term, namely

$$f(\tau) \approx f(\tau_s) + \frac{1}{2} f''(\tau_s)(\tau - \tau_s)^2 \tag{A.63}$$

accordingly,

$$\tilde{I}(\Omega) \approx \int_{-\infty}^{\infty} \mathrm{e}^{\,\mathrm{j}\Omega f''(\tau_s)(\tau-\tau_s)^2/2}\, \mathrm{d}\tau \tag{A.64}$$

On the other hand

$$\int_{-\infty}^{\infty} \mathrm{e}^{-ax^2}\, \mathrm{d}x = \sqrt{\frac{\pi}{a}} \tag{A.65}$$

and therefore

$$\tilde{I}(\Omega) \approx \sqrt{\frac{\pi}{-\mathrm{j}\Omega f''(\tau_s)/2}} = \sqrt{\frac{\mathrm{j}2\pi}{\Omega f''(\tau_s)}} = \sqrt{\frac{2\pi}{\Omega |f''(\tau_s)|}}\, \mathrm{e}^{\,\mathrm{j\,sgn}[f''(\tau_s)]\pi/4} \tag{A.66}$$

wherein

$$\mathrm{sgn}(x) = \begin{cases} +1 & \text{for } x > 0 \\ -1 & \text{for } x < 0 \end{cases} \tag{A.67}$$

is the sign of x function.

By substituting (A.66) into (A.61), it results

$$I(\Omega) \underset{\Omega\to\infty}{\approx} \sqrt{\frac{2\pi}{\Omega |f''(\tau_s)|}}\, F(\tau_s)\, \mathrm{e}^{\,\mathrm{j}\Omega f(\tau_s)}\, \mathrm{e}^{\,\mathrm{j\,sgn}[f''(\tau_s)]\pi/4} \tag{A.68}$$

A.4.2 Double integrals

Let

$$I(\Omega) = \int_{-\infty}^{\infty}\int_{-\infty}^{\infty} g(x,y)\, \mathrm{e}^{\,\mathrm{j}\Omega f(x,y)}\, \mathrm{d}x\, \mathrm{d}y \tag{A.69}$$

be a double integral, where Ω is a large positive parameter, $g(x,y)$ is a regular real or complex function of the real variables x, y and $f(x,y)$ is a regular real function. In the hypothesis that $g(x,y)$ is a slowly varying function of x, y and that $f(x,y)$ has an isolated first order stationary phase point at x_s, y_s, namely

$$\left.\frac{\partial f}{\partial x}\right|_{\substack{x=x_s\\y=y_s}} = f'_x(x_s, y_s) = 0; \qquad \left.\frac{\partial f}{\partial y}\right|_{\substack{x=x_s\\y=y_s}} = f'_y(x_s, y_s) = 0$$

but

$$\left.\frac{\partial^2 f}{\partial x^2}\right|_{\substack{x=x_s\\y=y_s}} = f''_{xx}(x_s, y_s) \neq 0; \quad \left.\frac{\partial^2 f}{\partial y^2}\right|_{\substack{x=x_s\\y=y_s}} = f''_{yy}(x_s, y_s) \neq 0; \quad \left.\frac{\partial^2 f}{\partial x \partial y}\right|_{\substack{x=x_s\\y=y_s}} = f''_{xy}(x_s, y_s) \neq 0$$

and the Hessian determinant

$$H[f(x_s, y_s)] = \begin{vmatrix} f''_{xx}(x_s, y_s) & f''_{xy}(x_s, y_s) \\ f''_{xy}(x_s, y_s) & f''_{yy}(x_s, y_s) \end{vmatrix} \neq 0,$$

such an integral can be asymptotically evaluated through the method of the stationary phase [18, 306].

It can be easily recognized that, as in the case of single integrals, only the integration region in the neighbourhood of the stationary phase point gives a significant contribution to the integral for $\Omega \to \infty$. Accordingly, $I(\Omega)$ can be approximated as:

$$I(\Omega) \underset{\Omega \to \infty}{\approx} g(x_s, y_s) \, \mathrm{e}^{\,\mathrm{j}\Omega f(x_s, y_s)} \, \tilde{I}(\Omega) \tag{A.70}$$

where

$$\tilde{I}(\Omega) = \int_{-\infty}^{\infty} \int_{-\infty}^{\infty} \mathrm{e}^{\,\mathrm{j}\Omega\,[f(x,y) - f(x_s, y_s)]} \, \mathrm{d}x \, \mathrm{d}y \tag{A.71}$$

To evaluate this last integral, it is convenient, in the neighbourhood of the stationary phase point x_s, y_s, to approximate $f(x, y)$ by a Taylor series truncated at the second order terms, namely

$$f(x,y) \approx f(x_s, y_s) + \frac{1}{2} f''_{xx}(x_s, y_s)(x - x_s)^2 + \frac{1}{2} f''_{yy}(x_s, y_s)(y - y_s)^2 + f''_{xy}(x_s, y_s)(x - x_s)(y - y_s) \tag{A.72}$$

From (A.72), by putting

$$A = \frac{1}{2} f''_{xx}(x_s, y_s); \qquad B = \frac{1}{2} f''_{yy}(x_s, y_s); \qquad C = f''_{xy}(x_s, y_s) \tag{A.73a}$$

$$\xi = (x - x_s); \qquad \eta = (y - y_s), \tag{A.73b}$$

it results

$$f(x,y) - f(x_s, y_s) = A\xi^2 + B\eta^2 + C\xi\eta \tag{A.74}$$

By substituting the approximation (A.72) into the integral (A.71), by taking into account the above positions, and, then, by applying the change of variables formula in the double integrals, it results

$$\tilde{I}(\Omega) = \int_{-\infty}^{\infty} \int_{-\infty}^{\infty} \mathrm{e}^{\,\mathrm{j}\Omega\,(A\xi^2 + B\eta^2 + C\xi\eta)} \, \mathrm{d}\xi \, \mathrm{d}\eta \tag{A.75}$$

since the Jacobian of the transformation is equal to 1, as can be easily verified.

To evaluate this last integral, it is convenient to rotate the axes ξ, η by a proper angle α, in such a way that it results

$$A\xi^2 + B\eta^2 + C\xi\eta = A'\xi'^2 + B'\eta'^2 \tag{A.76}$$

where the variables ξ, η are related to the ξ', η' ones by the relations

$$\begin{cases} \xi = \xi' \cos\alpha - \eta' \sin\alpha \\ \eta = \xi' \sin\alpha + \eta' \cos\alpha \end{cases} \tag{A.77}$$

By applying the change of variables formula to the integral (A.75), since also in such a case the Jacobian of the transformation is equal to 1, it follows that

$$\tilde{I}\,(\Omega) = \int_{-\infty}^{\infty}\int_{-\infty}^{\infty} \mathrm{e}^{\,\mathrm{j}\Omega\,\left(A'\xi'^2 + B'\eta'^2\right)} \mathrm{d}\xi'\mathrm{d}\eta' = \int_{-\infty}^{\infty} \mathrm{e}^{\,\mathrm{j}\Omega A'\xi'^2} \mathrm{d}\xi' \int_{-\infty}^{\infty} \mathrm{e}^{\,\mathrm{j}\Omega\, B'\eta'^2} \mathrm{d}\eta' \tag{A.78}$$

and therefore

$$\tilde{I}\,(\Omega) = \sqrt{\frac{\pi}{\Omega\,|A'|}}\,\mathrm{e}^{\,\mathrm{j}\,\mathrm{sgn}\,(A')\,\pi/4}\sqrt{\frac{\pi}{\Omega\,|B'|}}\,\mathrm{e}^{\,\mathrm{j}\,\mathrm{sgn}\,(B')\,\pi/4} = \frac{\mathrm{j}\pi\,\delta}{\Omega\sqrt{|A'||B'|}} \tag{A.79}$$

where

$$\delta = \begin{cases} 1 \ \text{if}\, A' > 0,\, B' > 0 \\ -1 \ \text{if}\, A' < 0,\, B' < 0 \\ -\mathrm{j} \ \text{if}\, A' \text{ and } B' \text{ have a different sign} \end{cases} \tag{A.80}$$

The expressions of the coefficients A', B' and the rotation angle α can be determined through the following steps. Relations (A.77) are substituted into (A.76), thus obtaining

$$A' = A\cos^2\alpha + B\sin^2\alpha + C\sin\alpha\cos\alpha \tag{A.81a}$$

$$B' = A\sin^2\alpha + B\cos^2\alpha - C\sin\alpha\cos\alpha \tag{A.81b}$$

Moreover, by imposing that the coefficient of the term in $\xi'\eta'$ is equal to 0, it results

$$\alpha = \frac{1}{2}\tan^{-1}\left(\frac{C}{A-B}\right) \tag{A.82}$$

Then, relations (A.81) are rewritten as function of $\cos 2\alpha$ and $\sin 2\alpha$, which in turn are expressed as function of $\tan 2\alpha$. At last, by taking into account that $\tan 2\alpha = C/\left(A - B\right)$, the following expressions for A' and B' are obtained.

$$A' = \frac{1}{2}\left[(A+B) + \sqrt{(A-B)^2 + C^2}\right]$$
$$= \frac{1}{2}\left[(A+B) + \sqrt{(A+B)^2 - (4AB - C^2)}\right] \tag{A.83a}$$

$$B' = \frac{1}{2}\left[(A+B) - \sqrt{(A-B)^2 + C^2}\right]$$
$$= \frac{1}{2}\left[(A+B) - \sqrt{(A+B)^2 - (4AB - C^2)}\right] \tag{A.83b}$$

Accordingly,

$$A' + B' = A + B \tag{A.84a}$$

$$A'B' = (4AB - C^2)/4 \tag{A.84b}$$

By taking into account relations (A.84), it can be easily verified that, if $(4AB - C^2) > 0$, then the coefficients A', B', A, and B have all the same sign, whereas if $(4AB - C^2) < 0$ then A' and B' have different sign. Therefore, the conditions (A.80) can be rewritten in the form

$$\delta = \begin{cases} 1 & \text{if } (4AB - C^2) > 0 \text{ and } A > 0 \\ -1 & \text{if } (4AB - C^2) > 0 \text{ and } A < 0 \\ -\mathrm{j} & \text{if } (4AB - C^2) < 0 \end{cases} \tag{A.85}$$

By taking into account relations (A.70), (A.79), and (A.84b), it follows that

$$I(\Omega) \underset{\Omega \to \infty}{\approx} g\left(x_s, y_s\right) \mathrm{e}^{\mathrm{j}\Omega f(x_s, y_s)} \frac{\mathrm{j}2\pi\delta}{\Omega\sqrt{|4AB - C^2|}} \tag{A.86}$$

On the other hand,

$$4AB - C^2 = f''_{xx}\left(x_s, y_s\right) f''_{yy}\left(x_s, y_s\right) - \left(f''_{xy}\left(x_s, y_s\right)\right)^2 = H\left[f\left(x_s, y_s\right)\right] \tag{A.87}$$

As a conclusion, by taking into account (A.87), relation (A.86) can be rewritten in the form

$$I(\Omega) \underset{\Omega \to \infty}{\approx} g(x_s, y_s) \mathrm{e}^{\mathrm{j}\Omega f(x_s, y_s)} \frac{\mathrm{j}2\pi\delta}{\Omega\sqrt{|H[f(x_s, y_s)]|}} \tag{A.88}$$

where

$$\delta = \begin{cases} 1 & \text{if } H[f(x_s, y_s)] > 0 \text{ and } f''_{xx}(x_s, y_s) > 0 \\ -1 & \text{if } H[f(x_s, y_s)] > 0 \text{ and } f''_{xx}(x_s, y_s) < 0 \\ -\mathrm{j} & \text{if } H[f(x_s, y_s)] < 0 \end{cases} \tag{A.89}$$

A.5 The fast Fourier transform

The fast Fourier transform (FFT) [300, 314] is an algorithm which allows one to efficiently compute the discrete Fourier transform (DFT) of the sequence of a sampled function or its inverse discrete Fourier transform (IDFT). It is briefly recalled in this appendix, since it can be conveniently employed to evaluate in a fast and accurate way the Fourier transform and the Fourier series of a function.

The DFT of the sequence $g(iT)$ of length N is defined [300] by:

$$G\left(\frac{n}{NT}\right) = \sum_{i=0}^{N-1} g(iT)\, \mathrm{e}^{-\mathrm{j}2\pi ni/N} \qquad n = 0, 1, \ldots, N-1 \tag{A.90}$$

whereas the IDFT is defined by:

$$g(iT) = \frac{1}{N} \sum_{n=0}^{N-1} G\left(\frac{n}{NT}\right) \mathrm{e}^{\mathrm{j}2\pi ni/N} \qquad i = 0, 1, \ldots, N-1 \tag{A.91}$$

As is well known, relations (A.90) and (A.91) can be efficiently computed via the FFT algorithm. Moreover, they require both the functions to be periodic, namely:

$$G\left(\frac{n+pN}{NT}\right)=G\left(\frac{n}{NT}\right) \qquad p=0,\pm 1,\pm 2,\ldots \tag{A.92a}$$

$$g\,[(i+pN)\,T]=g\,(iT) \qquad p=0,\pm 1,\pm 2,\ldots \tag{A.92b}$$

References

[1] Johnson R.C., Ecker H.A., Hollis J.S. 'Determination of far-field antenna patterns from near-field measurements'. *Proceedings of the IEEE*. 1973;**61**(12):1668–94.

[2] Yaghjian A.D. 'An overview of near-field antenna measurements'. *IEEE Transactions on Antennas and Propagation*. 1986;**AP-34**(1):30–45.

[3] Appel-Hansen J., Dyson J.D., Gillespie E.S, *et al.* 'Antenna measurements' in Rudge A.W., Milne K., Olver A.D., *et al* (eds.). *The Handbook of antenna design*. London, UK: Peter Peregrinus; 1986. pp. 584–694.

[4] Gillespie E.S. 'Special issue on near-field scanning techniques'. *IEEE Transactions on Antennas and Propagation*. 1988;**36**(6):727–901.

[5] Slater D. *Near-field antenna measurements*. Boston, MA, USA: Artech House; 1991.

[6] Gennarelli C., Riccio G., D'Agostino F, *et al. Near-field – far-field transformation techniques*. Vol. 1. Salerno, Italy: CUES; 2004.

[7] Gregson S., McCormick J., Parini C. *Principles of planar near-field antenna measurements*. London, UK: IET; 2007.

[8] Francis M.H., Wittmann R.C. 'Near-field scanning measurements: theory and practice' in Balanis C.A. (ed.). *Modern antenna Handbook*. Hoboken, NJ, USA: John Wiley & Sons, Inc; 2008. pp. 929–76.

[9] Francis M.H. (ed.) *1720-2012-IEEE Recommended practice for near-field antenna measurements*. IEEE; 2012.

[10] Gennarelli C., Capozzoli A., Foged L.J. 'Special issue on recentadvances in near-field to far-field transformation techniques'. *International Journal of Antennas and Propagation*. 2012;**2012**.

[11] Ferrara F., Gennarelli C., Guerriero R, *et al.* 'Near-field antenna measurement techniques' in Chen Z.N., Liu D., Nakano H. (eds.). *Handbook of antenna technologies*. Singapore: Springer; 2016. pp. 2107–63.

[12] Sierra Castañer M., Foged L.J. *Post-Processing techniques in antenna measurement*. London, UK: SciTech Publishing IET; 2019.

[13] Parini C., Gregson S., McCormick J, *et al. Theory and practice of modern antenna range measurements*. London, UK: SciTech Publishing IET; 2020.

[14] Rahmat-Samii Y., Lemanczyk J. 'Application of spherical near-field measurements to microwave holographic diagnosis of antennas'. *IEEE Transactions on Antennas and Propagation*. 1988;**36**(6):869–78.

[15] Yaccarino R.G., Rahmat-Samii Y., Williams L.I. 'The bi-polar planar near-field measurement technique, part II: near-field to far-field transformation and holographic imaging methods'. *IEEE Transactions on Antennas and Propagation*. 1994;**42**(2):196–204.

[16] Franceschetti G. *Electromagnetics*. New York, USA: Plenum Press; 1997.
[17] Balanis C.A. *Advanced engineering electromagnetics*. Hoboken, NJ, USA: John Wiley & sons, Inc; 2012.
[18] Balanis C.A. *Antenna theory analysis and design*. Hoboken, NJ, USA: John Wiley & sons, Inc; 2016.
[19] Silver S. *Microwave antenna theory and design*. London, UK: Peter Peregrinus; 1984.
[20] Hollis J.S., Lyon T.J., Clayton L. *Microwave antenna measurements*. Atlanta, GA, USA: Scientific Atlanta; 1985. p. 14.
[21] Hacker P.S., Schrank H.E. 'Range distance requirements for measuring low and ultralow sidelobe antenna patterns'. *IEEE Transactions on Antennas and Propagation*. 1982;**AP-30**(5):956–66.
[22] Hansen R.C. 'Measurement distance effects on low sidelobe patterns'. *IEEE Transactions on Antennas and Propagation*. 1984;**AP-32**(6):591–94.
[23] Corona P., Ferrara G., Gennarelli C. 'Measurement distance requirements for both symmetrical and antisymmetrical aperture antennas'. *IEEE Transactions on Antennas and Propagation*. 1989;**37**(8):990–95.
[24] Joy E.B., Paris D.T. 'Spatial sampling and filtering in near-field measurements'. *IEEE Transactions on Antennas and Propagation*. 1972;**AP-20**(3):253–61.
[25] Paris D.T., Leach W.M., Joy E.B. 'Basic theory of probe-compensated near-field measurements'. *IEEE Transactions on Antennas and Propagation*. 1978;**AP-26**(3):373–79.
[26] Joy E.B., Leach W.M., Rodrigue G.P., Paris D.T. 'Applications of probe-compensated near-field measurements'. *IEEE Transactions on Antennas and Propagation*. 1978;**AP-26**(3):379–89.
[27] Kerns D.M. 'Correction of near-field antenna measurements made with an arbitrary but known measuring antenna'. *Electronics Letters*. 1970;**6**(11):346–47.
[28] Kerns D.M. 'Plane-wave scattering-matrix theory of antennas and antenna-antenna interactions: formulation and applications'. *Journal of Research of the National Bureau of Standards, Section B*. 1976;**80B**(1):5.
[29] Kerns D.M. *Plane-wave scattering-matrix theory of antennas and antenna-antenna interactions. NBS Monograph no.162*. Washington, DC, USA: U.S. Government Printing Office; 1981.
[30] Wang J.J.H. 'An examination of the theory and practices of planar near-field measurement'. *IEEE Transactions on Antennas and Propagation*. 1988;**36**(6):746–53.
[31] Repjar A.G., Newell A.C., Francis M.H. 'Accurate determination of planar near-field correction parameters for linearly polarized probes'. *IEEE Transactions on Antennas and Propagation*. 1988;**36**(6):855–68.
[32] Rahmat-Samii Y., Mittra R. 'A plane-polar approach for far-field construction from near-field measurements'. *IEEE Transactions on Antennas and Propagation*. 1980;**AP-28**(2):216–30.

[33] Rahmat-Samii Y., Gatti M.S. 'Far-field patterns of spaceborne antennas from plane-polar near-field measurements'. *IEEE Transactions on Antennas and Propagation*. 1985;**AP-33**(6):638–48.

[34] Gatti M.S., Rahmat-Samii Y. 'FFT applications to plane-polar near-field antenna measurements'. *IEEE Transactions on Antennas and Propagation*. 1988;**36**(6):781–91.

[35] Bennett J.C., Farhat K.S. 'Near-field measurements on plane-polar facility'. *IEE Proceedings H Microwaves, Antennas and Propagation*. 1989;**136**(3):202.

[36] Yaghijian A.D. 'Antenna coupling and near-field sampling in plane-polar coordinates'. *IEEE Transactions on Antennas and Propagation*. 1992;**40**(3):304–12.

[37] Yaghjian A.D., Woodworth M.B. 'Sampling in plane-polar coordinates'. *IEEE Transactions on Antennas and Propagation*. 1996;**44**(5):696.

[38] Costanzo S., Di Massa G. 'Efficient near-field to far-field transformation from plane-polar samples'. *Microwave and Optical Technology Letters*. 2006;**48**(12):2433–36.

[39] Bucci O.M., Gennarelli C., Savarese C. 'Fast and accurate near-field-far-field transformation by sampling interpolation of plane-polar measurements'. *IEEE Transactions on Antennas and Propagation*. 1991;**39**(1):48–55.

[40] Bucci O.M., Gennarelli C., Riccio G, *et al*. 'Fast and accurate far-field evaluation from a non redundant, finite number of plane-polar measurements'. *Proceedings of IEEE Antennas and Propagation Society International Symposium*; Seattle, USA, 1994. pp. 540–43.

[41] Bucci O.M., D'Elia G., Migliore M.D. 'Advanced field interpolation from plane-polar samples: experimental verification'. *IEEE Transactions on Antennas and Propagation*. 1998;**46**(2):204–10.

[42] Bucci O.M., Gennarelli C., Riccio G., Savarese C. 'Near-field – far-field transformation from nonredundant plane-polar data: effective modellings of the source'. *IEE Proceedings - Microwaves, Antennas and Propagation*. 1998;**145**(1):33.

[43] Bucci O.M., D'Agostino F., Gennarelli C. 'NF–FF transformation with plane-polar scanning: ellipsoidal modelling of the antenna'. *Automatika*. 2000;**41**(3–4):159–64.

[44] D'Agostino F., Ferrara F., Gennarelli C., Guerriero R., Migliozzi M. 'Far-field pattern reconstruction from a nonredundant plane-polar near-field sampling arrangement: experimental testing'. *IEEE Antennas and Wireless Propagation Letters*. 2016;**15**:1345–48.

[45] D'Agostino F., Ferrara F., Gennarelli C, *et al*. 'Reconstruction of the antenna far-field pattern through a fast plane-polar scanning'. *Applied Computational Electromagnetics Society Journal*. 2016;**31**(12):1362–69.

[46] Williams L.I., Rahmat-Samii Y., Yaccarino R.G. 'The bi-polar planar near-field measurement technique, part I: implementation and measurement comparisons'. *IEEE Transactions on Antennas and Propagation*. 1994;**42**(2):184–95.

[47] Rahmat-Samii Y., Williams L.I., Yaccarino R.G. 'The UCLA bi-polar planar-near-field antenna-measurement and diagnostics range'. *IEEE Antennas and Propagation Magazine*. 1995;**37**(6):16–35.

[48] Costanzo S., Di Massa G. 'Direct far-field computation from bi-polar near-field samples'. *Journal of Electromagnetic Waves and Applications*. 2006;**20**(9):1137–48.

[49] D'Agostino F., Gennarelli C., Riccio G., Savarese C. 'Data reduction in the NF-FF transformation with bi-polar scanning'. *Microwave and Optical Technology Letters*. 2003;**36**(1):32–36.

[50] DAgostino F., Ferrara F., Gennarelli C., Guerriero R., Migliozzi M. 'Fast and accurate far-field prediction by using a reduced number of bi-polar measurements'. *IEEE Antennas and Wireless Propagation Letters*. 2017;**16**:2939–42.

[51] D'Agostino F., Ferrara F., Gennarelli C., Guerriero R., Migliozzi M. 'Laboratory tests on an near-field to far-field transformation technique from non-redundant bi-polar data'. *IET Microwaves, Antennas & Propagation*. 2018;**12**(5):712–17.

[52] Leach W.M., Paris D.T. 'Probe compensated near-field measurements on a cylinder'. *IEEE Transactions on Antennas and Propagation*. 1973;**AP-21**(4):435–45.

[53] Yaghjian A.D. *Near-field antenna measurement on a cylindrical surface: a source scattering matrix formulation. NBS Tech. note 696*. Washington, DC, USA: U.S. Government Printing Office; 1977.

[54] Borgiotti G.V. 'Integral equation formulation for probe corrected far-field reconstruction from measurements on a cylinder'. *IEEE Transactions on Antennas and Propagation*. 1978;**AP-26**(4):572–78.

[55] Appel-Hansen J. 'On cylindrical near-field scanning techniques'. *IEEE Transactions on Antennas and Propagation*. 1980;**AP-28**(2):231–34.

[56] Hansen T.B. 'Complex point sources in probe-corrected cylindrical near-field scanning'. *Wave Motion*. 2006;**43**(8):700–12.

[57] Hansen T.B. 'Probe-corrected near-field measurements on a truncated cylinder'. *The Journal of the Acoustical Society of America*. 2006;**119**(2):792.

[58] Bucci O.M., Gennarelli C. 'Use of sampling expansions in near-field-far-field transformation: the cylindrical case'. *IEEE Transactions on Antennas and Propagation*. 1988;**36**(6):830–35.

[59] Bucci O.M., Gennarelli C., Riccio G., Speranza V., Savarese C. 'Nonredundant representation of the electromagnetic fields over a cylinder with application to the near-field far-field transformation'. *Electromagnetics*. 1996;**16**(3):273–90.

[60] Bucci O.M., Gennarelli C., Riccio G., Savarese C. 'NF–FF transformation with cylindrical scanning: an effective technique for elongated antennas'. *IEE Proceedings - Microwaves, Antennas and Propagation*. 1998;**145**(5):369.

[61] D'Agostino F., Ferrara F., Gennarelli C., Riccio G., Savarese C. 'NF-FF transformation with cylindrical scanning from a minimum number of data'. *Microwave and Optical Technology Letters*. 2002;**35**(4):264–70.

[62] D'Agostino F., Ferrara F., Gennarelli C., Gennarelli G., Guerriero R., Migliozzi M. 'On the direct non-redundant near-field-to-far-field transformation in a cylindrical scanning geometry'. *IEEE Antennas and Propagation Magazine*. 2012;**54**(1):130–38.

[63] D'Agostino F., Ferrara F., Gennarelli C., Gennarelli G., Guerriero R., Migliozzi M. 'Antenna pattern reconstruction directly from nonredundant near-field measurements collected by a cylindrical facility'. *Progress In Electromagnetics Research M*. 2012;**24**:235–49.

[64] Hansen T.B. 'Complex-point dipole formulation of probe-corrected cylindrical and spherical near-field scanning of electromagnetic fields'. *IEEE Transactions on Antennas and Propagation*. 2009;**57**(3):728–41.

[65] Qureshi M.A., Schmidt C.H., Eibert T.F. 'Adaptive sampling in spherical and cylindrical near-field antenna measurements'. *IEEE Antennas and Propagation Magazine*. 2013;**55**(1):243–49.

[66] Jensen F. *Electromagnetic near-field far-field correlations*. Denmark, Lyngby, Denmark: Ph.D. dissertation,Technical Univ; 1970.

[67] Wacker P.F. *Non-Planar near-field measurements: spherical scanning*. NBSIR 75-809; 1975.

[68] Jensen F. 'On the probe compensation for near-field measurements on a sphere'. *Archiv Der Elektrischen Übertragung*. 1975;**29**:305–08.

[69] Larsen F.H. 'Probe correction of spherical near-field measurements'. *Electronics Letters*. 1977;**13**(14):393.

[70] Larsen F.H. 'Improved algorithm for probe-corrected spherical near-field/far-field transformation'. *Electronics Letters*. 1979;**15**(19):588.

[71] Jensen F. *Probe-corrected spherical near-field antenna measurement*. Denmark, Lyngby: Ph.D. dissertation, Technical Univ; 1980.

[72] Yaghjian A.D. 'Simplified approach to probe-corrected spherical near-field scanning'. *Electronics Letters*. 1984;**20**(5):195.

[73] Yaghjian A.D., Wittmann R.C. 'The receiving antenna as a linear differential operator: application to spherical near-field scanning'. *IEEE Transactions on Antennas and Propagation*. 1985;**AP-33**(11):1175–85.

[74] Lewis R., Wittmann R. 'Improved spherical and hemispherical scanning algorithms'. *IEEE Transactions on Antennas and Propagation*. 1987;**AP-35**(12):1381–88.

[75] Sarkar T.K., Petre P., Taaghol A., Harrington R.F. 'An alternative spherical near field to far field transformation'. *Progress In Electromagnetics Research*. 1997;**PIER 16**:269–84.

[76] Hald J., Hansen J.E., Jensen F, *et al. Spherical near-field antenna measurements*. London, UK: Peter Peregrinus; 1998.

[77] Rodriguez Varela F., Iraguen B.G., Sierra-Castaner M. 'Under-sampled spherical near-field antenna measurements with error estimation'. *IEEE Transactions on Antennas and Propagation*. 2020;**68**(8):6364–71.

[78] Laitinen T.A., Pivnenko S., Breinbjerg O. 'Iterative probe correction technique for spherical near-field antenna measurements'. *IEEE Antennas and Wireless Propagation Letters*. 2005;**4**:221–23.

[79] Laitinen T.A., Pivnenko S., Breinbjerg O. 'Odd-order probe correction technique for spherical near-field antenna measurements'. *Radio Science*. 2005;**40**:1–11.

[80] Laitinen T., Pivnenko S., Breinbjerg O. 'Application of the iterative probe correction technique for a high-order probe in spherical near-field antenna measurements'. *IEEE Antennas and Propagation Magazine*. 2006;**48**(4):179–85.

[81] Laitinen T., Pivnenko S. 'Probe correction technique for symmetric odd-order probes for spherical near-field antenna measurements'. *IEEE Antennas and Wireless Propagation Letters*. 2007;**6**:635–38.

[82] Laitinen T., Breinbjerg O. 'A first/third-order probe correction technique for spherical near-field antenna measurements using three probe orientations'. *IEEE Transactions on Antennas and Propagation*. 2008;**56**(5):1259–68.

[83] Saccardi F., Mioc F., Giacomini A., Iversen P.O., Foged L.J. 'Fully probe corrected spherical near field offset measurements with minimum sampling using the translated-SWE algorithm'. *Proc. 2018 AMTA*; Williamsburg, VA, USA, 2018. pp. 1–5.

[84] Laitinen T. 'Double ϕ-step θ-scanning technique for spherical near-field antenna measurements'. *IEEE Transactions on Antennas and Propagation*. 2008;**56**(6):1633–39.

[85] Laitinen T. 'Modified θ-scanning technique for first/third-order probes for spherical near-field antenna measurements'. *IEEE Transactions on Antennas and Propagation*. 2008;**57**(6):1590–96.

[86] Laitinen T., Pivnenko S., Nielsen J.M., Breinbjerg O. 'Theory and practice of the FFT/matrix inversion technique for probe-corrected spherical near-field antenna measurements with high-order probes'. *IEEE Transactions on Antennas and Propagation*. 2010;**58**(8):2623–31.

[87] Hansen T.B. 'Spherical near-field scanning with higher-order probes'. *IEEE Transactions on Antennas and Propagation*. 2011;**59**(11):4049–59.

[88] Hansen T.B. 'Numerical investigation of the system-matrix method for higher-order probe correction in spherical near-field antenna measurements'. *International Journal of Antennas and Propagation*. 2012;**2012**:1–8.

[89] Cornelius R., Heberling D. 'Spherical near-field scanning with point-wise probe correction'. *IEEE Transactions on Antennas and Propagation*. 2017;**65**(2):995–97.

[90] Foged L.J., Saccardi F., Mioc F., Iversen P.O. 'Spherical near field offset measurements using downsampled acquisition and advanced NF/FF transformation algorithm'. *Proc. 2016 EUCAP, Davos*; Davos, Switzerland, 2016. pp. 1–3.

[91] Cornelius R., Heberling D. 'Spherical wave expansion with arbitrary origin for near-field antenna measurements'. *IEEE Transactions on Antennas and Propagation*. 2017;**65**(8):4385–88.

[92] Rodriguez Varela F., Galocha Iraguen B., Sierra Castaner M. 'Fast spherical near-field to far-field transformation for offset-mounted

antenna measurements'. *IEEE Antennas and Wireless Propagation Letters*. 2020;**19**(12):2255–59.

[93] Bucci O.M., D'Agostino F., Gennarelli C., Riccio G., Savarese C. 'Data reduction in the NF-FF transformation technique with spherical scanning'. *Journal of Electromagnetic Waves and Applications*. 2001;**15**(6):755–75.

[94] Arena A., D'Agostino F., Gennarelli C, *et al.* 'Probe compensated NF–FF transformation with spherical scanning from a minimum number of data'. *Atti Della Fondazione Giorgio Ronchi*. 2004;**59**(3):312–26.

[95] D'Agostino F., Ferrara F., Gennarelli C., Guerriero R., Migliozzi M. 'Effective antenna modellings for NF-FF transformations with spherical scanning using the minimum number of data'. *International Journal of Antennas and Propagation*. 2011;**2011**:1–11.

[96] D'Agostino F., Ferrara F., Gennarelli C., Guerriero R., Migliozzi M. 'Experimental testing of nonredundant near-field to far-field transformations with spherical scanning using flexible modellings for nonvolumetric antennas'. *International Journal of Antennas and Propagation*. 2013;**2013**:1–10.

[97] D'Agostino F., Ferrara F., Gennarelli C., Guerriero R., Migliozzi M. 'Nonredundant spherical NF-FF transformations using ellipsoidal antenna modeling: experimental assessments [measurements corner]'. *IEEE Antennas and Propagation Magazine*. 2013;**55**(4):166–75.

[98] D'Agostino F., Ferrara F., Gennarelli C., Guerriero R., Migliozzi M. 'Nonredundant spherical near-field to far-field transformation for a volumetric antenna in offset configuration'. *The Open Electrical & Electronic Engineering Journal*. 2019;**13**(1):19–29.

[99] D'Agostino F., Ferrara F., Gennarelli C., Guerriero R., Migliozzi M. 'A nonredundant sampling representation requiring the same number of spherical near-field measurements for both onset and offset mountings of A quasi-planar antenna'. *International Journal on Communications Antenna and Propagation*. 2019;**9**(5):311–19.

[100] D'Agostino F., Ferrara F., Gennarelli C., Guerriero R., Migliozzi M. 'A nonredundant sampling representation managing an offset mounting of an elongated antenna in a spherical near-field facility'. *IEEE Antennas and Wireless Propagation Letters*. 2019;**18**(12):2671–75.

[101] D'Agostino F., Ferrara F., Gennarelli C., Guerriero R., Migliozzi M. 'A spherical near-to-far-field transformation using a non-redundant voltage representation optimized for non-centered mounted quasi-planar antennas'. *Electronics*. 2020;**9**(6):944.

[102] Bucci O.M., Franceschetti G. 'On the spatial bandwidth of scattered fields'. *IEEE Transactions on Antennas and Propagation*. 1987;**AP-35**(12):1445–55.

[103] Bucci O.M., Franceschetti G. 'On the degrees of freedom of scattered fields'. *IEEE Transactions on Antennas and Propagation*. 1989;**37**(7):918–26.

[104] Bucci O.M., Gennarelli C., Savarese C. 'Representation of electromagnetic fields over arbitrary surfaces by a finite and nonredundant number of samples'. *IEEE Transactions on Antennas and Propagation*. 1998;**46**(3):351–59.

[105] Bucci O.M., Gennarelli C. 'Application of nonredundant sampling representations of electromagnetic fields to NF-FF transformation techniques'. *International Journal of Antennas and Propagation*. 2012;**2012**:1–14.

[106] Bucci O.M., Gennarelli C., Savarese C. 'Optimal interpolation of radiated fields over a sphere'. *IEEE Transactions on Antennas and Propagation*. 1991;**39**(11):1633–43.

[107] Gennarelli C., Riccio G., Savarese C., Speranza V. 'Fast and accurate interpolation of radiated fields over a cylinder'. *Progress In Electromagnetics Research*. 1994;**PIER 8**:349–75.

[108] Ferrara F., Gennarelli C., Guerriero R., Riccio G., Savarese C. 'An efficient near-field to far-field transformation using the planar wide-mesh scanning'. *Journal of Electromagnetic Waves and Applications*. 2007;**21**(3):341–57.

[109] D'Agostino F., De Colibus I., Ferrara F., Gennarelli C., Guerriero R., Migliozzi M. 'Far-field pattern reconstruction from near-field data collected via a nonconventional plane-rectangular scanning: experimental testing'. *International Journal of Antennas and Propagation*. 2014;**2014**:1–9.

[110] Bolomey J.-C., Cown B.J., Fine G., *et al*. 'Rapid near-field antenna testing via arrays of modulated scattering probes'. *IEEE Transactions on Antennas and Propagation*. 1988;**36**(6):804–14.

[111] Bolomey J.C., Gardiol F.E. *Engineering applications of the modulated scatterer technique*. Boston, MA, USA: Artech House; 2001.

[112] Yaccarino R.G., Williams L.I., Rahmat-Samii Y. 'Linear spiral sampling for the bipolar planar near-field antenna measurement technique'. *IEEE Transactions on Antennas and Propagation*. 1996;**44**(7):1049–51.

[113] Bucci O.M., Gennarelli C., Riccio G., Savarese C. 'Probe compensated NF-FF transformation with helicoidal scanning'. *Journal of Electromagnetic Waves and Applications*. 2000;**14**(4):531–49.

[114] Bucci O.M., Gennarelli C., Riccio G., Savarese C. 'Nonredundant NF-FF transformation with helicoidal scanning'. *Journal of Electromagnetic Waves and Applications*. 2001;**15**(11):1507–19.

[115] D'Agostino F., Ferrara F., Gennarelli C., Guerriero R., Migliozzi M. 'Experimental results validating the near-field to far-field transformation technique with helicoidal scan'. *The Open Electrical & Electronic Engineering Journal*. 2010;**4**(1):10–15.

[116] D'Agostino F., Ferrara F., Gennarelli C., Guerriero R., Migliozzi M. 'NEAR-field-far-field transformation technique with helicoidal scanning for elongated antennas'. *Progress In Electromagnetics Research B*. 2008;**4**:249–61.

[117] D'Agostino F., Ferrara F., Gennarelli C., Guerriero R., Migliozzi M. 'Laboratory tests assessing the effectiveness of the NF-FF transformation with helicoidal scanning for electrically long antennas'. *Progress In Electromagnetics Research*. 2009;**PIER 98**:375–88.

[118] D'Agosti F., Ferrara F., Fordham J.A., *et al*. 'An effective near-field-far-field transformation technique for elongated antennas using a fast helicoidal scan [measurements corner]'. *IEEE Antennas and Propagation Magazine*. 2009;**51**(4):134–41.

[119] D'Agostino F., Ferrara F., Gennarelli C., Guerriero R., Migliozzi M. 'An innovative direct NF-FF transformation technique with helicoidal scanning'. *International Journal of Antennas and Propagation.* 2012;**2012**:1–9.

[120] D'Agostino F., Ferrara F., Gennarelli C., Guerriero R., Migliozzi M. 'Evaluation of the far field radiated by long antennas directly from data acquired through a fast helicoidal scanning'. *Progress In Electromagnetics Research M.* 2012;**26**:157–71.

[121] D'Agostino F., Ferrara F., Gennarelli C., Guerriero R., Migliozzi M. 'An efficient transformation technique for the direct recovering of the antenna far-field pattern from near-field measurements collected along a helix'. *The Open Electrical & Electronic Engineering Journal.* 2013;**7**(1):21–30.

[122] D'Agostino F., Gennarelli C., Riccio G., Savarese C. 'Theoretical foundations of near-field-far-field transformations with spiral scannings'. *Progress In Electromagnetics Research.* 2006;**PIER 61**:193–214.

[123] Gennarelli C., Riccio G., D'Agostino F, *et al. Near-field – far-field transformation techniques.* Vol. 2. Salerno, Italy: CUES; 2006.

[124] D'Agostino F., Ferrara F., Gennarelli C., Guerriero R., Migliozzi M. 'The unified theory of near-field-far-field transformations with spiral scannings for nonspherical antennas'. *Progress In Electromagnetics Research B.* 2009;**14**:449–77.

[125] F. D'Agostino F., F. Ferrara C., Gennarelli C, *et al. Near-field – far-field transformation techniques with spiral scannings.* Salerno, Italy: CUES; 2009.

[126] Cicchetti R., D'Agostino F., Ferrara F., Gennarelli C., Guerriero R., Migliozzi M. 'Near-field to far-field transformation techniques with spiral scannings: a comprehensive review'. *International Journal of Antennas and Propagation.* 2014;**2014**:1–13.

[127] Bucci O.M., D'Agostino F., Gennarelli C., Savarese C., Riccio G. 'Probe compensated far-field reconstruction by near-field planar spiral scanning'. *IEE Proceedings - Microwaves, Antennas and Propagation.* 2002;**149**(2):119–23.

[128] D'Agostino F., Ferrara F., Gennarelli C, *et al.* 'A planar NF–FF transformation for quasi-spherical antennas using the innovative spiral scanning'. *Applied Computational Electromagnetics Society Journal.* 2018;**33**(1):115–18.

[129] D'Agostino F., Ferrara F., Gennarelli C., Guerriero R., Migliozzi M. 'An effective NF-FF transformation technique with planar spiral scanning tailored for quasi-planar antennas'. *IEEE Transactions on Antennas and Propagation.* 2008;**56**(9):2981–87.

[130] D'Agostino F., Ferrara F., Gennarelli C., Guerriero R., McBride S., Migliozzi M. 'Fast and accurate antenna pattern evaluation from near-field data acquired via planar spiral scanning'. *IEEE Transactions on Antennas and Propagation.* 2016;**64**(8):3450–58.

[131] Bucci O.M., D'Agostino F., Gennarelli C, *et al.* 'NF–FF transformation with spherical spiral scanning'. *IEEE Antennas and Wireless Propagation Letters.* 2003;**2**:263–66.

[132] D'Agostino F., Ferrara F., Fordham J.A., Gennarelli C., Guerriero R., Migliozzi M. 'An experimental validation of the near-field-far-field transformation with spherical spiral scan [amta corner]'. *IEEE Antennas and Propagation Magazine*. 2013;**55**(3):228–35.

[133] D'Agostino F., Ferrara F., Gennarelli C., Guerriero R., Migliozzi M., Riccio G. 'A nonredundant near-field to far-field transformation with spherical spiral scanning for nonspherical antennas'. *The Open Electrical & Electronic Engineering Journal*. 2009;**3**(1):1–8.

[134] D'Agostino F., Ferrara F., Gennarelli C., Guerriero R., Migliozzi M. 'Far-field reconstruction from a minimum number of spherical spiral data using effective antenna modelings'. *Progress In Electromagnetics Research B*. 2012;**37**:43–58.

[135] D'Agostino F., Ferrara F., Gennarelli C., Guerriero R., Migliozzi M. 'Far-field reconstruction from near-field data acquired via a fast spherical spiral scan: experimental evidences'. *Progress In Electromagnetics Research*. 2013;**140**:719–32.

[136] D'Agostino F., Ferrara F., Gennarelli C., Guerriero R., Migliozzi M. 'Experimental assessment of an effective near-field–far-field transformation with spherical spiral scanning for quasi-planar antennas'. *IEEE Antennas and Wireless Propagation Letters*. 2013;**12**:670–73.

[137] D'Agostino F., Ferrara F., Gennarelli C., Guerriero R., Migliozzi M. 'Experimental testing on an effective technique to reconstruct the far- field pattern of a long antenna from near-field measurements acquired via spherical spiral scan'. *The Open Electrical & Electronic Engineering Journal*. 2014;**8**(1):1–9.

[138] D'Agostino F., Ferrara F., Gennarelli C., Guerriero R., Migliozzi M. 'Efficient reconstruction of the pattern radiated by a long antenna from data acquired via a spherical-spiral-scanning near-field facility [measurements corner]'. *IEEE Antennas and Propagation Magazine*. 2014;**56**(2):146–53.

[139] D'Agostino F., Ferrara F., Gennarelli C., Guerriero R., Migliozzi M. 'Reconstruction of the far field radiated by an offset mounted volumetric AUT from non-redundant spherical spiral near-field measurements'. *IET Microwaves, Antennas & Propagation*. 2020;**14**(14):1711–18.

[140] D'Agostino F., Ferrara F., Gennarelli C., Guerriero R., Migliozzi M., Riccio G. 'Pattern evaluation from NF spherical spiral data in the noncentered quasi-planar antennas case'. *IEEE Antennas and Wireless Propagation Letters*. 2020;**19**(12):2275–79.

[141] D'Agostino F., Ferrara F., Gennarelli C., Guerriero R., Migliozzi M., Riccio G. 'Optimization of the near-field spherical spiral scanning for the radiation pattern reconstruction of an offset quasi-planar antenna'. *International Journal of Electronics and Communications*. 2021;**132**:153607.

[142] Costanzo S., di Massa G. 'Far-field reconstruction from phaseless near-field data on a cylindrical helix'. *Journal of Electromagnetic Waves and Applications*. 2004;**18**(8):1057–71.

[143] Costanzo S., Di Massa G. 'NEAR-field to far-field transformation with planar spiral scanning'. *Progress In Electromagnetics Research.* 2007;**PIER 73**:49–59.

[144] Petre P., Sarkar T.K. 'Planar near-field to far-field transformation using an equivalent magnetic current approach'. *IEEE Transactions on Antennas and Propagation.* 1992;**40**(11):1348–56.

[145] Petre P., Sarkar T.K. 'Differences between modal expansion and intergral equation methods for planar near-field to far-field transformation'. *Progress In Electromagnetics Research.* 1996;**PIER 12**:37–56.

[146] Taaghol A., Sarkar T.K. 'Near-field to near/far-field transformation for arbitrary near-field geometry, utilizing an equivalent magnetic current'. *IEEE Transactions on Electromagnetic Compatibility.* 1996;**38**(3):536–42.

[147] Sarkar T.K., Taaghol A. 'Near-field to near/far-field transformation for arbitrary near-field geometry utilizing an equivalent electric current and mom'. *IEEE Transactions on Antennas and Propagation.* 1999;**47**(3):566–73.

[148] Las-Heras F., Sarkar T.K. 'Radial field retrieval in spherical scanning for current reconstruction and NF-FF transformation'. *IEEE Transactions on Antennas and Propagation.* 2002;**50**(6):866–74.

[149] Las-Heras F., Pino M.R., Loredo S., Alvarez Y., Sarkar T.K. 'Evaluating near-field radiation patterns of commercial antennas'. *IEEE Transactions on Antennas and Propagation.* 2006;**54**(8):2198–207.

[150] Alvarez Y., Las-Heras F., Pino M.R. 'Reconstruction of equivalent currents distribution over arbitrary three-dimensional surfaces based on integral equation algorithms'. *IEEE Transactions on Antennas and Propagation.* 2007;**55**(12):3460–68.

[151] Alvarez Y., Las-Heras F., Pino M.R. 'On the comparison between the spherical wave expansion and the sources reconstruction method'. *IEEE Transactions on Antennas and Propagation.* 2008;**56**(10):3337–41.

[152] Lopez Yu.A., Las-Heras Andres F., Pino M.R., Sarkar T.K. 'An improved super-resolution source reconstruction method'. *IEEE Transactions on Instrumentation and Measurement.* 2009;**58**(11):3855–66.

[153] Quijano J.L.A., Vecchi G. 'Improved-accuracy source reconstruction on arbitrary 3-D surfaces'. *IEEE Antennas and Wireless Propagation Letters.* 2009;**8**:1046–49.

[154] Araque Quijano J.L., Vecchi G. 'Near- and very near-field accuracy in 3-D source reconstruction'. *IEEE Antennas and Wireless Propagation Letters.* 2010;**9**:634–37.

[155] Lopez-Fernandez J.A., Lopez-Portugues M., Alvarez Lopez Y., GonzÃ¡lez C.G., MartÃnez D., Las-Heras F. 'Fast antenna characterization using the sources reconstruction method on graphics processors'. *Progress In Electromagnetics Research.* 2012;**126**:185–201.

[156] Eibert T.F., Schmidt C.H. 'Multilevel fast multipole accelerated inverse equivalent current method employing rao–wilton–glisson discretization of

electric and magnetic surface currents'. *IEEE Transactions on Antennas and Propagation*. 2009;**57**(4):1178–85.

[157] Eibert T.F., Kaliyaperumal E., Schmidt C.H, *et al.* ' inverse equivalent surface current method with hierarchical higher order basis functions, full probe correction and multilevel fast multipole acceleration '. *Progress In Electromagnetics Research*. 2010;**106**:377–94.

[158] Alvarez Y., Las-Heras F., Pino M.R. 'Probe-distortion correction for the sources reconstruction method'. *IEEE Antennas and Propagation Magazine*. 2008;**50**(6):117–24.

[159] Bennett J., Anderson A., McInnes P., Whitaker A. 'Microwave holographic metrology of large reflector antennas'. *IEEE Transactions on Antennas and Propagation*. 1976;**24**(3):295–303.

[160] Smith D., Leach M., Elsdon M., Foti S.J. 'Indirect holographic techniques for determining antenna radiation characteristics and imaging aperture fields'. *IEEE Antennas and Propagation Magazine*. 2007;**49**(1):54–67.

[161] Laviada J., Las-Heras F. 'Phaseless antenna measurement on non-redundant sample points via leith-upatnieks holography'. *IEEE Transactions on Antennas and Propagation*. 2013;**61**(8):4036–44.

[162] Laviada Martinez J., Arboleya-Arboleya A., Alvarez-Lopez Y., Garcia-Gonzalez C., Las-Heras F. 'Phaseless antenna diagnostics based on off-axis holography with synthetic reference wave'. *IEEE Antennas and Wireless Propagation Letters*. 2014;**13**:43–46.

[163] Laviada J., Alvarez-Lopez Y., Arboleya-Arboleya A., Garcia-Gonzalez C., Las-Heras F. 'Interferometric technique with nonredundant sampling for phaseless inverse scattering'. *IEEE Transactions on Antennas and Propagation*. 2014;**62**(2):739–46.

[164] Arboleya A., Laviada J., Ala-Laurinaho J., Alvarez Y., Las-Heras F., Raisanen A.V. 'Phaseless characterization of broadband antennas'. *IEEE Transactions on Antennas and Propagation*. 2016;**64**(2): 484–95.

[165] Costanzo S., Di Massa G. 'An integrated probe for phaseless plane-polar near-field measurements'. *Microwave and Optical Technology Letters*. 2001;**30**(5):293–95.

[166] Costanzo S., Di Massa G. 'An integrated probe for phaseless near-field measurements'. *Measurement*. 2002;**31**(2):123–29.

[167] Costanzo S., Di Massa G., Migliore M.D. 'A novel hybrid approach for far-field characterization from near-field amplitude-only measurements on arbitrary scanning surfaces'. *IEEE Transactions on Antennas and Propagation*. 2005;**53**(6):1866–74.

[168] Costanzo S., Di Massa G. 'Wideband phase retrieval technique for amplitude-only near-field data'. *Radioengineering*. 2008;**17**(4):8–12.

[169] Anderson A.P., Sali S. 'New possibilities for phaseless microwave diagnostics. part1: error reduction techniques'. *IEE Proceedings H Microwaves, Antennas and Propagation*. 1985;**132**(5):291.

[170] Bucci O.M., D'Elia G., Leone G., Pierri R. 'Far-field pattern determination from the near-field amplitude on two surfaces'. *IEEE Transactions on Antennas and Propagation.* 1990;**38**(11):1772–79.

[171] Bucci O.M., D'elia G., Pierri R., Leone G. 'Far-field computation from amplitude near-field data on two surfaces: cylindrical case'. *IEE Proceedings H Microwaves, Antennas and Propagation.* 1992;**139**(2):143.

[172] Isernia T., Leone G., Pierri R. 'New approach to antenna testing from near field phaseless data: the cylindrical scanning'. *IEE Proceedings H Microwaves, Antennas and Propagation.* 1992;**139**(4):363.

[173] Isernia T., Leone G., Pierri R. 'Radiation pattern evaluation from near-field intensities on planes'. *IEEE Transactions on Antennas and Propagation.* 1996;**44**(5):701.

[174] Bucci O.M., D'Elia G., Migliore M.D. 'An effective near-field far-field transformation technique from truncated and inaccurate amplitude-only data'. *IEEE Transactions on Antennas and Propagation.* 1999;**47**(9):1377–85.

[175] Yaccarino R.G., Rahmat-Samii Y. 'Phaseless bi-polar planar near-field measurements and diagnostics of array antennas'. *IEEE Transactions on Antennas and Propagation.* 1999;**47**(3):574–83.

[176] Migliore M.D., Soldovieri F., Pierri R. 'Far-field antenna pattern estimation from near-field data using a low-cost amplitude-only measurement setup'. *IEEE Transactions on Instrumentation and Measurement.* 2000;**49**(1):71–76.

[177] Capozzoli A., Curcio C., D'Elia G., Liseno A. 'Phaseless antenna characterization by effective aperture field and data representations'. *IEEE Transactions on Antennas and Propagation.* 2009;**57**(1):215–30.

[178] Capozzoli A., Curcio C., D'Elia G., *et al.* 'Photonic probes and advanced (also phaseless) near-field far-field techniques'. *IEEE Antennas and Propagation Magazine.* 2012;**52**(5):232–41.

[179] Rodriguez Varela F., Fernandez Alvarez J., Galocha Iraguen B., Sierra Castaner M., Breinbjerg O. 'Numerical and experimental investigation of phaseless spherical near-field antenna measurements'. *IEEE Transactions on Antennas and Propagation.* 2021;**69**(12):8830–41.

[180] Pierri R., D'Elia G., Soldovieri F. 'A two probes scanning phaseless near-field far-field transformation technique'. *IEEE Transactions on Antennas and Propagation.* 1999;**47**(5):792–802.

[181] Las-Heras F., Sarkar T.K. 'A direct optimization approach for source reconstruction and NF-FF transformation using amplitude-only data'. *IEEE Transactions on Antennas and Propagation.* 2002;**50**(4):500–10.

[182] Alvarez Y., Las-Heras F., Pino M.R. 'The sources reconstruction method for amplitude-only field measurements'. *IEEE Transactions on Antennas and Propagation.* 2010;**58**(8):2776–81.

[183] Alvarez Y., Las-Heras F., García C. 'Reconstruction method for antenna diagnostics and imaging applications' in Kishk A. (ed.). *Solutions and applications of scattering, propagation, radiation and*

emission of electromagnetic waves. London, UK: InTechOpen; 2012. pp. 159–86.

[184] Li P., Jiang L. 'An iterative source reconstruction method exploiting phaseless electric field data'. *Progress In Electromagnetics Research*. 2013;**134**:419–35.

[185] Brown T., Jeffrey I., Mojabi P. 'Multiplicatively regularized source reconstruction method for phaseless planar near-field antenna measurements'. *IEEE Transactions on Antennas and Propagation*. 2017;**65**(4):2020–31.

[186] Lopez Y.A., Garcia-Fernandez M., Las-Heras F. 'A portable cost-effective amplitude and phase antenna measurement system'. *IEEE Transactions on Instrumentation and Measurement*. 2020;**69**(9):7240–51.

[187] Bod M., Moradi G., Sarraf-shirazi R. 'Phaseless near-field to far-field transformation based on source current reconstruction and signal subspace optimization'. *International Journal of RF and Microwave Computer-Aided Engineering*. 2022;**32**(1):1–8.

[188] Schmidt C.H., Leibfritz M.M., Eibert T.F. 'Fully probe-corrected near-field far-field transformation employing plane wave expansion and diagonal translation operators'. *IEEE Transactions on Antennas and Propagation*. 2008;**56**(3):737–46.

[189] Schmidt C.H., Eibert T.F. 'Multilevel plane wave based near-field far-field transformation for electrically large antennas in free-space or above material halfspace'. *IEEE Transactions on Antennas and Propagation*. 2009;**57**(5):1382–90.

[190] Schmidt C.H., Eibert T.F. 'Assessment of irregular sampling near-field far-field transformation employing plane-wave field representation'. *IEEE Antennas and Propagation Magazine*. 2011;**53**(3):213–19.

[191] Coifman R., Rokhlin V., Wandzura S. 'The fast multipole method for the wave equation: a pedestrian prescription'. *IEEE Antennas and Propagation Magazine*. 1993;**35**(3):7–12.

[192] Chew W., Jin J., Michielssen E, *et al. Fast and efficient algorithms in computational electromagnetics*. Boston, MA, USA: Artech House; 2001.

[193] Capozzoli, A., Curcio C., Liseno A, *et al.* 'Field sampling and field reconstruction: a new perspective'. *Radio Science*. 2010;**45**:1–31.

[194] Capozzoli A., Curcio C., D'Elia G., Liseno A. 'Singular-value optimization in plane-polar near-field antenna characterization'. *IEEE Antennas and Propagation Magazine*. 2010;**52**(2):103–12.

[195] Capozzoli A., Curcio C., Liseno A. 'Multi-frequency planar near-field scanning by means of SVD optimization'. *IEEE Antennas and Propagation Magazine*. 2011;**53**(6):212–21.

[196] Capozzoli A., Curcio C., Liseno A. 'Truncation in "quasi-raster" near-field acquisitions'. *IEEE Antennas and Propagation Magazine*. 2012;**54**(5):174–83.

[197] Capozzoli A., Curcio C., D'Elia G., *et al.* 'A probe-compensated helicoidal NF-FF transformation for aperture antennas using a prolate spheroidal expansion'. *International Journal of Antennas and Propagation.* 2012;**2012**:1–13.

[198] Capozzoli A., Celentano L., Curcio C., Liseno A., Savarese S. 'Optimized trajectory tracking of a class of uncertain systems applied to optimized raster scanning in near-field measurements'. *IEEE Access : Practical Innovations, Open Solutions.* 2018;**6**(5):8666–81.

[199] Capozzoli A., Celentano L., Curcio C., Liseno A., Savarese S. 'Twice optimised near-field scanning system for antenna characterisation'. *IET Microwaves, Antennas & Propagation.* 2020;**14**(3):163–73.

[200] Foged L.J., Scialacqua L., Saccardi F., *et al.* 'Satellite near-field testing and verification using physics-based interpolation technique and optimum sampling [amta corner]'. *IEEE Antennas and Propagation Magazine.* 2013;**55**(6):261–70.

[201] Giordanengo G., Righero M., Vipiana F. 'Fast antenna testing with reduced near field sampling'. *IEEE Transactions on Antennas and Propagation.* 2014;**62**(5):2501–13.

[202] Massa A., Rocca P., Oliveri G. 'Compressive sensing in electromagnetics-a review'. *IEEE Antennas and Propagation Magazine.* 2015;**57**(1):224–38.

[203] Migliore M.D. 'On the sampling of the electromagnetic field radiated by sparse sources'. *IEEE Transactions on Antennas and Propagation.* 2015;**63**(2):553–64.

[204] Oliveri G., Salucci M., Anselmi N., Massa A. 'Compressive sensing as applied to inverse problems for imaging: theory, applications, current trends, and open challenges'. *IEEE Antennas and Propagation Magazine.* 2017;**59**(5):34–46.

[205] Migliore M.D. 'A simple introduction to compressed sensing/sparse recovery with applications in antenna measurements'. *IEEE Antennas and Propagation Magazine.* 2014;**56**(2):14–26.

[206] Fuchs B., Coq L.L., Migliore M.D. 'Fast antenna array diagnosis from a small number of far-field measurements'. *IEEE Transactions on Antennas and Propagation.* 2016;**64**(6):2227–35.

[207] Oliveri G., Rocca P., Massa A. 'Reliable diagnosis of large linear arrays—a Bayesian compressive sensing approach'. *IEEE Transactions on Antennas and Propagation.* 2011;**60**(10):4627–36.

[208] Costanzo S., Borgia A., Di Massa G., Pinchera D., Migliore M.D. 'Radar array diagnosis from undersampled data using a compressed sensing/sparse recovery technique'. *Journal of Electrical and Computer Engineering.* 2013;**2013**:1–5.

[209] Migliore M.D. 'A compressed sensing approach for array diagnosis from a small set of near-field measurements'. *IEEE Transactions on Antennas and Propagation.* 2011;**59**(6):2127–33.

[210] Fuchs B., Le Coq L., Rondineau S., Migliore M.D. 'Fast antenna far-field characterization via sparse spherical harmonic expansion'. *IEEE Transactions on Antennas and Propagation*. 2017;**65**(10):5503–10.

[211] Cornelius R., Heberling D., Koep N., Behboodi A., Mathar R. 'Compressed sensing applied to spherical near-field to far-field transformation'. *Proc. 2016 EUCAP*; Davos, Switzerland, 2016. pp. 1–4.

[212] Culotta-López C., Heberling D., Bangun A, *et al.* 'A compressed sampling for spherical near-field measurements'. *Proc. 2018 AMTA*; Williamsburg, VA, USA, 2018. pp. 1–6.

[213] Culotta-Lopez C., Walkenhorst B., Ton Q., Heberling D. 'Practical considerations in compressed spherical near-field measurements'. *Proc. 2019 AMTA*; San Diego, CA, USA, 2019. pp. 1–6.

[214] Hofmann B., Neitz O., Eibert T.F. 'On the minimum number of samples for sparse recovery in spherical antenna near-field measurements'. *IEEE Transactions on Antennas and Propagation*. 2019;**67**(12): 7597–610.

[215] Joy E.B., Wilson R.E. 'A simplified technique for probe position error compensation in planar surface near-field measurements'. *Proc. 1982 AMTA*; USA, 1982. pp. 14.

[216] Jory V.V., Joy E.B., Leach W.M. 'Current antenna near field measurement research at the georgia institute of technology'. *Proc. 13th European Microwave Conference*; Nuremberg, West Germany, 1983. pp. 823–28.

[217] Corey L.E., Joy E.B. 'On computation of electromagnetic fields on planar surfaces from fields specified on nearby surfaces'. *IEEE Transactions on Antennas and Propagation*. 1981;**AP-29**(2):402–04.

[218] Mahmoud S.F. 'Correction of probe position errors in antenna near field to far field transformation'. *Journal of Electromagnetic Waves and Applications*. 1988;**2**(5–6):545–53.

[219] Bucci O.M., Schirinzi G., Leone G. 'A compensation technique for positioning errors in planar near-field measurements'. *IEEE Transactions on Antennas and Propagation*. 1988;**36**(8):1167–72.

[220] Muth L.A., Lewis R.L. 'A general technique to correct probe position errors in planar near-field measurements to arbitrary accuracy'. *IEEE Transactions on Antennas and Propagation*. 1990;**38**(12):1925–32.

[221] Muth L.A. 'General analytic correction for probe-position errors in spherical near-field measurements'. *Journal of Research of the National Institute of Standards and Technology*. 1991;**96**(4):391–410.

[222] Dutt A., Rokhlin V. 'Fast Fourier transforms for nonequispaced data'. *SIAM Journal on Scientific Computing*. 1993;**14**(6):1368–93.

[223] Wittmann R.C., Alpert B.K., Francis M.H. 'Near-field antenna measurements using nonideal measurement locations'. *IEEE Transactions on Antennas and Propagation*. 1998;**46**(5):716–22.

[224] Wittmann R.C., Alpert B.K., Francis M.H. 'Near-field, spherical-scanning antenna measurements with nonideal probe locations'. *IEEE Transactions on Antennas and Propagation*. 2004;**52**(8):2184–87.

[225] Dehghanian V., Okhovvat M., Hakkak M. 'A new interpolation technique for the reconstruction of uniformly spaced samples from non-uniformly spaced ones in plane-rectangular near-field antenna measurements'. *Progress In Electromagnetics Research*. 2007;**PIER 72**:47–59.

[226] Yen J. 'On nonuniform sampling of bandwidth-limited signals'. *IRE Transactions on Circuit Theory*. 1956;**CT-3**(4):251–57.

[227] Bucci O.M., Gennarelli C., Savarese C. 'Interpolation of electromagnetic radiated fields over a plane by nonuniform samples'. *IEEE Transactions on Antennas and Propagation*. 1993;**41**(11):1501–08.

[228] Bucci O.M., Gennarelli C., Riccio G. 'Electromagnetic fields interpolation from nonuniform samples over spherical and cylindrical surfaces'. *IEE Proceedings - Microwaves, Antennas and Propagation*. 1994;**141**(2):77.

[229] Golub G.H., Van Loan C.F. *Matrix computations*. Baltimore, MD, USA: J.Hopkins University Press; 2007.

[230] Ferrara F., Savarese C. 'Far-field reconstruction from nonuniform plane-polar data: a SVD-based approach'. *Electromagnetics*. 2003;**23**(5):417–29.

[231] D'Agostino F., Ferrara F., Gennarelli C., Guerriero R., Migliozzi M. 'Two efficient procedures to correct the positioning errors in the plane-polar scanning'. *IET Microwaves, Antennas & Propagation*. 2016;**10**(13):1453–58.

[232] D'Agostino F., Ferrara F., Gennarelli C, *et al*. 'Near to far-field plane-polar transformation from probe positioning error affected data'. *Applied Computational Electromagnetics Society Journal*. 2018;**33**(4):419–29.

[233] Ferrara F., Gennarelli C., Iacone M, *et al*. 'NF-FF transformation with bi-polar scanning from nonuniformly spaced data'. *Applied Computational Electromagnetics Society Journal*. 2005;**20**(1):35–42.

[234] D'Agostino F., Ferrara F., Gennarelli C, *et al*. 'On the correction of the probe positioning errors in a non-redundant bi-polar near to far-field transformation'. *Applied Computational Electromagnetics Society Journal*. 2018;**33**(12):1374–82.

[235] D'Agostino F., Ferrara F., Gennarelli C., Guerriero R., Migliozzi M. 'Laboratory testing of an SVD-based approach to recover the non-redundant bi-polar NF data from the positioning error affected ones'. *IET Microwaves, Antennas & Propagation*. 2019;**13**(6):736–41.

[236] D'Agostino F., Ferrara F., Gennarelli C., Guerriero R., Migliozzi M., Riccio G. 'A singular value decomposition based approach for far-field reconstruction from irregularly spaced planar wide-mesh scanning data'. *Microwave and Optical Technology Letters*. 2007;**49**(7):1768–72.

[237] D'Agostino F., Ferrara F., Gennarelli C., Guerriero R., Migliozzi M. 'An SVD-based approach to reconstruct the planar wide-mesh scanning NF data from inaccurately probe-positioned samples'. *IEEE Antennas and Wireless Propagation Letters*. 2018;**17**(4):641–44.

[238] D'Agostino F., Ferrara F., Gennarelli C, *et al*. 'An effective iterative algorithm to correct the probe positioning errors in a non-redundant plane-rectangular near-field to far-field transformation'. *Applied Computational Electromagnetics Society Journal*. 2019;**34**(11):1662–70.

[239] D'Agostino F., Ferrara F., Gennarelli C., Guerriero R., Migliozzi M. 'Two approaches to get efficient NF/FF transformations from inaccurately positioned planar wide-mesh scanning samples'. *International Journal of Electronics and Communications*. 2020;**119**:153169.

[240] Ferrara F., Gennarelli C., Riccio G., Savarese C. 'NF-FF transformation with cylindrical scanning from nonuniformly distributed data'. *Microwave and Optical Technology Letters*. 2003;**39**(1):4–8.

[241] D'Agostino F., Ferrara F., Gennarelli C., Guerriero R., Migliozzi M. 'On the compensation of probe positioning errors when using a nonredundant cylindrical NF-FF transformation'. *Progress In Electromagnetics Research B*. 2010;**20**:321–35.

[242] Ferrara F., Gennarelli C., Guerriero R., Migliozzi M. 'Experimental validation of the NF-FF transformation with cylindrical scan from nonuniformly distributed data'. *Microwave and Optical Technology Letters*. 2011;**53**(4):915–20.

[243] D'Agostino F., Ferrara F., Gennarelli C., Guerriero R., Migliozzi M. 'Near-field/far-field transformation with helicoidal scanning from irregularly spaced data'. *International Journal of Antennas and Propagation*. 2010;**2010**:1–8.

[244] D'Agostino F., Ferrara F., Gennarelli C., Guerriero R., Migliozzi M. 'An iterative technique to compensate for positioning errors in the NF-FF transformation with helicoidal scanning for long antennas'. *Progress In Electromagnetics Research C*. 2011;**18**:73–86.

[245] D'Agostin F., Ferrara F., Gennarelli C., Guerriero R., Migliozzi M. 'Far-field pattern reconstruction from positioning errors affected near-field data acquired via helicoidal scanning'. *Journal of Electromagnetic Analysis and Applications*. 2012;**04**(2):60–68.

[246] Gennarelli C. 'Two techniques for compensating the probe positioning errors in the spherical NF-FF transformation for elongated antennas'. *The Open Electrical & Electronic Engineering Journal*. 2011;**5**(1):29–36.

[247] Agostino F.D., Ferrara F., Gennarelli C., Guerriero R., Migliozzi M. 'Spherical near-field-far-field transformation for quasi-planar antennas from irregularly spaced data'. *Journal of Electromagnetic Analysis and Applications*. 2012;**04**(4):147–55.

[248] D'Agostino F., Ferrara F., Gennarelli C., Guerriero R., Migliozzi M. 'Two effective approaches to correct the positioning errors in a spherical near-field–far-field transformation'. *Electromagnetics*. 2016;**36**(2): 78–93.

[249] D'Agostino F., Ferrara F., Gennarelli C., Guerriero R., Migliozzi M. 'Far-field pattern evaluation from data acquired on a spherical surface by an inaccurately positioned probe'. *IEEE Antennas and Wireless Propagation Letters*. 2016;**15**:402–05.

[250] D'Agostino F., Ferrara F., Gennarelli C, *et al*. 'Probe position errors corrected near-field-far-field transformation with spherical scanning'. *Applied Computational Electromagnetics Society Journal*. 2016;**31**(2):106–17.

[251] Cicchetti R., D'Agostino F., Ferrara F., Gennarelli C., Guerriero R., Migliozzi M. 'Correction of known position errors in a spherical near to far-field transformation for long antennas'. *The Open Electrical & Electronic Engineering Journal*. 2017;**11**(1):141–53.

[252] Newell A.C. 'Error analysis techniques for planar near-field measurements'. *IEEE Transactions on Antennas and Propagation*. 1988;**36**(6):754–68.

[253] Gregson S.F., Newell A.C., Hindman G.E, *et al.* 'Extension of the mathematical absorber reflection suppression technique to the planar near-field geometry'. *Proc. 2010 AMTA*; Atlanta, GA, USA, 2010. pp. 1–7.

[254] Gregson S.F., Newell A.C., Hindman G.E, *et al.* 'Range multipath reduction in plane-polar near-field antenna measurements'. *Proc. 2012 AMTA*; Seattle, WA, USA, 2012. pp. 1–7.

[255] Gregson S.F., Newell A.C., Hindman G.E. 'Reflection suppression in cylindrical near-field antenna measurement systems – cylindrical mars'. *Proc. 2009 AMTA*; Salt Lake City, UT, USA, 2009. pp. 1–7.

[256] Gregson S.F., Newell A.C., Hindman G.E, *et al.* 'Comparison of cylindrical and spherical mathematical absorber reflection suppression'. *Proc. 2010 LAPC*; Loughborough UK, 2010. pp. 181–84.

[257] Hindman G.E., Newell A.C. 'Reflection suppression in a large spherical near-field range'. *Proc. 2005 AMTA*; Newport, RI, USA, 2005. pp. 1–6.

[258] Hindman G.E., Newell A.C. 'Reflection suppression to improve anechoic chamber performance'. *Proc. 2006 AMTA Europe*; Munich, Germany, 2006. pp. 297–302.

[259] Gregson S.F., Dupuy J., Parini C.G., Newell A.C., Hindman G.E. 'Application of mathematicalabsorber reflection suppression to direct far-field antenna range measurements'. *Proc. 2011 AMTA*; Denver, CO, USA, 2011. pp. 1–5.

[260] Gregson S.F., Newell A.C., Hindman G.E. 'Examination of far-field mathematical absorber reflection suppression through computational electromagnetic simulation'. *International Journal of Antennas and Propagation*. 2012;**2012**:1–10.

[261] Foged L.J., Scialacqua L., Mioc F., *et al.* 'Echo suppression by spatial-filtering techniques in advanced planar and spherical near-field antenna measurements [AMTA corner]'. *IEEE Antennas and Propagation Magazine*. 2013;**55**(5):235–42.

[262] Quijano J.L.A., Scialacqua L., Zackrisson J., Foged L.J., Sabbadini M., Vecchi G. 'Suppression of undesired radiated fields based on equivalent currents reconstruction from measured data'. *IEEE Antennas and Wireless Propagation Letters*. 2011;**10**:314–17.

[263] Bucci O.M., D'Elia G., Migliore M.D. 'A general and effective clutter filtering strategy in near-field antenna measurements'. *IEE Proceedings - Microwaves, Antennas and Propagation*. 2004;**151**(3):227.

[264] Bucci O.M., D'Elia G., Migliore M.D. 'A new strategy to reduce the truncation error in near-field/far-field transformations'. *Radio Science*. 2000;**35**(1):3–17.

[265] Bolomey J.-C., Bucci O.M., Casavola L., D'Elia G., Migliore M.D., Ziyyat A. 'Reduction of truncation error in near-field measurements of antennas of base-station mobile communication systems'. *IEEE Transactions on Antennas and Propagation*. 2004;**52**(2):593–602.

[266] Ferrara F., Gennarelli C., Riccio G, *et al.* 'Extrapolation of near-field data in sampling representations: an efficient SVD-based approach'. *Proc. 2002 APMC*; Kyoto, Japan, 2002. pp. 1–4.

[267] Ferrara F., Gennarelli C., Riccio G, *et al.* 'A SVD–based method for extrapolating outside data in near-field measurements'. *Proc. 2003 ISPACS*; Awaji Island, Japan, 2003. pp. 1–4.

[268] D'Agostino F., Ferrara F., Gennarelli C., Guerriero R., Riccio G. 'An effective technique for reducing the truncation error in the near-field-far-field transformation with plane-polar scanning'. *Progress In Electromagnetics Research*. 2007;**PIER 73**:213–38.

[269] D'Agostino F., Ferrara F., Gennarelli C., Guerriero R., Riccio G., Savarese C. 'An efficient technique to lower the error due to the truncation of the scanning region in a bipolar facility'. *Microwave and Optical Technology Letters*. 2007;**49**(12):3033–37.

[270] Ferrara F., Gennarelli C., Guerriero R, *et al.* 'SVD-based approach for estimating the data external to the measurement region in the planar wide-mesh scanning'. *Proc. 2005 ICECOM*; Dubrovnik, Croatia, 2005. pp. 533–36.

[271] Ferrara F., Gennarelli C., Guerriero R., Riccio G., Savarese C. 'Extrapolation of the outside near-field data in the cylindrical scanning'. *Electromagnetics*. 2008;**28**(5):333–45.

[272] D'Agostino F., Gennarelli C., Ferrara F., Guerriero R., Riccio G., Savarese C. 'An efficient approach to extrapolate the data falling in the zone not covered by measurements in a near-field spherical facility'. *The Open Electrical & Electronic Engineering Journal*. 2007;**1**:42–50.

[273] Martini E., Breinbjerg O., Maci S. 'Reduction of truncation errors in planar near-field aperture antenna measurements using the gerchberg-papoulis algorithm'. *IEEE Transactions on Antennas and Propagation*. 2008;**56**(11):3485–93.

[274] Gerchberg R.W. 'Super-resolution through error energy reduction'. *Optica Acta*. 1974;**21**(9):709–20.

[275] Papoulis A. 'A new algorithm in spectral analysis and band-limited extrapolation'. *IEEE Transactions on Circuits and Systems*. 1975;**22**(9): 735–42.

[276] Cano-Fácila F.J., Pivnenko S., Sierra-Castañer M. 'Reduction of truncation errors in planar, cylindrical, and partial spherical near-field antenna measurements'. *International Journal of Antennas and Propagation*. 2012;**2012**:1–19.

[277] Wittmann R.C., Stubenrauch C.F., Francis M.H. 'Using truncated data sets in spherical-scanning antenna measurements'. *International Journal of Antennas and Propagation*. 2012;**2012**:1–6.

[278] Hansen T.B., Yaghjian A.D. 'Planar near-field scanning in the time domain.1. formulation'. *IEEE Transactions on Antennas and Propagation.* 1994;**42**(9):1280–91.

[279] Hansen T.B., Yaghjian A.D. 'Planar near-field scanning in the time domain.2. sampling theorems and computation schemes'. *IEEE Transactions on Antennas and Propagation.* 1994;**42**(9):1292–300.

[280] Hansen T.B., Yaghjian A.D. 'Formulation of probe-corrected planar near-field scanning in the time domain'. *IEEE Transactions on Antennas and Propagation.* 1995;**43**(6):569–84.

[281] Hansen T.B., Yaghjian A.D. *Plane-wave theory of time-domain fields.* Piscataway, NJ, USA: IEEE Press; 1999.

[282] Hansen T.B. 'Formulation of spherical near-field scanning for electromagnetic fields in the time domain'. *IEEE Transactions on Antennas and Propagation.* 1997;**45**(4):620–30.

[283] Serhir M. 'On the near-field sampling and truncation errors in planar time-domain near-field to far-field transformation'. *Progress In Electromagnetics Research B.* 2015;**62**:181–93.

[284] Serhir M. 'Transient UWB antenna near-field and far-field assessment from time domain planar near-field characterization: simulation and measurement investigations'. *IEEE Transactions on Antennas and Propagation.* 2015;**63**(11):4868–76.

[285] Rammal R., Lalande M., Martinod E., *et al.* 'Far-field reconstruction from transient near-field measurement using cylindrical modal development'. *International Journal of Antennas and Propagation.* 2009;**2009**:1–7.

[286] Blech M.D., Leibfritz M.M., Hellinger R., *et al.* 'A time domain spherical near-field measurement facility for UWB antennas employing a hardware gating technique'. *Advances in Radio Science.* 2010;**8**:243–50.

[287] Blech M.D., Leibfritz M.M., Hellinger R., *et al.* 'Time-domain spherical near-field antenna measurement system employing a switched continuous-wave hardware gating technique'. *IEEE Transactions on Instrumentation and Measurement.* 2010;**59**(2):387–95.

[288] Bucci O.M., D'Elia G., Migliore D. 'Optimal time-domain field interpolation from plane-polar samples'. *IEEE Transactions on Antennas and Propagation.* 1997;**45**(6):989–94.

[289] Bucci O.M., D'Elia G., Migliore M.D. 'Near-field far-field transformation in time domain from optimal plane-polar samples'. *IEEE Transactions on Antennas and Propagation.* 1998;**46**(7):1084–88.

[290] D'Agostino F., Ferrara F., Gennarelli C., Guerriero R., Migliozzi M. 'Optimal sampling interpolation over a plane from transient bi-polar near-field data'. *IET Microwaves, Antennas & Propagation.* 2016;**10**(13):1445–52.

[291] D'Agostino F., Ferrara F., Gennarelli C., Guerriero R., Migliozzi M. 'TD optimal sampling interpolation over a plane from NF data collected through a non-conventional plane-rectangular scanning'. *IET Microwaves, Antennas & Propagation.* 2017;**11**(12):1681–86.

[292] Ricciardi G.F., Stutzman W.L. A near-field to far-field transformation for spheroidal geometry utilizing an eigenfunction expansion. *IEEE Transactions on Antennas and Propagation*. 2004;**52**(12):3337–49.

[293] Milne-Thomson L.M. 'Elliptic integrals' in Abramowitz M., Stegun I.A. (eds.). *Handbook of mathematical functions*. New York, USA: Dover Publications, Inc; 1970. pp. 587–626.

[294] Whittaker E.T. 'XVIII.—on the functions which are represented by the expansions of the interpolation-theory'. *Proceedings of the Royal Society of Edinburgh*. 1915;**35**:181–94.

[295] Knab J.J. 'The sampling window'. *IEEE Transactions on Information Theory*. 1983;**IT-29**(1):157–59.

[296] Bucci O.M., Gennarelli C., Savarese C. *Interpolation of radiated fields over a sphere*. Vol. LVII. I.U.N., Naples, Italy: Annali della Facoltà di Scienze Nautiche; 1990. pp. 1–114.

[297] Bucci O.M., D'agostino F., Ferrara F., Gennarelli C. 'Ellipsoidal source modelings for optimal far-field interpolation'. *Microwave and Optical Technology Letters*. 2001;**29**(3):181–85.

[298] Collin R.E. *Antennas and radiowave propagation*. New York, USA: MacGraw-Hill; 1985.

[299] Papoulis A. *Signal analysis*. New York, USA: McGraw-Hill; 1977.

[300] Brigham E.O. *The fast Fourier transform*. NJ, USA: Prentice-Hall, Englewood Cliffs; 1974.

[301] Yaghjian A.D. *Upper-bound errors in far-field antenna parameters determined from planar near-field measurements, part 1: analysis ', Nbs tech. note 667*. Washington, DC, USA: U.S. Government Printing Office; 1975.

[302] Newell A.C., Crawford M. *Planar near-field measurements on high performance array antennas*. NBSIR; 1974. pp. 74–380.

[303] Yaghjian A.D. 'Approximate formulas for the far field and gain of open-ended rectangular waveguide'. *IEEE Transactions on Antennas and Propagation*. 1984;**AP-32**(4):378–84.

[304] D'Agostino F., Ferrara F., Gennarelli C, *et al.* 'A planar NF-FF transformation for quasi-spherical antennas using the innovative spiral scanning'. *Sensors*. 2021;**21**.

[305] Clemmow P.C. *The plane wave spectrum representation of electromagnetic fields*. Oxford, UK: Pergamon Press; 1966 - Piscataway, NJ, USA: IEEE Press; 1996.

[306] James G.L. *Geometrical theory of diffraction for electromagnetic waves*. London, UK: Peter Peregrinus; 1986.

[307] Stark H., Wengrovitz M. 'Comments and corrections on the use of polar sampling theorems in CT'. *IEEE Transactions on Acoustics, Speech, and Signal Processing*. 1983;**31**(5):1329–31.

[308] Harrington R.F. *Time-harmonic electromagnetic fields*. New York, USA: McGraw-Hill; 2001.

[309] Stratton J.A. *Electromagnetic theory*. McGraw-Hill, New York, USA, 1941, Wiley-IEEE Press, New York, USA; 2007.

[310] Belousov S.L. *Tables of normalized associated Legendre polynomials.* Oxford, UK: Pergamon Press; 1962.

[311] Edmonds A.R. *Angular momentum in quantum mechanics.* Princeton, New Jersey, USA: Princeton University Press; 1974.

[312] Bruning J., Yuen Lo T. 'Multiple scattering of em waves by spheres part I – multipole expansion and ray-optical solutions'. *IEEE Transactions on Antennas and Propagation.* 1971;**AP-19**(3):378–90.

[313] Antosiewicz H.A. 'Bessel functions of fractional order' in Abramowitz M., Stegun I.A. (eds.). *Handbook of mathematical functions.* New York, USA: Dover Publications, Inc; 1970. pp. 435–78.

[314] Cooley J.W., Tukey J.W. 'An algorithm for the machine calculation of complex Fourier series'. *Mathematics of Computation.* 1965;**19**(90):297–301.

Index